CELLULAR PEPTIDASES IN IMMUNE FUNCTIONS AND DISEASES

Recent Volumes in this Series

CELLULAR PEPTIDASES IN IMMUNE FUNCTIONS AND DISEASES

Edited by

Siegfried Ansorge
Otto von Guericke University
Magdeburg, Germany

and

Jürgen Langner
Martin Luther University
Halle, Germany

SPRINGER SCIENCE+BUSINESS MEDIA, LLC

Library of Congress Cataloging-in-Publication Data

Cellular peptidases in immune functions and diseases / edited by Siegfried Ansorge and Jürgen Langner.
p. cm. -- (Advances in experimental medicine and biology ; v. 421)
Includes bibliographical references and index.

DOI 10.1007/978-1-4757-9613-1

1. Peptidase--Immunology. 2. Peptidase--Pathophysiology. I. Ansorge, Siegfried. II. Langner, Jürgen. III. Series.
QP609.P45C45 1997
616.07'9--dc21 97-21524
CIP

Proceedings of an International Conference on Cellular Peptidases in Immune Functions and Diseases, held November 3 – 5, 1996, in Magdeburg, Germany

Originally published by Plenum Press, New York in 1997
MyCopy version of the original edition 1997

http://www.plenum.com

10 9 8 7 6 5 4 3 2 1

PREFACE

This book stems from a conference held at Magdeburg-Herrenkrug, Germany, in November 1996 on "Cellular Peptidases in Immune Functions and Diseases." This symposium was designed to bring together scientists from diverse areas of expertise to discuss issues of newly identified relevance of proteolytic processes and their role as molecular regulators in the immune system and in diseases. The meeting was organized by the Sonderforschungsbereich 387 of the Deutsche Forschungsgemeinschaft "Zelluläre Proteasen, Bedeutung für Immunmechanismen und entzündliche Erkrankungen," which links research groups from the Otto von Guericke University Magdeburg and the Martin Luther University Halle.

It has become clear during the last decade that proteolysis, the processing and degradation of peptides and proteins, has to be considered as a special level of epigenetic control of practically all processes of life and—an understanding dating from the last three years or so—also of cell death or apoptosis. There is also increasing evidence that proteolysis does play a crucial role in all areas of immune functions as well as in inflammatory and neoplastic diseases. Cellular proteases have central functions in natural (nonspecific) as well as acquired (specific) immunity. They are involved in the cognitive phase (antigen processing and presentation), in the activation phase (e.g., generation and processing of cytokines, function of CD26, CD13, processing of transcription factors) as well as in the effector phase of the immune response (e.g., complement system, granzymes, elastase, proteinase 3, matrix metallo proteases). Consequently, most inflammatory processes are based on or are at least connected to the action of proteases. Thus, proteolysis must be considered as an integral event in immune response and in diseases with an immunological background. In the context of the topic of this meeting the question arises: Where are the roots of the knowledge on the mechanisms of proteolysis? It is often not easy to establish when something was first discovered and when something starts for the first time.[1] It is to some degree sure, however, that the development of research on the molecular mechanisms of protein and peptide degradation and processing started not very far from the place of this meeting, in Berlin. On January 6, 1906, just 91 years ago, **Emil Fischer** (Figure 1) held his famous lecture in the German Chemical Society "Investigations on Amino Acids, Polypeptides and Proteins" in which he described for the first time the architectural principles of protein structure, the fundamentals of the knowledge of proteins and peptides of today.[2] At the same time, there worked in his laboratory, the Kaiser-Wilhelm-Institut in Berlin, a young scientist from Switzerland, **Emil Abderhalden** (Figure 2). He was particularly interested in the mechanisms of physiological protein and peptide degradation. To our knowledge, he was the first who gave us a description of the molecular mechanisms of protein degradation, i.e., the hydrolysis of peptide bonds and the release of free amino acids and peptides. The first review he wrote, interestingly enough, was published in his textbook

Figure 1. Emil Fischer, 1852–1919.

on physiological chemistry,[3] which appeared in 1906 (Figure 3). Here, one can find a prototyped scheme on the different ways of proteolysis, the catabolic pathway to single, free amino acids on the one hand, and the proteolytic processing or limited proteolysis generating peptides on the other hand (Figure 4). Moreover, in this book one can find the prophecy that "proteolytic enzymes also act in tissues and cells, and many observations speak clearly to have us believe that degradation in them occurs in a very similar way as described already for trypsin. Not only the animal but also the plant cell has such enzymes." In 1911 Emil Abderhalden left Berlin. He was appointed as professor of physiology at the University of Halle/Saale, where he worked until 1945 mainly in the field of proteolysis. His pupil, **Horst Hanson** (Figure 5), nurtured this idea for more than 3 further decades until 1977 in the same institute in Halle, the Institute of Physiological Chemistry, under complicated and politically difficult conditions. He and his group were one of the first to bring together scientists from all over the world working with proteolytic enzymes in the Reinhardsbrunn Meetings of the 1970s and 1980s. Many of the colleagues who attended the recent meeting here in Magdeburg-Herrenkrug knew Horst Hanson and some of them were his pupils (Siegfried Ansorge, Heidrun Kirschke, Jürgen Langner, Bernd Wiederanders).

This, at least partially, may explain why this conference on "Cellular Peptidases in Immune Functions and Diseases" was organized in this region. We hope that this historical background will be a good justification to hold further meetings covering this growing scientific field here in the near future.

Figure 2. Emil Abderhalden, 1877–1950.

LEHRBUCH

DER

PHYSIOLOGISCHEN CHEMIE

IN DREISSIG VORLESUNGEN.

VON

EMIL ABDERHALDEN,

MIT 3 FIGUREN.

URBAN & SCHWARZENBERG

BERLIN WIEN

1906

Figure 3. The first page of the textbook on physiological chemistry, published by Emil Fischer in 1906.

ACKNOWLEDGMENTS

The devoted secretarial and excellent editorial assistance of Barbara Schotte, Gudrun Plexnies, and Christel Walcker is gratefully acknowledged.

Financial support for the conference was provided in part by: The Deutsche Forschungsgemeinschaft (Germany), the Fonds der Chemischen Industrie (Germany), the Ministerium für Kultur des Landes Sachsen-Anhalt and the companies Biermann GmbH/Bad Nauheim, Laboserv GmbH/Giessen, Inter-Ärzte-Versicherungen/Magdeburg, Becton Dickinson GmbH/Heidelberg, Mucos Pharma GmbH/Geretsried, PAA Laboratories/Cölbe,

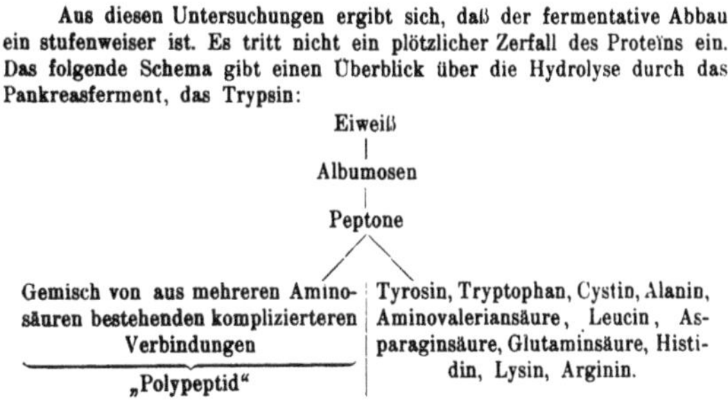

Aus diesen Untersuchungen ergibt sich, daß der fermentative Abbau ein stufenweiser ist. Es tritt nicht ein plötzlicher Zerfall des Proteïns ein. Das folgende Schema gibt einen Überblick über die Hydrolyse durch das Pankreasferment, das Trypsin:

Eiweiß

Albumosen

Peptone

Gemisch von aus mehreren Aminosäuren bestehenden komplizierteren Verbindungen — **„Polypeptid"**

Tyrosin, Tryptophan, Cystin, Alanin, Aminovaleriansäure, Leucin, Asparaginsäure, Glutaminsäure, Histidin, Lysin, Arginin.

Figure 4. Emil Abderhalden's view of enzymatic protein degradation. "From these investigations results can be seen that enzymatic degradation takes place in a stepwise manner. A sudden decay of the proteins does not occur. The following scheme shows a survey of hydrolysis by the pancreatic enzyme trypsin" (above). "Mixture of complicated compounds consisting of several amino acids" (left hand, below).

Figure 5. Horst Hanson, 1911–1978.

Pharmingen/Hamburg, Recker GmbH/Berlin, Carl Zeiss/Jena, Coulter-Immunotech/Hamburg, Immuno GmbH/Heidelberg, Bio-Rad/München, Stadtsparkasse Halle, Faust GmbH/Engelsdorf, Perkin Elmer/Markleeberg, Bayr. Vereinsbank/Halle, Pharmica Biotec/Freiburg i. Br.

S. Ansorge
Otto-von-Guericke-University Magdeburg
J. Langner
Martin-Luther-University Halle
1997

REFERENCES

1. Bond, J. S. , and A. J. Barrett, 1993, Proteolysis and protein turnover. Proceeding of the 9th ICOP Meeting, Williamsburg, Virginia, U. S. A. , Portland Press Proceedings. page XIV.
2. Hoesch, Kurt, 1921, Emil Fischer, sein Leben und sein Werk, Verlag Chemie, G.m.b.H. . Berlin und Leipzig, 480 pages.
3. Abderhalden, Emil, 1906, Lehrbuch der Physiologischen Chemie, Urbahn und Schwarzenberg, Berlin, Wien, 787 pages.

CONTENTS

Part I: Membrane Peptidases in Immune Functions and Inflammation

Part III: Peptidases Involved in Cytokine Actions and in the Pathogenesis of Disease

1

MEMBRANE METALLOENDOPEPTIDASES IN IMMUNE FUNCTION AND DISEASE

Judith S. Bond and Weiping Jiang

Department of Biochemistry and Molecular Biology
The Pennsylvania State University College of Medicine
Hershey, Pennsylvania 17033-0850

The enzymes that compose the 'Metallopeptidases' are a diverse group[1]. Forty-seven distinct evolutionary families of metallopeptidases have been identified in the last eight years; more than for any other protease classes, i.e., the serine/threonine, cysteine, or aspartic classes of proteases (see the Peptidase World Wide Web sites: http://www.qmw.ac.uk/~ugca000/iupac/enzyme/ and htpp://prolysis.phys.univ-tours.fr/Prolysis). In 1987, the primary amino acid sequence of very few metallopeptidases and of only one mammalian metalloendopeptidase (human fibroblast collagenase) were known, and the 3-dimensional structures of very few metallopeptidases (thermolysin, carboxypeptidase A and B) were solved[2,3]. Now hundreds of sequences of members of this Class are known, and many x-ray structures have been determined to high resolution (see, for example, reference 4). Thus our information about this class is expanding rapidly.

Proteases are broadly grouped into the categories of exopeptidases and endopeptidases. The action of exopeptidases is directed by the ends of a polypeptide chain, and they remove mono-, di- or tripeptides (amino- or carboxypeptidases), or they remove blocked or substituted NH_2- or COOH-terminal amino acid residues (for example, omega peptidases remove N-acyl amino acids). Endopeptidases cleave internally in a polypeptide chain. The exo- and endometalloproteases all have a metal atom (almost always zinc) tightly associated with the protein at the active site, and this zinc enhances the reactivity of a water molecule or a hydroxyl ion that cleaves a peptide bond in the substrate. In general, metalloproteases act optimally at neutral or alkaline pH values. The *Leishmania* cell surface metalloprotease (EC 3.4.24.36) is an exception to this rule; the parasite lives in lysosomes of host macrophages and this protease has an acid pH optimum[5]. There are metalloproteases in subcellular compartments, such as the cytosol (e.g., thimet oligopeptidase EC 3.4.24.15), and mitochondria (e.g., mitochondrial intermediate peptidase EC 3.4.24.59). However, the great majority of metallopeptidases described are secreted (e.g., thermolysin EC 3.4.24.27, collagenases, snake venom enzymes) or localized to the cell surface (e.g., membrane alanyl aminopeptidase EC 3.4.11.2, angiotensin I-converting enzyme EC 3.4.15.1, meprins EC 3.4.24.18 and 3.4.24.63, neprilysin EC 3.4.24.11).

Cellular Peptidases in Immune Functions and Diseases, edited by Ansorge and Langner
Plenum Press, New York, 1997

The integral membrane metalloproteases are thought to be critical for a variety of physiological and pathological processes such as growth factor activation and release (shedding), processing of secreted proteins, modification of receptors and adhesion molecules, degradation of biologically active peptides, hormones, and extracellular proteins, formation of amino acids for absorption, and generation of toxins (e.g., see reference 6, 7, 8). These processes are determinative in growth, differentiation, transformation, cell-cell interactions, and fertilization, as well as metastasis, Alzheimer's disease, and inflammatory processes. Some cell surface events involving metalloproteases are relatively well-defined in terms of the proteases involved, such as blood-pressure control (involving angiotensin I-converting enzyme)[3]. However, the specific proteases involved in many of these processes such as release of growth factors, ß-amyloid formation, activation of secreted proteins, release of selectin, CDs, receptors, and adhesion molecules are unknown. The structure, activities, and regulation of the proteases at the cell surface are only just being elucidated, however, it is clear that these proteases are highly regulated at the transcriptional and posttranscriptional level, and contain multidomains that that are critical for latency/activation, and interactions with proteins and substrates.

Several membrane-associated metalloendopeptidases have been cloned and sequenced in the 1990s, and examples of some of the best characterized enzymes in this group are presented in Table I. The enzymes generally have subunits that are integral membrane proteins of type I (transmembrane domain in the COOH-terminal region) or type II (transmembrane domain in the NH_2-terminal region), that span the membrane once. Leishmanolysin represents a protein that is anchored by glycosylphosphatidylinositol (GPI). Neprilysin (EC 3.4.24.11, also called common acute lymphocytic leukemia antigen or CALLA, endopeptidase 24.11, neutral endopeptidase, NEP, enkephalinase) and endothelin-converting enzyme 1 (ECE, EC 3.4.24.71) are members of Clan MA, or are referred to as Gluzincins[1, 18]. A Clan is a group of families that contain distinct evolutionary relationships in the catalytic residues and tertiary structures of the catalytic domain. Members of Clan MA or Gluzincins contain metalloprotease families that have **HEXXH...E** as

Table I. Representative integral membrane metalloendopeptidases

Enzyme	Anchor	Substrate/specificity	Subunit	Reference
Neprilysin	type II	Peptides (e.g., enkephalins) on the N-terminal side of hydrophobic residues; does not cleave proteins	94 kDa	9, 10
Endothelin-converting enzyme 1	type II	Cleaves Trp-Val bond in big endothelin to produce endothelin	S-S linked homodimer; each subunit: 130 kDa	11
Leishmanolysin	GPI	Proteins (e.g., gelatin, fibrinogen, casein) and peptides	63 kDa	5
Meprin A	type I	Proteins and peptides, e.g. azocasein, gelatin, α-melanocyte-stimulating hormone, parathyroid hormone, and bradykinin	Homo- or heterotetramer of S-S linked dimers; α subunit: 80-100 kDaß subunit: 80-110 kDa	12, 13
Meprin B	type I	Proteins such as azocasein and gelatin, and clips off the carboxyl terminal tail of the catalytic subunit of secreted protein kinase A	Homotetramer of S-S linked dimers; ß subunits: 80-110 kDa	14, 15
MT-MMP-1	type I	Activates pro-gelatinase A	63 kDa	16
Membrane-associated metalloproteinase	type I	Proteins such as myelin basic protein	58 kDa	17

the zinc-binding motif. Neprilysin cleaves many oligopeptides such as enkephalins, and does not hydrolyze proteins. It is thought to metabolize biologically active peptides at the cell surface and have a role in lymphocytic function. ECE converts big endothelin to endothelin, a potent vasoactive peptide that functions in blood pressure control.

The type I metalloendopeptidases are all members of a superfamily termed the 'Metzincins,'[4] or 'Clan MB.'[1] The Metzincins can be further subdivided into four families: astacins, reprolysins, serralysins, matrixins[1, 18, 19]. There have been some exciting advances recently indicating that these metalloendopeptidases are involved in the 'shedding' of selectins and tumor necrosis factor-α (TNFα), molecules implicated in inflammatory processes[20], in processing of procollagen[21], and in the activation of secreted matrix metalloproteases[22]. The active sites of the four families are related in terms of zinc binding ligands (all use three histidines in the motif **HE**XX**H**XXXXX**H**), and tertiary structure (ß sheet structure, and a Met-turn near the active site). The meprins are the only members of the astacin family that contain membrane-spanning regions, though it has been noted that several other secreted members of this family act in association with plasma membranes[23]. For example, bone morphogenetic protein-1 in sea urchins and procollagen C-endopeptidase in mammalian tissues are secreted enzymes of the astacin family but are found at the cell surface where they function. Meprins are implicated in degradation of parathyroid hormone (PTH), α-melanocyte-stimulating hormone (MSH), transforming growth factor-α, and collagen/gelatin, as well as in processing of urinary proteins in rodents, and inactivation of secreted protein kinases[12, 15, 24–28]. Several membrane-type matrix metalloproteinases (MT-MMPs) have been identified, and these enzymes are proposed to be important in the activation of secreted matrixins (matrix metalloproteinases)[16, 29]. Perhaps the complexes that the MT-MMPs form with secreted enzymes are

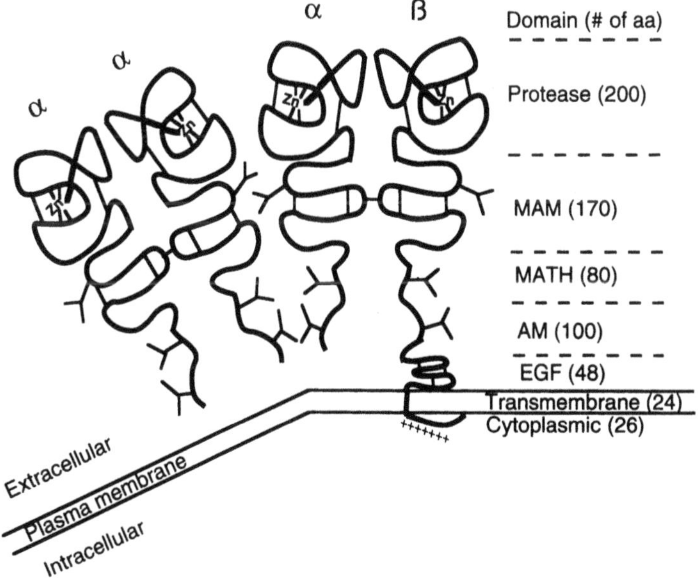

Figure 1. Diagram of a meprin heterotetramer. A heterotetrameric form of meprin A ($\alpha_3\beta_1$) is shown. The ß subunit spans the membrane; the α subunits are linked to the membrane via S-S bridges or adhesion to the anchored dimer. The signal and prosequences for the subunits, as well as COOH-terminal domains of the α subunit, have been cleaved off during maturation. Vertical lines connecting portions of the primary amino acid sequence indicate S-S bridges, Y-shaped structures indicate oligosaccharides, pluses near the cytoplasmic domain indicate positively charged amino acids.

important for physiological function of this family of enzymes at the cell surface. The disintegrins, including enzymes termed membrane-associated metalloproeinase (Table I), KUZ, fertilin (PH-30), MS2 antigen, meltrin, are membrane-bound members of the reprolysin family;[17,30–34] a family that contains a large number of secreted snake venom metalloendopeptidases. These enzymes have been implicated in fertilization, release of TNFα from the cell surface, and degradation of myelin basic protein.

The membrane metalloendopeptidases have not been directly implicated in immune system functions. However, there are several properties/characteristics of these enzymes, and particularly meprins, that relate these enzymes to the immune system, as discussed below.

Meprins are composed of two evolutionarily-related subunits, the α and ß subunits, and exist as homo- or hetero-oligomeric complexes.[14,23,35,36] The cDNAs for the two subunits predict similar domain structures, and both are initially synthesized as membrane-spanning type I proteins[37, 38]. However, proteolytic processing during biosynthesis differs for the two subunits. Both subunits are synthesized with the following domains: signal (S), prosequence (Pro), catalytic (Protease), adhesion (Meprin, A-5 protein, receptor protein-tyrosine phosphatase μ, or MAM[39]), interaction domains (Meprin And TRAF Homology or MATH[40]; AfterMATH or AM; epidermal growth factor-like or EGF-like), transmembrane (TM), cytoplasmic (C). During biosynthesis the signal sequence is removed, and in some instances the prosequence is removed. The latter process (removal of the prosequence) is cell-specific, and determines whether the protein will be catalytically active or not. Proteolytic processing in the COOH-terminus differs for the two subunits. The α subunit is synthesized with a 56-amino acid sequence, called the I or Inserted domain, between the AM and EGF-like domains that is essential for COOH-terminal processing[38]. The subunit is proteolytically processed, in or near the I domain, in the endoplasmic reticulum, and secreted as a homooligomer if expressed in the absence of the meprin ß subunit. In the presence of meprin ß, the α subunit is retained at the cell surface due to disulfide bridging to ß or adhesion of the α homodimer to a ß-containing dimer. The subunits are highly glycosylated (25 - 30% of the subunit estimated to be carbohydrate). This is a unique structure for a protease, and reminiscent of many cell surface molecules such as integrins, receptors, and IgG/HMC molecules.

The meprin α subunit contains a seven amino acid sequence (YNCTATH) that is a signature sequence for IgG/MHC molecules[35, 41]. In addition, this subunit is linked to the major histocompatibility complex (MHC). The original localization of the meprin α gene resulted from observations of differential expression of this subunit in kidneys of inbred mice, mice with defined MHC genotypes[42]. Subsequent physical mapping of the structural genes for meprin subunits showed that the meprin α gene is on mouse chromosome 17 telomeric to the H-2 complex, and on human chromosome 6p21.2-p21.1[43]. The meprin ß structural gene is on chromosome 18 of both the mouse and human genome[44].

Expression of meprin subunits is particularly high in proximal tubule brush border membranes of kidney in rodents, and in brush border membranes of intestine in rodents and humans[45]. In the rodent kidney, meprins have been implicated in the processing of urinary peptides and proteins, and these polypeptides may be important for pheromone-binding [27, 45, 46]. This may have consequences in mating, marking territories, and identification of young. It has been known that mating preferences are linked to the MHC in rodents[47], and meprins may be a link. In the intestine, meprins are more highly expressed in ileum than duodenum[48]. The ileum is associated with the intestinal immune system; perhaps these cell surface proteases play a role in the hydrolysis and formation of peptides that are then presented to the antigen producing cells.

There is no direct evidence that meprins participate in immune system processes, or evidence for expression of the subunits in immune system cells. However, the similarities of the meprin structures to IgG/MHC molecules, the linkage with genes of the MHC, the cell surface location, affinity of the isoforms for proteins and polypeptides, and the possibilities that the enzymes are involved in processes of self/nonself recognition are intriguing possible links with the immune system.

REFERENCES

1. Rawlings ND, and Barrett AJ (1995) Evolutionary families of metallopeptidases. Methods Enzymol 248:183–228
2. Bond JS, and Butler PE (1987) Intracellular proteases. Annu Rev Biochem 56:333–364
3. Bond JS, and Beynon RJ (1987) Proteolysis and physiological regulation. Molec Aspects Med 9:173–287
4. Stöcker W, Grams F, Baumann U, Reinemer P, Gomis-Rüth FX, McKay DB, and Bode W (1995) The metzincins - Topological and sequential relations between the astacins, adamalysins, serralysins, and matrixins (collagenases) define a superfamily of zinc-peptidases. Protein Sci 4:823–840
5. Bouvier J, Schneider P, and Etges R (1995) Leishmanolysin: Surface metalloproteinase of Leishmania. Methods Enzymol 248:614–633
6. Ehlers MRW, and Riordan JF (1991)Membrane prtoeins with soluble counterparts: Role of proteolysis in the release of transmembrane proteins. Biochemistry 30:10065–10074
7. Massagué J, and Pandiella A (1993) Membrane-anchored growth factors. Annu Rev Biochem 62:515–541
8. Rose-John S, and Heinrich PC (1994) Soluble receptors for cytokines and growth factors: Generation and biological function. Biochem J 300:281–290
9. Beaumont A, Fournie-Zaluski MC, and Roques BP (1996) Neutral endopeptidase-24.11: structure and design and clinical use of inhibitors. In: Zinc Metalloproteases in Health and Disease, NM Hooper, ed, Taylor & Francis Ltd, London, pp. 105–129
10. Li C, and Hersh LB (1995) Neprilysin: Assay methods, purification, and characterization. Methods Enzymol 248:253–263
11. Turner AJ, and Tanzawa K (1996) Endothelin converting enzyme: structure and localization. In: Zinc Metalloproteases in Health and Disease, NM Hooper, ed, Taylor & Francis Ltd, London, pp. 311–331
12. Wolz RL, and Bond JS (1995) Meprins A and B. Methods Enzymol 248:325–345
13. Marchand P, and Bond JS (1996) Structure and membrane association of mouse and rat meprins. In: Intracelluar Protein Catabolism, K Suzuki and JS Bond, eds, Plenum Press, New York, pp. 13–22
14. Gorbea CM, Marchand P, Jiang W, Copeland NG, Gilbert DJ, Jenkins NA, and Bond JS (1993) Cloning, expression, and chromosomal localization of the mouse meprin ß subunit. J Biol Chem 268:21035–21043
15. Chestukhin A, Muradov K, Litovchick L, and Shaltiel S (1996) The cleavage of protein kinase A by the kinase-splitting membranal proteinase is reproduced by meprin ß. J Biol Chem 271:30272–30280
16. Sato H, Takino T, Okada Y, Cao J, Shinagawa A, Yamamoto E, and Seiki M (1994) A matrix metalloproteinase expressed on the surface of invasive tumour cells. Nature 370:61–65
17. Chantry A, Gregson NA, and Glynn P (1989) A novel metalloproteinase associated with brain myelin membranes. J Biol Chem 264:21603–21607
18. Hooper NM (1994) Families of zinc metalloproteases. FEBS Lett 354:1–6
19. Jiang W, and Bond JS (1992) Families of metalloendopeptidases and their relationships. FEBS Lett 312:110–114
20. Bennett TA, Lynam EB, Sklar LA, and Roge S (1996) Hydroxamate-based metalloprotease inhibitor blocks shedding of L-selectin adhesion molecule from leukocytes: Functional consequences for neutrophil aggregation. J Immun 156:3093–3097
21. Kessler E, Takahara K, Biniaminov L, Brusel M, and Greenspan DS (1996) Bone morphogenetic protein-1: The type I procollagen C-proteinase. Science 271:360–362
22. Strongin AY, Collier I, Bannikov G, Marmer BL, Grant GA, and Goldberg GI (1995) Mechanism of cell surface activation of 72-kDa type IV collagenase. J Biol Chem 270:5331–5338
23. Bond JS, and Beynon RJ (1995) The astacin family of metalloendopeptidases. Protein Sci 4:1247–1261
24. Choudry Y, and Kenny AJ (1991) Hydrolysis of transforming growth factor-α by cell-surface peptidases *in vitro*. Biochem J 280:57–60

25. Stephenson SL, and Kenny AJ (1988) The metabolism of neuropeptides: Hydrolysis of peptides by the phosphoramidon-insensitive rat kidney enzyme 'endopeptidase-2' and by rat microvillar membranes. Biochem J 255:45–51
26. Kaushal GP, Walker PD, and Shah SV (1994) An old enzyme with a new function: Purification and characterization of a distinct matrix-degrading metalloproteinase in rat kidney cortex and its identification as meprin. J Cell Bio. 126:1319–1327
27. Bond JS, and Beynon RJ (1986) Meprin: A membrane-bound metalloendo-peptidase. Curr Top Cell Regul 28:263–290
28. Yamaguchi T, Kido H, Fukase M, Fujita T, and Katunuma N (1991) A membrane-bound metallo-endopeptidase from rat kidney hydrolyzing parathyroid hormone. Eur J Biochem 200:563–571
29. Takino T, Sato H, Shinagawa A, and Seiki M (1995) Identification of the second membrane-type matrix metalloproteinase (MT-MMP) gene from a human placenta cDNA library. J Biol Chem 270:23013–23020
30. Fox JW, and Bjarnason JB (1996) The reprolysins: a family of metalloproteinases defined by snake venom and mammalian metalloproteinases. In: Zinc Metalloproteases in Health and Disease, NM Hooper, ed. Taylor & Francis Ltd, London, pp. 47–81
31. Rooke J, Pan D, Xu T, and Rubin GM (1996) KUZ, a conserved metalloprotease-disintegrin protein with two roles in *Drosophila* neurogenesis. Science 273:1227–1231
32. Howard L, and Glynn P (1995) Membrane-associated metalloproteinase recognized by characteristic cleavage of myelin basic protein: Assay and isolation. Methods Enzymol 248:388–395
33. Yagami-Hiromasa T, Sato T, Kurisaki T, Kamijo K, Nabeshima Y, and Fujisawa-Sehara (1995) A metalloprotease-disintegrin participating in myoblast fusion. Nature 377:652–656
34. Wolfsberg TG, Straight PD, Gerena RL, Huovila APJ, Primakoff P, Myles DG, and White JM (1995) ADAM, a widely distributed and developmentally regulated gene family encoding membrane protein with a disintegrin and metalloprotease domain. Dev Biol 169:378–383
35. Jiang W, Gorbea CM, Flannery AV, Beynon RJ, Grant GA, and Bond JS (1992) The α subunit of meprin A: Molecular cloning and sequencing, differential expression in inbred mouse strains and evidence for divergent evolution of the α and ß subunits. J Biol Chem 267:9185–9193
36. Gorbea CM, Flannery AV, and Bond JS (1991) Homo- and heterotetrameric forms of the membrane-bound metalloendopeptidases meprin A and B. Arch Biochem Biophys 290:549–553
37. Marchand P, Tang J, and Bond JS (1994) Membrane association and oligomeric organization of the α and ß subunits of mouse meprin A. J Biol Chem 269:15388–15393
38. Marchand P, Tang J, Johnson GD, and Bond JS (1995) COOH-terminal proteolytic processing of secreted and membrane foms of the α subunit of the metalloprotease meprin A: Requirement of the I domain for processing in the endoplasmic reticulum. J Biol Chem 270:5449–5456
39. Beckmann G, and Bork P (1993) An adhesive domain detected in functionally diverse receptors. Trends in Biochem Sci 18:40–41
40. Uren AG, and Vaux DL (1996) TRAF proteins and meprins share a conserved domain. Trends in Biochem Sci 21:244–245
41. Bairoch A (1990) PROSITE: A Dictionary of Protein Sites and Patterns, 6th release, University of Geneva Switzerland
42. Bond JS, Beynon RJ, Reckelhoff JF, and David CS (1984) Mep-1 gene controlling a kidney metalloendopeptidase is linked to the major histocompatibility complex in mice. Proc Natl Acad Sci USA 81:5542–5545
43. Jiang W, Dewald G, Brundage E, Mücher G, Schildhaus H-U, Zerres K, and Bond JS (1995) Fine mapping of MEP1A, the gene encoding the α subunit of the metalloendopeptidase meprin, to human chromosome 6p21. Biochem Biophys Res Comm 216:630–635
44. Bond JS, Rojas K, Overhauser J, Zoghbi HY, and Jiang W (1995) The structural genes MEP1A and MEP1B, for the α and ß subunits of the metalloendopeptidase meprin map to human chromosome 6p and 18q, respectively. Genomics 25:300–303
45. Craig SS, Reckelhoff JF, and Bond JS (1987) Distribution of meprin in kidneys from mice with high- and low-meprin activity. Am J Physiol 253:C535-C540
46. Beynon RJ, Oliver S, and Robertson DHL (1996) Characterization of the soluble, secreted form of urinary meprin. Biochem J 315:461–466
47. Boyse EA, Yanakazi K, Yamaguchi M, and Thomas L (1980) Sensory communication amoung mice according to their MHC types. In: The Immune System: Functions and Therapy of Dysfunction, G Doria and A Kshkol, eds, Academic Press, New York, pp 45–53
48. Bankus JM, and Bond JS (1996) Expression and distribution of meprin protease subunits in mouse intestine. Arch Biochem Biophys 331:87–94

2

STRUCTURAL STUDIES OF AMINOPEPTIDASE P

A Novel Cellular Peptidase

Anthony J. Turner,* Ralph J. Hyde, Jaeseung Lim, and Nigel M. Hooper

Department of Biochemistry and Molecular Biology
University of Leeds
Leeds LS2 9JT, United Kingdom

1. INTRODUCTION

The plasma membrane of many cell types contains a cohort of peptidases that serve to modulate the activity of circulating regulatory peptides. Many of these are zinc metallopeptidases and these enzymes are particularly abundant in the brush border membranes of renal and intestinal microvilli. Some of them such as neprilysin (NEP; EC 3.4.24.11) and aminopeptidase N (AP-N; EC 3.4.11.2) also exist as cluster differentiation antigens (CD10 and CD13 respectively) on the surface of leukocytes and may play a role in regulation of the immune system[1]. Both NEP and AP-N also have roles in the metabolism of certain cardiovascular and neuropeptides, e.g. natriuretic peptides[2] and enkephalins[3]. NEP and AP-N are type II integral membrane proteins and possess the typical HExxH zincin motif[4] characteristic of many zinc peptidases. Recently, we have focused on another brush border zinc peptidase, aminopeptidase P (AP-P; X-Pro aminopeptidase; EC 3.4.11.9) which we showed to be unusual among the membrane peptidases in being anchored to the membrane by a glycosyl-phosphatidylinositol (GPI) moiety[5]. AP-P also shows a number of other atypical features which led us to attempt the molecular cloning of the pig kidney enzyme.

2. PROLINE-SPECIFIC PEPTIDASES

The Xaa-Pro sequence is found at the N-terminus of more than 100 naturally occurring peptides and proteins, including peptide hormones, kinins, neurotransmitters, toxins and mediators of the immune response[6]. The presence of the Xaa-Pro motif confers upon a peptide considerable resistance to non-specific proteolytic breakdown because the conformational

* Corresponding author: Professor A. J. Turner, Department of Biochemistry and Molecular Biology, University of Leeds, Leeds, LS2 9JT U.K. Tel: (0113) 233 3131; fax: (0113) 242 3187; E-mail: a.j.turner@leeds.ac.uk.

Cellular Peptidases in Immune Functions and Diseases, edited by Ansorge and Langner
Plenum Press, New York, 1997

constraint induced by the cyclic imino acid restricts proteolysis. Nature has, however, designed a small number of highly specific enzymes that can cleave such bonds[7,8]. Three distinct serine peptidases are able to cleave on the C-terminal side of a prolyl residue: the exopeptidases dipeptidyl peptidase IV (CD26; EC 3.4.14.5) and lysosomal prolyl carboxypeptidase (EC 3.4.16.2) and the unique mammalian endopeptidase, prolyl oligopeptidase (EC 3.4.21.26). There are three proline-specific metallopeptidases: AP-P, membrane carboxypeptidase P and the dipeptidase, prolidase, which all share the feature that they are activated by manganese ions. AP-P, dipeptidyl peptidase IV and carboxypeptidase P are all integral membrane proteins existing as ectoenzymes. This collection of proline-specific enzymes is therefore able to act in concert to initiate the hydrolysis of proline-containing dietary or biologically active peptides in a controlled fashion at the cell-surface.

3. AMINOPEPTIDASE P

3.1 Location and Physiological Roles

AP-P is a ubiquitous enzyme found in organisms as diverse as bacteria, yeast, fish and mammals. In mammals, AP-P is principally located in the brush border membranes of kidney and intestine[9,10]. A membrane-associated AP-P activity has also been identified in the lungs of several species[11,12]. A soluble form of AP-P has been characterized in serum[13] which is most likely derived by cleavage of the membrane bound form. A cytosolic, intracellular form of AP-P has been identified in a number of tissues, especially brain[14], and also in human platelets[15] and activated lymphocytes[16,17]. The cytosolic AP-P appears to be closely related in properties to the membrane form but is likely to represent a distinct gene product.

Mammalian AP-P was first identified as an activity in pig kidney able to hydrolyse peptides of the type Xaa-Pro- and a synthetic peptide substrate (glycyl-prolyl-hydroxyproline; Gly-Pro-Hyp) was routinely used to monitor activity[18]. This specificity led to the suggestion that AP-P could have a role in the degradation of collagen, because of its high degree of repeated Gly-Pro-Y sequences[18]. The association of AP-P deficiency with an inherited metabolic disease characterized by massive urinary excretion of imino-oligopeptides such as Gly-Pro-Hyp-Gly[19] provides some support for AP-P participation in collagen metabolism. However, exposure of susceptible internal Gly-Pro-Y sequences to AP-P hydrolysis would require the initial action of an appropriate endopeptidase. In recent years the relationship between AP-P and the turnover of collagen has been little explored and attention has rather focused on the role of the enzyme in the processing and metabolism of regulatory peptides, especially the vasodilator bradykinin.

Bradykinin is susceptible to hydrolysis by a number of peptidases *in vitro*, especially angiotensin converting enzyme (ACE) and AP-P (Fig. 1). In an attempt to assess the relative contributions of these and other activities, Prechel et al[20] have used a newly developed inhibitor of AP-P, apstatin (N-{(2S,3R)-3-amino-2-hydroxy-4-phenylbutanoyl}-L-prolyl-L-alaninamide), which exhibits a K_i of 2.6 μM towards the rat lung enzyme and is relatively inactive against other cell-surface peptidases. (^{3}H)-bradykinin was perfused through the isolated rat lung in the presence or absence of apstatin and/or an ACE inhibitor. Analysis of the perfusate by HPLC revealed that the bradykinin-degrading activity in the rat pulmonary vascular bed can be fully accounted for by a combination of AP-P (30%) and ACE (70%). Thus, inhibitors of AP-P may have therapeutic potential in the treatment of certain cardiovascular disorders although it may be premature to extrapolate from bradykinin metabolism in the rat to the human

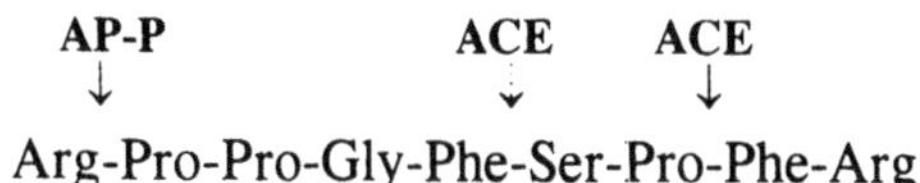

Figure 1. The amino acid sequence of bradykinin and sites of hydrolysis by AP-P and angiotensin converting enzyme. Full arrow indicates primary sites of cleavage. Broken arrow indicates secondary site of cleavage by angiotensin converting enzyme.

clinical situation. Other regulatory peptides susceptible to N-terminal processing by AP-P include members of the pancreatic polypeptide family (neuropeptide Y, peptide YY)[21,22]. Substance P can also be hydrolysed by AP-P *in vitro*[23] although there is no direct evidence for a role in inactivation of this neuropeptide *in vivo*; rather this appears to be due principally to the action of NEP[24]. An AP-P activity in human leukocytes has been characterised by Rusu and Yaron[25]. They were able to demonstrate a possible involvement of AP-P in the inflammatory process and immune response, through the degradation of interleukin-6. Furthermore, AP-P activity increased upon activation of the immune response with the mitogen, phytohaemagglutinin (PHA), as did DPP IV activity. We have similarly observed a marked increase in the transcription of AP-P and DPP IV messages upon activation of human T cells by PHA together with a phorbol ester over a period of 48 h (U. Lendeckel, R. J. Hyde, S. Ansorge and A.J. Turner, unpublished observations). Thus, like a number of other cell-surface peptidases, AP-P may have a role in modulation of the immune response. The large number of cytokines with an N-terminal Xaa-Pro- sequence reinforces this suggestion although most of these have not yet been demonstrated as substrates for AP-P. The development of more potent and selective AP-P inhibitors than apstatin will be required to evaluate the significance of the suggested physiological roles for AP-P.

3.2 Characterization of AP-P

Initial indications that AP-P was a metallopeptidase were provided by its sensitivity to inhibition by chelating agents, such as EDTA and 1,10-phenanthroline[11,26,27]. Subsequently, pig kidney AP-P was shown to contain one mol of zinc per mol of protein[28]. However, AP-P also displays a requirement for additional metal ions. The initial studies on the enzyme by Dehm and Nordwig[18] indicated that the addition of Mn^{2+} ions was required for maximal activity. Assays using Gly-Pro-Hyp as a substrate have therefore generally included Mn^{2+} in the assay buffer, a concentration of 4–10 mM proving optimal. A more detailed investigation of the pig kidney enzyme requirements indicated that Mn^{2+} and Co^{2+}, and to a lesser extent Ca^{2+} and Mg^{2+}, stimulated enzyme activity, whereas other cations e.g. Cu^{2+}, Ni^{2+} and Zn^{2+} were inhibitory[27]. The substrate specificity of AP-P was also examined in some detail in the thorough study of Dehm & Nordwig[18]. The hydrolysis of Gly-Pro-Y showed the fastest reaction rates of those peptides tested of the X-Pro-Y type. Studies on the lung enzyme[11,29] further established the requirement for proline in the P1′ position, although a much slower rate of hydrolysis could be detected when Pro was replaced with Ala. These studies indicated that Arg-Pro-Pro was cleaved most rapidly and that Gly-Pro-Hyp was in fact a relatively poor substrate. However, Mn^{2+} was not included in the assay buffer and this could account for the discrepancies seen between these studies and the original observations of Dehm and Nordwig[18]. The effect of Mn^{2+} on enzyme activity appears to be a substrate-dependent phenomenon as subsequently demonstrated in two recent studies[23,30]. Hydrolysis of certain substrates is substantially activated by Mn^{2+} (Gly-Pro-Hyp, substance P) whereas the hydrolysis of certain other substrates (e.g.

bradykinin, Arg-Pro-Pro) shows no activation at any concentration of the cation and concentrations above 10^{-5}M were inhibitory. There appears to be no obvious relationship between the N-terminal amino acid sequence of a particular substrate and the effect of Mn^{2+} on catalysis.

AP-P also demonstrates an unusual inhibitor profile which again is influenced by the presence of Mn^{2+}. Unlike many other aminopeptidases, AP-P is not inhibited by actinonin, amastatin, bestatin or puromycin, although certain ACE inhibitors (e.g. enalaprilat and ramiprilat) at micromolar concentrations can inhibit the hydrolysis of Gly-Pro-Hyp by AP-P[28]. The degradation of bradykinin by AP-P is partially inhibited by the ACE inhibitor enalaprilat but only when Mn^{2+} is included in the assay buffer[23,30]. These various anomalous effects on catalysis together with the distinct kinetic behaviours of bradykinin and Gly-Pro-Hyp[23], suggested the possibility that AP-P might possess two active sites. However, other explanations relating to the unique properties of the Xaa-Pro bond may be of significance, for example the isomeric state of the bond[8]. The various proline-specific peptidases will only cleave when the peptide bond preceding the prolyl residue is in the *trans* conformation and *cis-trans* isomerisation may therefore be a determining factor.

3.3 Mode of Anchorage of AP-P

Bovine lung[11] and rat intestinal[10] AP-P were both shown to have the characteristics of an integral membrane protein. Subsequently, we were able to demonstrate the efficient solubilisation of AP-P activity from human and pig kidney membranes by the action of phosphatidylinositol-specific phospholipase C (PI-PLC) from both *Staphylococcus aureus* and *Bacillus thuringiensis* which suggested the presence of a glycosyl-phosphatidylinositol (GPI) anchor on the protein[5]. The presence of a GPI-anchor on pig kidney AP-P was also indicated by the differential solubilisation of the enzyme by a range of detergents. Brush border GPI-anchored proteins are selectively solubilized only by detergents of high critical micellar concentration (CMC), such as octyl glucoside, and this phenomenon was also observed with AP-P[5,31]. The release of AP-P by phospholipase D (PLD) activity in plasma[32] also helped to establish that anchorage to the lipid bilayer is through a GPI-moiety. The release of AP-P from pig kidney membranes in a hydrophilic form by PI-PLC proved a useful starting point for isolation of the enzyme which was first achieved in 1990[27]. The release of AP-P by PI-PLC has subsequently been used in the purification of the enzyme from bovine lung[29], rat lung[12] and intestine[33] and guinea pig kidney[34]. PI-PLC-cleaved and purified AP-P expresses the anti-CRD epitope typical of GPI-anchored proteins[35]. AP-P is therefore one of only three recognised GPI-anchored cell-surface peptidases, the others being membrane dipeptidase and carboxypeptidase M[36].

4. MOLECULAR CLONING AND EXPRESSION OF AP-P

We have now isolated a cDNA clone encoding AP-P from a porcine kidney cortex cDNA library[37]. Degenerate PCR primers were used to amplify fragments from single stranded cDNA synthesised from RNA isolated from pig kidney cortex. Cloning and sequencing of these products provided sequence information enabling the design of specific PCR primers. Specific PCR reactions were used to identify sub-libraries enriched in AP-P clones. Plaque hybridisation of one of these sub-libraries resulted in the isolation of a full length clone of 3563 bp which was sequenced completely. The full sequence of the clone (GenBank/EMBL/DDBJ Nucleotide Sequence Database accession number U55039) comprises

345 nucleotides of 5′- untranslated sequence, an open reading frame of 2019 nucleotides, and 1199 nucleotides of 3′- untranslated sequence, including 24 nucleotides of the poly(A) tail. The open reading frame encodes a protein of 673 amino acids with a calculated M_r of 75 755 which compares with the estimated M_r of 91 000 for the purified protein. Thus, approximately 17% of the mass of the mature protein is due to glycosylation, consistent with the presence of six possible N-linked glycosylation sites. A hydropathy plot of the predicted amino acid sequence using the Kyte and Doolittle algorithm[38] identifies strong hydrophobic regions at both termini of the protein. The region at the N-terminus is typical of a cleavable signal sequence, responsible for targeting the protein to the cell surface. Analysis of this N-terminal region by using the weight-matrix method of von Heijne[39] identifies two possible cleavage sites. If Lys24 is considered as the N-terminus, the highest 'score' (6.70) is generated. However, this position puts Pro23 at the -1 position which is unusual in eukaryotic signal sequences. When His22 is considered as the N-terminal residue a lower 'score' is seen (6.15) but the important positions of -1 and -3 are occupied by more favourable amino acids. At present, therefore, we cannot distinguish between these two possibilities for the N-terminus particularly since we find the N-terminus of the purified protein to be blocked.

The C-terminal hydrophobic domain is consistent with a signal region for attachment of a GPI-anchor. The signal sequence for pig kidney AP-P shows the hydrophobic domain separated from the point of GPI-anchor attachment by a spacer region. The cleavage prediction criteria established by Udenfriend and Kodukula[40] would predict that Ala649 is the residue to which the pre-formed GPI-moiety is attached. This would therefore place Arg650 and Ala651 at the $\omega + 1$ and $\omega + 2$ positions respectively. Fig. 2 shows an alignment of several GPI attachment sites from a range of proteins, including that for pig kidney AP-P.

Expression of the cloned pig kidney AP-P in COS-1 cells[37] has been achieved. The expressed enzyme can hydrolyse both Gly-Pro-Hyp and bradykinin to the correct products. The

Enzyme	Species/Tissue	C-terminal sequence
Alkaline phosphatase	Human placenta	. . . AGTT**D**AAHPGRSVVPALLPLLAGTLLLLETATAP
Membrane dipeptidase	Human kidney	. . . YGYS**S**GASSLHRHWGLLLASLAPLVLCLSLL
Membrane dipeptidase	Pig kidney	. . . YGY**S**.AAPSLHLPPGSLLASLVPLLL.LSLP
5'-Nucleotidase	Rat liver	. . . RIKF**S**AASHYQGSFPLILLSFWAVILVLYQ
5'-Nucleotidase	Human placenta	. . . RIKF**S**TGSHCHGSFSLIFLSLWAVIFVLYQ
Aminopeptidase P	Pig kidney	. . . EPLS**A**RAAPTTSLGSLMTVSALAILGWSV

Figure 2. Alignment of the deduced C-terminal sequences of some brush border GPI-anchored hydrolases. The bold amino acids indicate the positions that have been identified as the site of GPI-anchor addition (ω residue; Udenfriend & Kodukula[40]). The underlined sections represent the C-terminal hydrophobic sequences in each case. The data are assembled from refs. 37, 48 and 49. The porcine and human membrane dipeptidase sequences have been aligned to give maximum homology by the introduction of two gaps in the porcine sequence as in[49].

hydrolysis of Gly-Pro-Hyp was stimulated by Mn^{2+} and inhibited by the ACE inhibitor enalaprilat. Inhibition of substrate hydrolysis was observed with the specific AP-P inhibitor, apstatin, and the metal chelator EDTA. Expression of AP-P at the cell surface of the COS-1 cells was revealed by immunocytochemistry. The presence of a GPI-anchor on the expressed protein was demonstrated by showing release from the membrane with PI-PLC treatment. Western blotting using a polyclonal antibody raised against pig kidney AP-P recognises a single polypeptide of approximate M_r 91 000 in membranes prepared from COS-1 cells expressing the AP-P clone. In conclusion, a fully functional AP-P was expressed in COS-1 cells that showed the expected kinetic and immunological properties of the previously purified enzyme.

5. AP-P IS A MEMBER OF THE PROLINE PEPTIDASE FAMILY

Prior to the successful cloning of pig kidney AP-P, a limited partial sequence of guinea pig AP-P had been obtained by Denslow et al.[41] which suggested that the mammalian enzyme may be related to members of the so-called 'proline peptidase' family. This classification derived from an earlier study[42] involving comparisons of the sequences of *E. coli* AP-P and methionine aminopeptidase, together with prolidase, creatinase, and agropine synthase. Although not all these proteins have peptidase activity, they share a common structural feature referred to as the 'pita-bread fold' as well as some other conserved regions. The assignment of mammalian AP-P to the proline peptidase family was confirmed when Vergas Romero et al.[43] were able to obtain a practically complete amino acid sequence of AP-P purified from pig kidney by a combination of Edman degradation and mass spectrometry. The sequence similarities between pig kidney AP-P and the other members of the proline peptidase family are restricted to the C-terminal half of the protein. This suggests that the active site of AP-P lies within this region. A partial alignment of some members of the proline peptidase family is shown in Fig. 3.

Although pig kidney AP-P has been shown to contain 1 mol of zinc per mol of enzyme[28], no recognisable zinc binding motif is apparent in the primary amino acid sequence and the enzyme therefore represents a novel zinc peptidase. The requirement of Mn^{2+} ions for optimal activity with some substrates suggests a dual ion dependence of the enzyme. Intriguingly, other members of the proline peptidase family also show dual metal ion character. It has been proposed that two metal ions are required at the active site of prolidase[44]. *E.coli* methionine aminopeptidase contains a pair of closely linked metal ions (Co^{2+} ions are reported to give optimal activity)[45]. The eukaryotic methionyl aminopeptidase, although only distantly related to the prokaryotic enzyme, is also cobalt dependent[46]. However, there is no Co^{2+} present in pig kidney AP-P[23]. The nature of any additional metal ion is therefore unknown at present.

Elucidation of active site residues in AP-P will require a combination of chemical modification and site-directed mutagenesis. As a start in this direction, we have examined modification of purified pig kidney AP-P with diethylpyrocarbonate which indicates the presence of at least two critical histidine residues in the protein[47]. Protection of the enzymic activity was achieved by preincubation with the substrates, bradykinin (1–5) and Gly-Pro-Hyp, and with the inhibitor apstatin. The C-terminal half of pig kidney AP-P (residues 376–625), in which the active site is likely to reside, has seven histidine residues, only four of which are conserved with other members of the proline peptidase family (residues 429, 519, 523 and 532; numbering refers to that of the pig kidney AP-P sequence[37]) and these therefore represent primary targets for mutagenesis. Preliminary data using the cysteine modifying reagent p-hydroxymercuriphenylsulfonic acid appear to indicate the additional presence of a critical

```
PIGAPP   376   IRYLAWLEKNVPTGTVDEFS.GAKRVEEFRGEEEFFSG....
APPECO   184   LRRAGEITAMAHTRAMEKCRPG...MFEYHLEGEIHHE.FNR
PROHUM   194   LRYTNKISSEAHREVMKAVKVG...MKEYGLESLFEHYCYSR
PROECO   170   MREAQKMAVNGHRAAEEAFRSG...MSEFDINIAYLTATGHR
CREAT    167   IRHGARIADIGGAAVVEAL..GD.QVPEYEVALHATQAMVRA

PIGAPP   413   .....PSFETISA.....SGLNAALAHYSPT.KELHRKLSSD
APPECO   222   HGARYPSYNTIVG.....SGENGCILHYTE....NECEMRDG
PROHUM   233   GGMRHSSYTCICG.....SGENSAVLHYGHAGAPNDRTIQNG
PROECO   209   DTD..VPYSNIVALNE.....HAAVLHYTKLDHQAPEEMRS.
CREAT    206   IADTFEDVELMDTWTWFQSGINTDGAHNPVT....TRKVNKG

PIGAPP   444   EMYLLDSGGQYWDGTTDITRTVHWGTPSAF...QKEAYTRVL
APPECP   255   DLVLIDAGCEYKGYAGDITRTFP..VNGKFTQAQREIYDIVL
PROHUM   270   DMCLFDMGGEYYSVASDITCSFP..RNGKFTADQKAVYEAVL
PROECO   243   ..FLLDAGAEYNGYAADLTRTWSAKSDNDYAQLVKDVNDEQL
CREAT    244   DILSLNCFPMIAGYYTALERTLFL...DHCSDDHLRLWQVNV

PIGAPP   483   IGNIDLSRLVFPAATSGRVVEAFA.RKALWDVGLN.......
APPECO   295   ESLETSLRLYRPGTSILEVTGEVV.RIMVSGLVK.LGILKGD
PROHUM   310   LSSRAVMGAMKPGDWWPDIDRLAD.RIHLEELA.HMGILSGS
PROECO   283   ....ALIATMKAGVSYVDYHIQFHQRIAKLLR.KHQIITDMS
CREAT    283   EVHEAGLKLIKPGARCSDIA.....RELNEIFLKHDVL....

PIGAPP   517   ...........YGHGTGHGIGNFLCVHEWPVGFQYGNIPM
APPECO   335   VDELIAQNAHRPFFMHGLSHWLGL..DVHD..VGVYGQDRSR
PROHUM   350   VDAMVQAHLGAVFMPHGLGHFLGI..DVHD..VGGYPEGVER
PROECO   320   EEAMVENDLTGPFMPHGIGHPLGL..QVHD..AGFMQDDSGT
CREAT    316   .......QYRTFGYGHSFG......TLSHY..YGREAGLELR

PIGAPP   546   AEGMFTSIEPGYYQDGEFGIRLEDVALVVEAKTKYPGTYLTF
APPECO   373   I...................LE............PGMVLTV
PROHUM   388   IDEPG.....LRSLRTARH..LQ............PGMVLTV
PROECO   358   HLAAPAKYPYLRCTRI.....LQ............PGMVLTI
CREAT    343   EDIDTV..............LE............PGMVVSM

PIGAPP   588   EVVSLVPYDRKLIDVSLLSPEQLQYLN.......RYYQAIR.
APPECO   383   EPGLYIAP................DAEVPEQYRGI...GIRI
PROHUM   411   EPGIYFID..HLLDEALADPARASFLNREVLQRFRGFGGVRI
PROECO   383   EPGIYFIES..LLA.PWREGQFSKHFNWQKIEALKPFGGIRI
CREAT    358   EPMI.......MLPEGLPGA.................GGYR.

PIGAPP   622   EK..VG
APPECO   406   EDDIVI
PROHUM   451   EEDVVV
PROECO   422   EDNVVI
CREAT    375   EHDILI
```

Figure 3. Sequence alignment of the C-terminal region of some members of the proline peptidase family. Conserved amino acid residues within the alignment are represented in bold type. The sequence of pig AP-P is that deduced from the cDNA clone[37]; the sole cysteine residue (Cys530) in this region of the protein is underlined. The other sequences are as aligned in[43]. Sequence abbreviations indicate: PIGAPP, AP-P; APPECO, *E. coli* AP-P; PROHUM, human prolidase; PROECO, *E.coli* prolidase; CREAT, *P. putida* creatinase.

cysteine residue in AP-P. Comparisons between the predicted amino acid sequence of pig kidney AP-P from the isolated cDNA clone[37] and the amino acid sequence of the purified enzyme[43] indicate one protein sequencing error. The cDNA clone predicts a cysteine residue at position 530 (See Fig. 3) whereas this was reported incorrectly as a glutamine in Vergas Romero et al[43]. This cysteine residue is the only one that occurs in the C-terminal half of pig kidney AP-P. A partial sequence of a human kidney AP-P cDNA clone (see below) indicates that this cysteine residue is conserved in the human. However an equivalent residue is not found in other members of the proline peptidase family. Its relevance to the thiol sensitivity of mammalian AP-P must therefore await mutagenesis studies.

6. HUMAN HOMOLOGUES OF AP-P

Taking advantage of the information available from the pig kidney AP-P cDNA clone we have initiated experiments to isolate human homologues of AP-P. By using the specific PCR primers for the pig sequence, fragments were amplified from a human kidney cortex cDNA library constructed in λgt10 and products of sizes equivalent to those expected from the pig sequence were purified. These products have been subcloned and are in the process of being sequenced. The initial sequence data indicate that the deduced amino acid sequence of the human fragments differs from the pig sequence at only one amino acid position. The fragments span the amino acid positions 189–230 and 431–610 in the pig sequence[37]. The sequence homology is maintained in both fragments, not only in the C-terminal domain as expected for the proline peptidase family of proteins, but also in the N-terminal domain which is usually more variable between family members. The C-terminal fragment shows the conservation of three of the four conserved histidine residues (519, 523 and 532) and Cys^{530}, the only cysteine residue in the C-terminal half of the protein. We are now in the process of isolating a full length functional clone of the human AP-P, although with the extremely high degree of sequence homology between the pig and the human sequences the enzymic properties of the two proteins will probably be highly similar. We also have obtained an expressed sequence tag (EST) which has homology to the C-terminal region of the pig AP-P sequence, with 58% sequence similarity to the pig AP-P sequence over a 103 amino acid region. The EST possesses a stop signal but does not encode a GPI-attachment sequence. It may, therefore, represent the cytosolic form of AP-P which has been previously described[14]. The isolation of different isoforms of AP-P and the characterisation of their enzymic properties, combined with site-directed mutagenesis and chemical modification studies should help to unravel the nature of the active site and mechanism of mammalian AP-P.

ACKNOWLEDGMENTS

We should like to thank the British Heart foundation for support of this work.

REFERENCES

1. Turner, A.J. (1993) Membrane peptidases of the nervous and immune systems. *Adv. Neuroimmunol.* **3**, 163–170.

2. Kenny, A.J. & Stephenson, S.L. (1988) Role of endopeptidase-24.11 in the inactivation of atrial natriuretic peptide. FEBS *Lett.* **232**, 1–8.
3. Turner, A.J. (1987) Metabolism of enkephalins. *ISI Atlas of Sci. Pharmacol.* **1**, 74–77.
4. Hooper, N.M. (1994) Families of zinc metalloproteases. *FEBS Lett.* **354**, 1–6.
5. Hooper, N.M. & Turner, A.J. (1988) Ectoenzymes of the kidney microvillar membrane. Aminopeptidase P is anchored by a glycosyl-phosphatidylinositol moiety. *FEBS Lett.* **229**, 340–344.
6. Yaron, A. (1987) The role of proline in the proteolytic regulation of biologically active peptides. *Biopolymers* **26**, S215-S222.
7. Mentlein, R. (1988) Proline residues in the maturation and degradation of peptide hormones and neuropeptides. *FEBS Lett.* **234**, 251–256.
8. Vanhoof, G., Goossens, F., De Meester, I., Hendriks, D. & Scharpé, S. (1995) Proline motifs in peptides and their biological processing. *FASEB J.* **9**, 736–744.
9. Kenny, A.J., Booth, A.G. & Macnair, R.D. (1977) Peptidases of the kidney microvillus membrane. *Acta Biol. Med. Germ.* **36**, 1575–1585.
10. Lasch, J., Koelsch, R., Ladhoff, A.-M., & Hartrodt, B. (1986) Is the proline-specific aminopeptidase P of the intestinal brush border an integral membrane enzyme? *Biomed. Biochim. Acta* **45**, 833–843.
11. Orawski, A.T., Susz, J.P. & Simmons, W.H. (1989) Aminopeptidase P from bovine lung: solubilisation, properties and potential role in bradykinin degradation. *Mol. Cell. Biochem.* **75**, 123–132.
12. Orawski, A.T. & Simmons, W.H. (1992) Purification and properties of membrane-bound aminopeptidase P from rat lung. *Biochemistry* **34**, 11227–11236.
13. Holtzmann, E.J., Pillay, G., Rosenthal, G. & Yaron, A. (1987) Aminopeptidase P activity in rat organs and human serum. *Analyt. Biochem.* **162**, 476–484.
14. Harbeck, H.-T. & Mentlein, R. (1991) Aminopeptidase P from rat brain. Purification and action on bioactive peptides. *Eur. J. Biochem.* **198**, 451–458.
15. Vanhoof, G., De Meester, I., Goossens, F., Hendriks, D., Scharpé, S. and Yaron, A. (1992) Kininase activity in human platelets: cleavage of the Arg^1-Pro^2 bond of bradykinin by aminopeptidase P. *Biochem. Pharmacol.* **44**, 479–487.
16. Kohno, H. & Kanno, T. (1985) Properties and activities of aminopeptidases in normal and mitogen-stimulated human lymphocytes. *Biochem. J.* **226**, 59–65.
17. Hendriks, D., De Meester, I., Umiel, T., Vanhoof, G., van Sande, M., Scharpé, S. & Yaron, A. (1991) Aminopeptidase P and dipeptidyl peptidase IV activity in human leukocytes and in stimulated lymphocytes. *Clin. Chim. Acta* **196**, 87–96.
18. Dehm, P. & Nordwig, A. (1970) The cleavage of prolyl peptides by kidney peptidases. Partial purification of a "X-prolyl-aminopeptidase" from swine kidney microsomes. *Eur. J. Biochem.* **17**, 364–371.
19. Blau, N., Niederwieser, A. & Shmerling, D.H. (1988) Peptiduria presumably caused by aminopeptidase P deficiency. A new inborn error of metabolism. *J. Inher. Metab. Dis.* **11**, 240–242.
20. Prechel, M.M., Orawski, A.T., Maggiora, L.L. & Simmons, W.H. (1995) Effect of a new aminopeptidase P inhibitor, apstatin, on bradykinin degradation in the rat lung. *J. Pharmacol. Exp. Therap.* **275**, 1136–1142.
21. Medeiros, M.S. & Turner, A.J. (1994) Processing and metabolism of peptide YY: pivotal roles for dipeptidyl peptidase IV, aminopeptidase P and endopeptidase-24.11. *Endocrinology* **134**, 2088–2094.
22. Mentlein, R., Dahms, P., Grandt, D. & Krüger, R. (1993) Proteolytic processing of neuropeptide Y and peptide YY by dipeptidyl peptidase IV. *Regul. Peptides* **49**, 133–134.
23. Lloyd, G.S., Hryszko, J., Hooper, N.M. & Turner, A.J. (1996) Inhibition and mteal ion activation of pig kidney aminopeptidase P. dependence on nature of substrate. *Biochem. Pharmacol.* **52**, 229–236.
24. Matsas, R., Fulcher, I.S., Kenny, A.J. & Turner, A.J. (1983) Substance P and [Leu]enkephalin are hydrolyzed by an enzyme in pig caudate synaptic membranes that is identical with the endopeptidase of kidney microvilli. *Proc. Natl. Acad. Sci. U.S.A.* **80**, 3111–3115.
25. Rusu, I. & Yaron, A. (1992) Aminopeptidase P from human leukocytes. *Eur. J. Biochem.* **210**, 93–100.
26. Lasch, J., Koelsch, R., Steinmetzer, J., Neumann, V. & Demuth, H.-V. (1988) Enzymic properties of intestinal aminopeptidase P: a new continuous assay. *FEBS Lett.* **227**, 171–174.
27. Hooper, N.M. & Turner, A.J. (1990) Purification and characterization of pig kidney aminopeptidase P. A glycosyl-phosphatidylinositol-anchored ectoenzyme. *Biochem. J.* **267**, 509–515.
28. Hooper, N.M., Hryszko, J., Oppong, S.Y. & Turner, A.J. (1992) Inhibition by converting enzyme inhibitors of pig kidney aminopeptidase P. *Hypertension* **19**, 281–285.
29. Simmons, W.H. & Orawski, A.T. (1992) Membrane bound aminopeptidase P from bovine lung. Its purification, properties and degradation of bradykinin. *J. Biol. Chem.* **267**, 4897–4903.
30. Orawski, A.T. & Simmons, W.H. (1995) Purification and properties of membrane-bound aminopeptidase P from rat lung. *Biochemistry* **34**, 11227–11236.

31. Hooper, N.M.. & Turner, A.J. (1988) Ectoenzymes of the kidney microvillar membrane. Differential solubilization by detergents can predict a glycosyl-phosphatidylinositol membrane anchor. *Biochem. J.* **250**, 865–869.
32. Hooper, N.M.. & Turner, A.J. (1989) Hydrolysis of the glycosyl-phosphatidylinositol anchors of renal microvillar peptidases by a plasma phospholipase D. *Biochem. Soc. Trans.* **17**, 885–886.
33. Koelsch, R., Gottwald, S. & Lasch, J. (1994) Release of GPI-anchored membrane aminopeptidase P by enzymes and detergents has some peculiarities. *Biochim. Biophys Acta* **1190**, 170–172.
34. Ryan, J.W., Denslow, N.D., Greenwald, J.A. & Rogoff, M.A. (1994) Immunoaffinity purifications of aminopeptidase P from guinea pig lungs, kidney and serum. *Biochem. Biophys. Res Commun.* **205**, 1796–1802.
35. Hooper, N.M., Broomfield, S.J. & Turner, A.J. (1991) Characterization of antibodies to the glycosyl-phosphatidylinositol membrane anchors of mammalian proteins. *Biochem. J.* **273**, 301–306.
36. Turner, A.J. & Hooper, N.M. (1997) The glycolipid-anchored peptidases. In: Cell-surface peptidases (ed. Kenny, A.J. & Boustead, C.M.), Bios, Oxford, in press.
37. Hyde, R.J., Hooper, N.M. & Turner, A.J. (1996) Molecular cloning and expression in COS-1 cells of pig kidney aminopeptidase P. *Biochem. J.* **319**, 197–201.
38. Kyte, J. & Doolittle, R.F. (1982) A simple method for displaying the hydropathic character of a protein. *J. Mol. Biol.* **157**, 105–132.
39. von Heijne, G. (1986) A new method for predicting signal sequence cleavage sites. *Nucleic Acids Res.* **14**, 4683–4690.
40. Udenfriend, S. & Kodukula, K. (1995) Prediction of ω site in nascent precursor of a glycosylphosphatidylinositol protein. *Meth. Enzymol.* **250**, 571–582.
41. Denslow, N.D., Ryan, J.W. & Nguyen, H.P. (1994) Guinea pig membrane-bound aminopeptidase P is a member of the proline peptidase family. *Biochem. Biophys. Res. Commun.* **205**, 1790–1795.
42. Bazan, J.F., Weaver, L.H., Roderick, S.L., Hüber, R. & Matthews, B.W. (1994) Sequence and structure comparison suggest that methionine aminopeptidase, prolidase, aminopeptidase P and creatinase share a common fold. *Proc. Natl. Acad. Sci. U.S.A.* **91**, 2473–2477.
43. Vergas Romero, C., Neudorfer, I., Mann, K. & Schäfer, W. (1995) Purification and amino acid sequence of aminopeptidase P from pig kidney. *Eur. J. Biochem.* **229**, 262–269.
44. Mock, W.L. & Liu, Y. (1995) Hydrolysis of picolinylprolines by prolidase. A general mechanism for dual-metal ion containing aminopeptidases. *J. Biol. Chem.* **270**, 18437–18446.
45. Roderick, S.L. & Matthews, B.W. (1993) Structure of the cobalt-dependent methionine aminopeptidase from *E. coli*: a new type of proteolytic enzyme. *Biochemistry* **32**, 3907–3912.
46. Arfin, S.M., Kendall, R.L., Hall, H., Weaver, L.H., Stewart, A.E., Matthews, B.W. & Bradshaw, R.A. (1995) Eukaryotic methionyl aminopeptidases: two classes of cobalt-dependent enzymes. *Proc. Natl. Acad. Sci. U.S.A.* **92**, 7714–7718.
47. Lim, J. & Turner, A.J. (1996) Chemical modification of porcine kidney aminopeptidase P indicates the involvement of two critical histidine residues. *FEBS Lett.* **381**, 188–190.
48. Turner, A.J., Medeiros, M.S. & Hooper, N.M. (1991) The molecular biology of GPI-anchored brush border hydrolases. *Cell Biol. Int. Reports* **15**, 1083–1099.
49. Brewis, I.A., Ferguson, M.A.J., Mehlert, A., Turner, A.J. & Hooper, N.M. (1995) Structures of the glycosylphosphatidylinositol anchors of porcine and human renal membrane dipeptidase. Comprehensive structural studies on the porcine anchor and interspecies comparison of the glycan core structures. *J. Biol. Chem.* **270**, 22946–22956.

AMINOPEPTIDASE P

A Cell-Surface Antigen of Endothelial and Lymphoid Cells

Jürgen Lasch, Sylke Moschner, and Regine Koelsch

Institute of Physiological Chemistry
Medical Faculty
Martin-Luther-University
Halle, Germany

1. INTRODUCTION

Many mammalian cell-surface proteins are linked posttranslationally to complex glycosyl phosphatidylinositol (GPI) anchors which associate these proteins with the outer leaflet of the plasma membrane (type V membrane proteins). At least 50 of such proteins are known today in mammals. GPI-proteins include complement regulatory factors [1], cell adhesion molecules [2], receptors of the Fc part of immunoglobulins [2], ADP-ribosyltransferases [3] and ectoenzymes [4], e.g. aminopeptidase P [5,6,7]. They have been found on a variety of cells including endothelial cells [8]. They differ from peptide-anchored transmembrane proteins (type I - IV membrane proteins) in cell-surface distribution, lateral mobility and lipid environment [9].

Some GPI-linked proteins, mainly complement regulatory factors, seem to transfer easily between lipid bilayers [10]. To what extent promiscuous transfer occurs is unknown.

We discovered in 1986 that aminopeptidase P (APP) of the small intestine of rat is membrane-anchored [11]. It is an exopeptidase liberating the N-terminal amino acid from peptides of the type X- Pro-Y....., i.e. it catalyses the hydrolysis of an imide peptide bond instead of the usual amide peptide bond.

The physiological function of the GPI-anchored aminopeptidase P variant is still elusive. Most researchers suppose that this enzyme inactivates biologically active peptides like bradykinin, neuropeptide tyrosine (NPY) and others [12]. Recent evidence suggests that at least some membrane-bound ectoproteases at the cell surface are multifunctional proteins with quite a number of non-catalytic functions [13]. APP is one of the candidates.

We show the cell surface localisation of APP by immunohistological methods, by measuring the enzyme activity of lymphoid cell membranes and intact endothelial cell (EC) monolayers and by a chemiluminescence assay of APP on endothelial cells.

Cellular Peptidases in Immune Functions and Diseases, edited by Ansorge and Langner
Plenum Press, New York, 1997

2. MATERIALS AND METHODS

Human endothelial cells (HUVECs) were isolated and cultured by conventional methods. The other cell lines were HL 60, U 937, K 562, HuT 78, Jurkat, MOLT-4, Reh, MSAB and Raij).*

2.1 Generation of Polyclonal Anti-APP Antibodies

After PI-PLC (phosphatidylinositol-specific phospholipase C) treatment of brush border membrane vesicles the aminopeptidase P of rat small intestine was partially purified in a series of conventional chromatographic steps (200 fold enrichment). The IgG elicited in rabbits against partially purified APP showed several precipitates in two-dimensional immunoelectrophoresis against this antigen. The antigen-antibody-precipitate with enzymic APP activity was excised and used as antigen in further immunisation steps. The resulting IgG fraction produced only one precipitate with APP-activity.

2.2 Luminescence Immunoassay[14]

The quantification of APP at the outer leaflet of the plasma membrane (ectoenzyme) was done by the peroxidase immunotechnique. The variant of antibody-peroxidase-anti-peroxidase (PAP) method we used was the peroxydase-catalysed oxidation of luminol to its chemiluminescent compound (the excited 3-aminophthalate dianion). The number of photons emitted per second is proportional to the number of antibody molecules disposed on the surface of the cells. The luminescence was measured with the luminometer Lumat LB 9501 (Berthold, Wildbad, Germany). Light emission was measured after 30, 60 and sometimes after 90 seconds. The maximal value was always reached after 60 seconds. It was expressed as **R**elative **L**ight **U**nits per second (RLU/sec).

Table 1. Chemiluminescence immunoassay of aminopeptidase P on endothelial cells

Sample	max.RLU [a]/s/10[4] cells (average)	APP-antibody dilution	HRP [b]Ab dilution
APP antibody omitted	76	-	1:500
APP antibody added	3665	1:100	1:500
APP antibody added	2003	1:200	1:500

[a] RLU: Relative Light Units, [b] HRP: Horse Radish Peroxidase

* Lymphoid and myeloid cell lines were a gift of Dr. Riemann, Institute of Medical Immunology of the Martin-Luther-University, Halle.

2.3 Enzymic Activity of APP

The following substrates were used: (i) Ala-Pro-Pro-pNA in combination with dipeptidase IV as auxiliary enzyme for photometric measurements [11], (ii) Gly-Pro-Pro-ßNA (II) and Fast Garnet salt for gel staining and Lys(Abz)-Pro-Pro-pNA (III) for fluorimetric measurements (λ_{ex}=310nm, λ_{em}=410 nm).[†] In all cases APP was preincubated with Mn^{2+}, 1 mM, at pH 7.4 for 1 hour at room temperature.

Enzymic activity at the cell surface of vital lymphoid and endothelial cells was measured by incubation with substrate (I). The reaction was stopped after 15 min with 10 mM EDTA, the sample centrifuged and the product measured at 405 nm photometrically. Total activity of these cells was obtained after solubilisation with 2% Triton X-100.

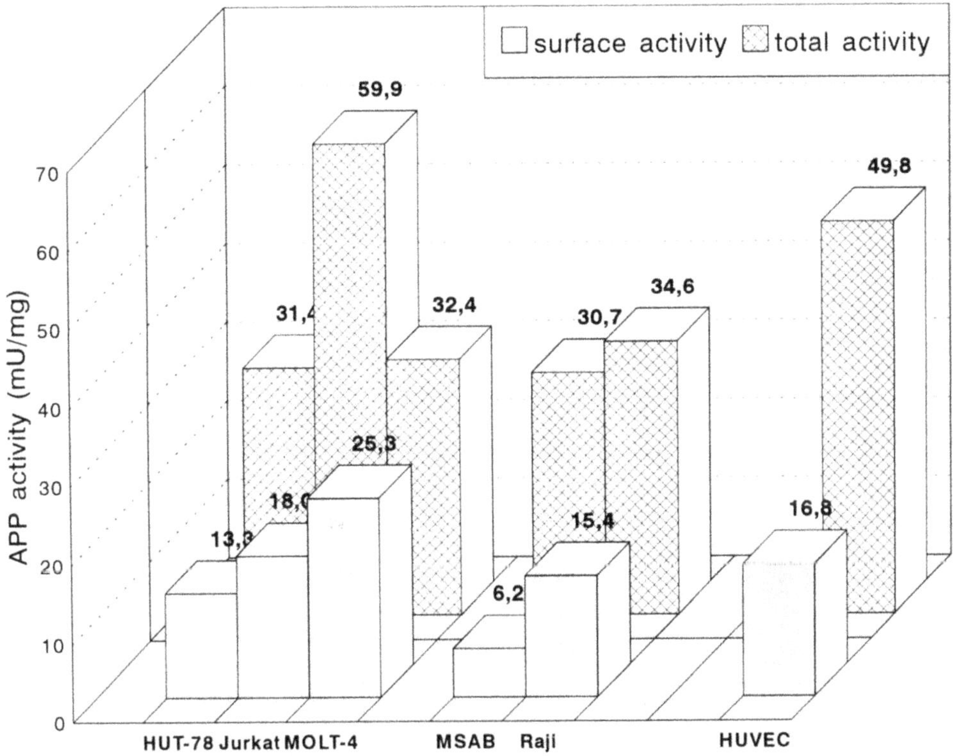

Figure 1. APP activities of various lymphoid and myeloid cell lines as well as of endothelial cells (HUVECs). Activities on the top of the columns are in mU/mg protein.

[†] We thank Mrs. A.Stöckel and Dr.B.Stiebitz of the Fachbereich Biochemie/Biotechnologie of the Martin-Luther-University for the fluorogenic APP substrate Lys(Abz)-Pro-Pro-pNA.

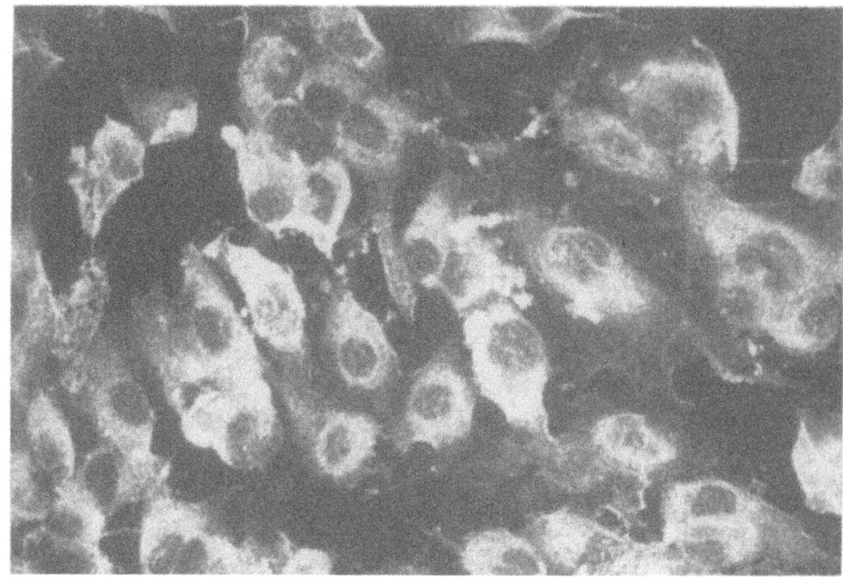

A

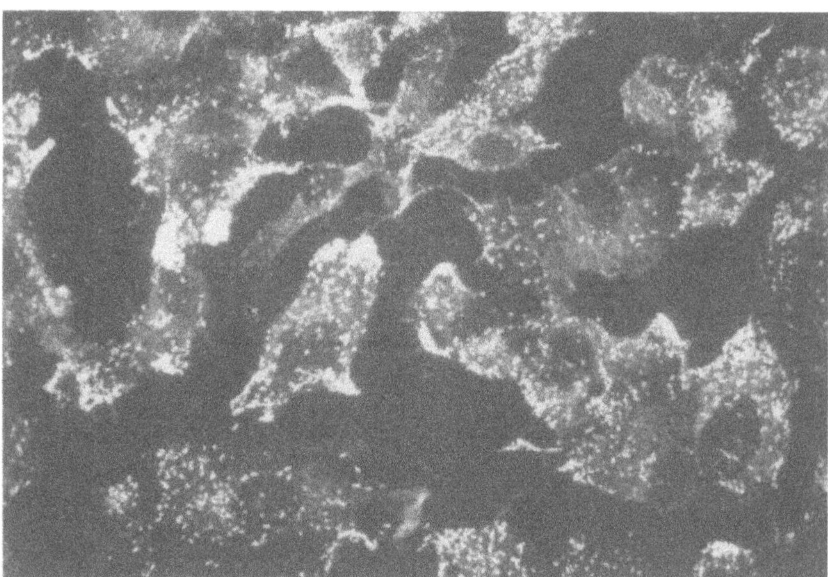

B

Figure 2. Immunofluorescence. Fluoromicrographs of ECs cultured on microscopic slides. (A) First antibody: anti-APP, second antibody: FITC-labeled anti-rabbit-IgG (B) First antibody: anti-human-von-Willebrand-factor, second antibody: FITC-labeled anti-rabbit-IgG. (C) First antibody: omitted, second antibody: FITC-labeled anti-rabbit-IgG.

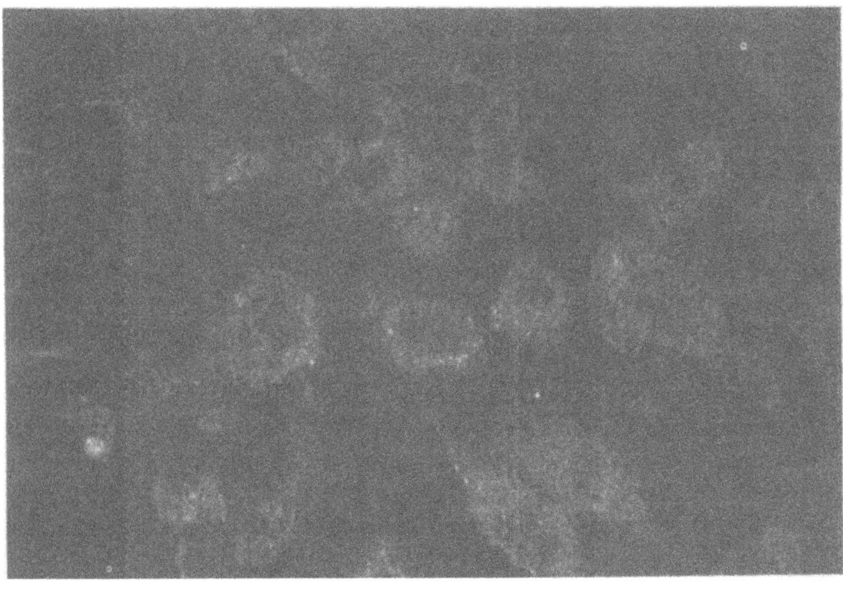

C

Figure 2. (*Continued*).

2.4 Measurement of the Enzymic Activity at the Surface of Cultured Endothelial Cells

HUVECs, grown for 2–3 days on microscopic coverslips, were introduced into the cuvette of the Hitachi F-2000 fluorimeter. The monitored volume was adjusted in front of the monolayer and the generation of the fluorescent product measured under stirring. At the end of the measurement (up to one hour) the intactness of the cells was microscopically controlled. The cell number in the cuvette was measured by the PicoGreen DNA assay after solubilisation of the cells with 1% Triton X-100.

2.5 Immunofluorescence of Endothelial Cells

HUVECs were seeded on coverslips pretreated with 0.5% gelatin. At confluence cells were rinsed three times with PBS and fixed with methanol at -20 C for 10 minutes. After washing three times with PBS the respective antibodies were added. All staining steps were carried out at room temperature. The first antibodies used were anti-APP (home-made) and anti-human-von-Willebrand-factor-IgG raised in rabbits, the second antibody was FITC-labeled anti-rabbit-IgG (Sigma ImmunoChemicals).

2.6 Measurement of the Transfer of GPI-Anchored APP to Other Lipid Bilayers

According to Lundahl [15] liposomes of various composition (egg yolk PC, 10 mg/ml and human stratum corneum lipid, 5 mg/ml) were sterically immobilised in Superdex 200 beads by freezing (-80 C) and thawing. Immobilised lipid was measured by the phosphate content after ashing. After extensive washing rat small intesine brush border vesicles were added. After 16 hrs at room temperature the enzymic activities of APP (GPI-anchored) and dipeptidase IV (transmembrane peptide anchor) were measured. The ratio of these activities was used as measure of enzyme transfer.

3. RESULTS

The enzyme activity on lymphoid cells is given in Fig.1.

The chemiluminescene immunoassay with endothelial cells gave a clear indication of the surface localisation of APP (TABLE 1.)

This was substantiated by extensive immunofluorescence studies including appropriate controls (Fig.2).

'Monolayer kinetics' with the fluorogenic substrate are shown in Fig.3. It proved essential to use a very sensitive and reliable method (PicoGreen method) to 'count' the cells on the coverslip.

There was no change of the ratio of enzymic activities of type II and type V membrane proteins after prolonged contact with beads of immobilised liposomes. That is, no transfer of APP between lipid bilayers could be shown under the experimental conditions used.

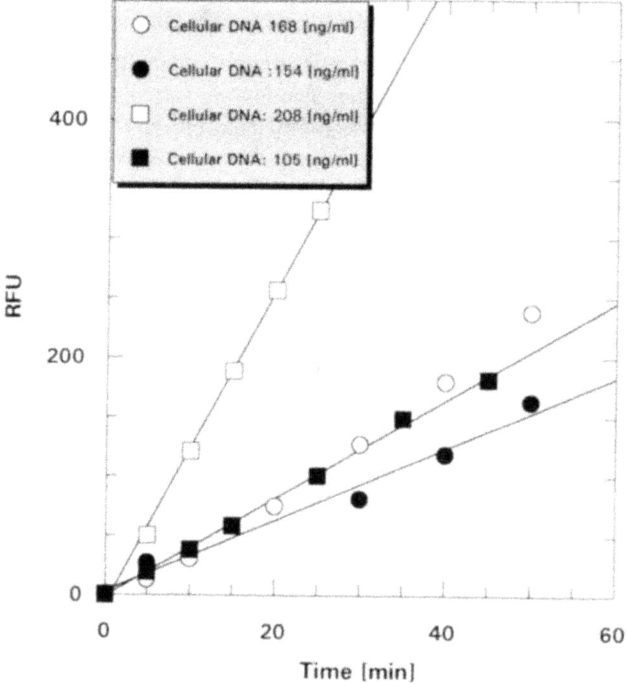

Figure 3. Product generation by endothelial cell monolayers on microscopic slides. Kinetic runs with the fluorogenic APP substrate Lys(Abz)-Pro-Pro-pNA.

4. CONCLUSIONS

Polyclonal antibodies raised in rabbits against APP of rat small intestine cross react with APP from human small intestine as well as with human endothelial APP (HUVECs).

APP is an ectoenzyme (cell surface antigen) of T and B lymphocytes as evidenced by a membrane-linked enzymic activity (substrate (I), Ala-Pro-Pro-pNA). Ryan et al.[16] found 0.67 U/mg protein with the substrate Arg-Pro-Pro[^{3}H]benzylamide on human aortic endothelial cells (HAECs).

Chemiluminescent enzyme immunoassays and immunohistochemical studies clearly show that APP is also a surface antigen of endothelial cells.

ACKNOWLEDGMENT

This work was supported by the Deutsche Forschungsgemeinschaft (Ko 1292/1–2) and the Fonds der Chemischen Industrie.

5. REFERENCES

1. Hatanaka,M., Seya, T., Matsumoto, M., Hara, T., Nonaka, M., Inoue, N., Takeda, J. and Shimizu, A. (1996), Mechanisms by which the surface expression of the glycosylphosphatidylinositol-anchored complement regulatory proteins decay-accelerating factor (CD 55) and CD 59 is lost in human leukaemia cell lines, Biochem.J. *314*, 969–976.
2. Edwards, S.W. (1996), Cell signalling by integrins and immunoglobulin receptors in primed neutrophiles, TIBS, *29*, 362–367.
3. Nemoto,E., Stohlman,S., Dennert, G. (1996), Release of a glycosyl phosphatidylinositol-anchored ADP-ribosyltransferase from cytotoxic T cells upon activation, J.Immunol.*156(1)*, 85–92.
4. Lasch,J., Koelsch,R., Ladhoff,A.-M., and Hartrodt, B. (1986), Is the proline-specific aminopeptidase P of the intestinal brush border an integral membrane enzyme? Biomed.Biochim.Acta *45(7)*, 833–843.
5. Hooper,N.M., Hryszko, J. and Turner,A.M. (1990), Purification and characterization of pig kidney aminopeptidase P. A glycosyl-phospatidyl- inositol-anchored ectoenzyme, Biochem.J. *267*, 509–515.
6. Koelsch,R., Gottwald,S. and Lasch,J. (1994), Release of GPI-anchored membrane aminopeptidase P by enzymes and detergents has some peculiarities, Biochim. Biophys. Acta *1190*, 170 - 172.
7. Orawski,A.T., Suzs, J.P. and Simmons,W.H. (1987), Aminopeptidase P from bovine lung: solubilization, properties, and potential role in bradykinin degradation, Mol.Cell.Biochem. *75*, 123–132.
8. Kooyman,D.L., Byrne,G.W., McClellan, S., Nielsen, D., Fone, M., Waldmann, H., Coffman, T.M., McCurry, K.R., Platt, J.L., Logan, J.S. (1995), In Vivo Transfer of GPI-Linked Complement Restriction Factors from Erythrocytes to the Endothelium, Science *269*, 89–92.
9. Brown,D.A. (1992), Interactions between GPI-anchored proteins and membrane lipids, Trends Cell Biol. *2*, 338–343.
10. Ilangumara,S., Robinson, P.J. and Hoessli,D.C. (1996), Transfer of exogenous glycosylphosphatidylinositol (GPI)-linked molecules to plasma membranes, Trends in Cell Biol.*6*, 163–167.
11. Lasch, J., Koelsch,R., Steinmetzer, T., Neumann, U.Demuth, H.-U.(1988), Enzymic properties of intestinal aminopeptidase P: a new continuous assay FEBS Letters *227*, 171 - 174.
12. Vanhoof,G., Gossens,F. De Meester, I., Hendriks, D. and Scharpé,S. (1995), Proline motifs in peptides and their biological processing, FASEB J. *9*, 736–744.
13. edo,A., Mandys,V., K epela,E. (1996), Cell Membrane-Bound Proteases Not "Only" Proteolysis, Physiol.Res. *45*, 169–176.
14. Kricka,L.J. and Thorpe, G.H.G. (1990), Bioluminescent and Chemiluminescent Detection of Horseradish Peroxidase Labels in Ligand Binding Assays, In: Luminescence Immunoassays and Molecular Applications, Van Dyke,K. and Van Dyke, R.(eds.), CRC Press, Boca Raton, Ann Arbor, Boston, pp.77–98.
15. Yang,Q., and Lundahl,P. (1994), Steric Immobilization of Liposomes in Chromatographic Gel Beads and Incorporation of Integral Membrane Proteins into Their Lipid Bilayers, Anal.Biochem. *218*, 210–221.
16. Ryan,W.J., Papapetropoulos,A., Ju,H., Denslow,N.D., Antonov,A., Virmani,R., Kolodgie,F.D., Gerrity,R.G., and Catravas,J.D.(1996), Aminopeptidase P is disposed on human endothelial cells, Immunopharmacology *32*, 149–152.

4

HUMAN LYMPHOCYTE X-PROLYL AMINOPEPTIDASE (AMINOPEPTIDASE P)-LIKE PROTEIN

A New Member of the Proline Peptidase Family?

G. Vanhoof,[1] F. Goossens,[1] M. A. Juliano,[2] L. Juliano,[2] I. De Meester,[1] D. Hendriks,[1] K. Schatteman,[1] and S. Scharpé[1]

[1]Department of Clinical Biochemistry
University of Antwerp
Universiteitsplein 1 S-6, B-2610 Wilrijk, Belgium
[2]Biophysics Department
Escola Paulista de Medicina
04044 Sao Paulo, Brasil

1. INTRODUCTION

Peptidases are grouped according to their mechanism of catalysis in serine-type, cysteine-type, aspartic-type and metallo-type peptidases. Amongst these groups, the metallopeptidases represent the most diverse group, comprising 25 different families (1,2). Most of the Zn-binding metallopeptidases, named zincins, have the HEXXH motif for Zn-binding. Some other Zn-binding motifs have been defined, as there are the HXXEH, the HXXE, and the HXH motif (3). X-prolyl aminopeptidase (aminopeptidase P, EC 3.4.11.9), is an aminopeptidase that has been reported to bind zinc, and is activated by manganese ions, but does not contain any of these Zn-binding motifs. Aminopeptidase P from *Escherichia coli* was classified in the peptidase family M24, a family of metallopeptidases in which the ligands for metal ion binding are predominantly carboxylic acids (2). Apart from *E. coli* aminopeptidase P, this family comprises *E. coli* methionyl aminopeptidase (EC 3.4.11.18) and *E. coli* and human proline dipeptidase (EC 3.4.13.9). We have been studying the soluble form of aminopeptidase P in human lymphocytes and platelets (4,5), and were interested in determining the nucleotide sequence that would provide useful information for the development of potent and specific inhibitors. Aminopeptidase P is a proline-specific metallo-aminopeptidase that catalyses specifically the removal of any unsubstituted N-terminal amino acid that is adjacent to a penultimate proline residue. Due to its specificity towards proline, it has been suggested that aminopeptidase P is important

Cellular Peptidases in Immune Functions and Diseases, edited by Ansorge and Langner
Plenum Press, New York, 1997

in the maturation and degradation of peptide hormones, neuropeptides, and tachykinins, and it may be important in the digestion of otherwise resistant dietary protein fragments, thereby complementing the pancreatic peptidases (6–8). There is evidence for the participation of aminopeptidase P in the activation and proliferation of T-lymphocytes (4). Recent studies using inhibitors for aminopeptidase P in rats demonstrate that aminopeptidase P plays a crucial role in the degradation of bradykinin, in this way regulating blood pressure responses induced by this peptide (8). The sequences of the aminopeptidase P genes (*pepP*) from *E. coli, Mycoplasma genitalium, Streptomyces lividans,* and *H. influenzae* have been reported (9–12). The complete amino acid sequence and the cDNA-sequence of the membrane-bound pig kidney aminopeptidase P have recently been published (13–14). Using the aminoacid sequence of the pig kidney aminopeptidase P, we cloned, by RT-PCR of lymphocyte mRNA, a cDNA encoding a protein exhibiting 42% sequence identity and 60% sequence similarity to the membrane-bound pig kidney aminopeptidase P. Here we show by sequence comparison that this cDNA codes for a protein that has the characteristics to be a new member of the metalloprotease family M24.

2. EXPERIMENTAL

2.1. mRNA Isolation and cDNA Synthesis

mRNA was isolated from phytohemagglutinin stimulated lymphocytes (10^6 cells) by oligo(dT)-Sepharose affinity chromatography using the Quickprep mRNA purification kit (Pharmacia). Total cDNA was synthesised by use of a combination of oligo(dT)-primers and random primers (Superscript Preamplification system, Gibco BRL Gaithersburg, MD, USA). Specific cDNA was enriched by PCR using specific oligonucleotides.

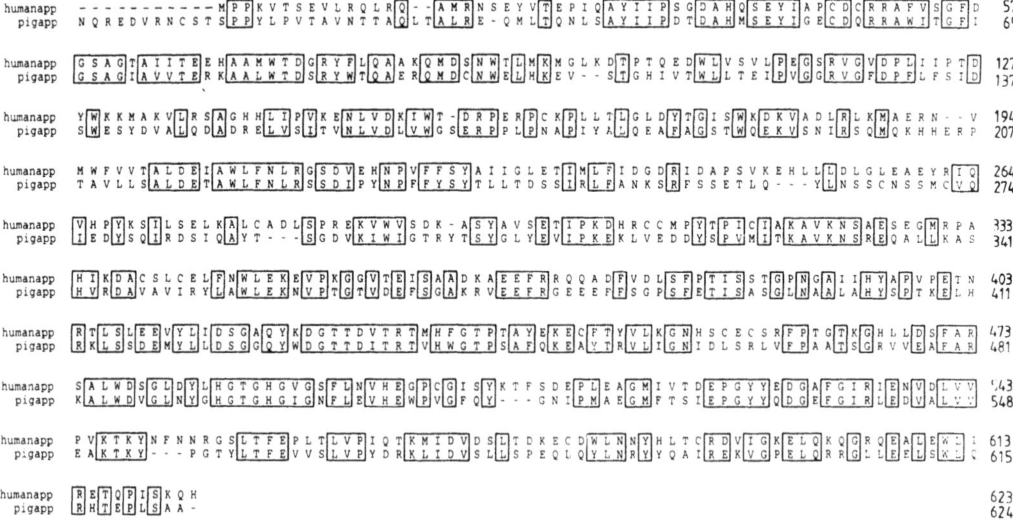

Figure 1. Amino acid sequence alignment of the protein encoded by XPNPEPL (humanapp) with the pig kidney X-prolyl aminopeptidase (pigapp). Identical and highly homologous amino acids are boxed.

2.2. Cloning and Sequence Analysis

PCR products were subcloned in the *SmaI* site of a pUC18 vector using the Pharmacia SureClone ligation kit. Transformation was performed in *E. coli* DH5α cells. The sequence of the insert was determined by the dideoxy-chain termination method.

2.3. Database Searches and Sequence Alignments

Amino acid sequence of the pig kidney X-prolyl aminopeptidase and nucleotide sequences obtained from the XPNPEPL gene were compared with available nucleotide and amino acid sequences using the BLAST algorithm at NCBI (15). Alignment of the amino acid sequence of the human XPNPEPL sequence with those of related proteins was performed using the PileUp multiple-sequence alignment software included in the genetics computer group (GCG) package (16).

```
                                 o                           *           *
XPNPEPL   380  SFPTISSTGPNGAIIHY  396      413  LIDSGAQ-YKDGTTDVTRT 430
Ecoli     229  SYNTIVGSGENGCILHY  245      259  LIDAGCE-YKGYAGDITRT 276
Strli1    276  GYGSICAAGPHACTLHW  292      305  LLDAGVETHTYYTADVTRT 323
Strli2    256  GYGTICAAGEHATIMHW  272      286  LLDAGVETRSLYTADVTRT 304
Bacsub    179  SFDMIVASGLRSSLPHG  195      208  TLDFG-AYYKGYCSDITRT 226
Haeminf   222  SYNSIVAGGSNACILHY  238      252  LIDAGCE-FAMYAGDITRT 269
Mycge     181  SFDPIVATGKNGANPHH  197      223  TCDFG-TIYNGYCSDITRT 240

                 o  *      o                       *
XPNPEPL   483  YLHGTGHGVGSFLNVHE  499      513  PLEAGMIVTDEPGYYE 528
Ecoli     349  FMHGLSHWLG--LDVHD  363      474  ILEPGMVLTVEPGLYI 489
Strli1    397  TLHGTGHMLG--MDVHD  411      423  TLEPGMVLTVEPGLYF 438
Strli2    378  TMAGTGHMLG--LDVHD  392      404  VLEPGMCLTVEPGLYF 419
Bacsub    281  FGHSTGHGLG--MEVHE  299      306  ILEPGMVVTVEPGIYI 321
Haeminf   342  YMHGLGHWLG--LDVHD  356      367  ILEIGMVITVEPGIYI 382
Mycge     287  FVHSTGHGVG--IDIHE  301      311  LLCENGVVTIEPGIYI 326

                    *
XPNPEPL   533  GIRIENVVLV   542
Ecoli     503  GIRIEDDIVI   512
Strli1    453  GVRIEDDILV   462
Strli2    434  GVRIEDDLVV   443
Bacsub    324  GVRIEDDIVI   333
Haeminf   396  GVRIEDNLLM   405
Mycge     331  GIRIEDMVLV   340
```

Figure 2. Amino acid sequence alignment of the regions conserved among the protein encoded by XPNPEPL and the procaryotic *pepP* gene products. Ecoli: *Escherichia coli*; Strli1:*Streptomyces lividans* gene1; Strli2: *Streptomyces lividans* gene2; Bacsub: *Bacillus subtilis*; Haeminf: *Haemophylus influenzae*; Mycge: *Mycoplasma genitalium*. Residues aligning with those residues involved in cobalt binding in methionine aminopeptidase are marked with an asterisk. Other residues possibly involved in catalytic activity are marked with a dot.

3. RESULTS AND DISCUSSION

Using combinations of primers designed on the basis of sequence information of the pig aminopeptidase P and human unidentified EST-sequences with high sequence-similarity to this protein, we cloned a 1872 bp cDNA coding for a 623 amino acid protein with high sequence similarity to the pig kidney aminopeptidase P (Figure 1). The clone was designated X-prolyl aminopeptidase (aminopeptidase P)-like (XPNPEPL).

The putative protein encoded by the XPNPEPL cDNA, not only shows sequence homology to the pig kidney aminopeptidase P, but also reveals highly conserved sequences with the prokaryotic *pepP* gene products (Figure 2). The amino acids involved in cobalt binding in methionyl aminopeptidase, a member of the M24 metalloprotease family, are strictly conserved in the XPNPEPL-protein. These aminoacids align with Asp-415, Asp-426, His 489, Glu 523 and Glu-537 in the human lymphocyte sequence. Significant sequence homology was also found with the third member of the M24 metalloprotease family, proline dipeptidase (EC 3.4.13.9), and the C-terminal part of the human XPNPEPL encoded sequence (Figure 3).

Remarkably, the human XPNPEPL gene translation product exhibits also a high sequence homology to the *Schizosaccharomyces pombe* chromosome I hypothetical protein C22G7.01c, and to the *Saccharomyces cerevisiae* ORF yll029w (results not shown).

```
                *    **** +   + *  +  *   +  *
XPNPEPL    351  EVPKGGVTEISAADKAEEFRRQQADFVDLS  380
prolidase  162  EALRNGVTERAVVSQIEYQLKLQKGVMQTS  191

                * **   * * *  *  *   +  *+   *
XPNPEPL    381  FPTISSTGPNGAIIHYAPVPETNRTLSLEE  410
prolidase  192  FDTIVQAGKNAANPHQGP---SMNTVQPNE  218

                + * * *  ++   +* +**+ +* **
XPNPEPL    411  VYLIDSGAQYKDGTTDVTRTMHFGTPTAYE  440
prolidase  219  LVLFDLGTMHEGYASDSSRTVAYGEPTDKM  248

                +*    * +            *     **
XPNPEPL    441  KECFTYVLKGNHSCECSRFPTGTKGHLLDS  470
prolidase  249  REI-YEVNRTAQQAAIDAAKPGMTASELDG  277

                 **  + *+*    ++*  ***+*    **
XPNPEPL    471  FARSALWDSG--LDYLHGTGHGVGSFLNVH  498
prolidase  278  VARKIITDAGYGEYFIHRLGHGIGME--VH  305

                * *   *    +*  ** **  + *** *
XPNPEPL    499  EGPCGISYKTFSDEPLEAGMIVTDEPGYYE  528
prolidase  306  EFP---SIANGNDVVLEEGMCFSIEPGIYI  333

                 *  *+***+  ++     *        *
XPNPEPL    529  DGAFGIRIENVDLVVPVKTKYNFNNRGSLT  558
prolidase  334  PGFAGVRIEDCGVLTKDGFKPFTHTSKELK  363
```

Figure 3. Sequence alignment of the C-terminal part of the human XPNPEPL amino acid sequence with *Lactobacillus delbrueckii* proline dipeptidase (EC 3.4.13.9; sp P46545). Identical amino acids are marked with an asterisk (*), homologous amino acids are marked with a plus sign (+). Gaps are indicated by a minus sign (-).

Bazan et al., showed that the C-terminal region of *E. coli, Streptomyces lividans, and Mycobacterium tuberculosis* aminopeptidase P share a common fold with methionyl aminopeptidase, proline dipeptidase, creatinase, agropine synthase, and rat p67 protein (17). His-395 of the human sequence, corresponding to the creatinase key residue, is also strictly conserved among the different aminopeptidase P sequences (Figure 2). Recently the human homologue of the rat p67 protein has been cloned and sequenced (18). This human sequence did show a very high homology to the rat p67 protein (92%), but did not show high identity with the *yeast* or *E. coli* methionyl aminopeptidase (22%). Expression of this cDNA revealed that the human homologue of the rat p67 protein displayed methionyl aminopeptidase activity (19). It remains to be shown however, whether the XPNPEPL protein shows protease activity, whether this activity is the same or resembles the aminopeptidase P activity. Expression and site-directed mutagenesis of the XPNPEPL cDNA clone will further elucidate this questions and reveal whether this protein represents another member of the peptidase family M24.

4. REFERENCES

1. Rawlings, N.D., Barrett, A.J. Evolutionary families of peptidases. Biochem J. 290, 205–218 (1993)
2. Rawlings, N.D., Barrett, A.J. Evolutionary families of metallopeptidases. In: Methods in Enzymology 248,183–210 (1995)
3. Hooper, N.M. Families of zinc metalloproteases. FEBS Letters, 354, 1–6 (1994)
4. Hendriks, D., De Meester, I., Umiel, T., Vanhoof, G., van Sande, M., Scharpé, S., Yaron, A. Aminopeptidase P and dipeptidyl peptidase IV activity in human leukocytes and in stimulated lymphocytes. Clinica Chimica Acta, 196, 87–96 (1991)
5. Vanhoof, G., De Meester, I., Goossens, F., Hendriks, D., Scharpé, S., Yaron, A. Kininase activity in human platelets: cleavage of the Arg[1]-Pro[2] bond of bradykinin by aminopeptidase P. Biochem Pharmacol 44, 479–487 (1992b)
6. Vanhoof, G., Goossens, F., De Meester, I., Hendriks, D., Scharpé, S. Proline motifs in peptides and their biological processing. FASEB J 9, 736–744 (1995)
7. Yaron, A., and Naider, F. Proline dependent structural and biological properties of peptides and proteins. Crit Rev Biochem Molec Biol 28,31–81 (1993)
8. Kitamura, S., Carbini, L.A., Carretero, O.A., Simmons, W.H., Scicli, A.G. Potentiation by aminopeptidase P of blood pressure response to bradykinin. Brit J Pharmacol 114, 6–7 (1995)
9. Yoshimoto, T., Tone, H., Honda, T., Osatomi, K., Kobayashi, R., Tsuru, D. Sequencing and high expression of aminopeptidase P gene from *Escherichia coli* HB101. J Biochem (Tokyo) 105, 412–416 (1989)
10. Butler, M.J., Aphale, J.S., DiZonno, M.A., Krygsman, P., Walczyk, E., Malek, L.T.: Intracellular aminopeptidases in *Streptomyces lividans* 66. J Ind Microbiol 13, 24–29 (1994)
11. Fleischmann, R.D., Adams, M.D., White, O., Clayton, R.A., Kirkness, E.F., Kerlavage, A.R., Bult, C.J., et al. Whole-genome random sequencing and assembly of *Haemophilus influenzae* RD. Science 269, 496–512 (1995)
12. Fraser, M.C., Gocayne, J.D., White, O., Adams, M.D., Clayton, R.A., Fleischmann, R.D., Bult, C.J. et al. The minimal gene complement of *Mycoplasma genitalium*. Science 270, 397–403 (1995)
13. Vergas Romero, C., Neudorfer, I., Mann, K., Schäfer, W. Purification and amino acid sequence of aminopeptidase P from pig kidney. Eur J Biochem 279, 262–269 (1995)
14. Hyde, R.J., Hooper, N.M., Turner, A.J. Molecular cloning and expression in COS-cells of pig kidney aminopeptidase P. Biochem. J. 319, 197–201 (1996)
15. Altschul, S.F., Gisch, W., Miller, W., Myers, E.W., and Lipman, D.J. Basic local alignment search tool. J Mol Biol 215, 403–410 (1990)
16. Rost, B., and Sander, C. Improved prediction of protein secondary structure by use of sequence profiles and neural networks, Proc Natl Acad Sci, USA 90, 7758–7762 (1993)
17. Bazan, J.F., Weaver, L.H., Roderick, S.L., Huber, R., Matthews, B.W. Sequencee and structure comparison suggest that methionine aminopeptidase, prolidase, aminopeptidase P, and creatinase share a common fold. Proc Natl Acad Sci, USA 91, 2473–2477 (1994)
18. Li, X., Chang, Y.H. Molecular cloning of a human complementary DNA encoding an initiation factor 2-associated protein (p67). Biochim Biophys Acta, 1260, 333–336 (1995)
19. LI, X., Chang, Y.H. Evidence that the human homologue of a rat initiation factor-2 associated protein (p67) is a methionine aminopeptidase. Biochem. Biophys. Res. Comm. 227,157–159 (1996)

5

SPECIFIC INHIBITORS OF AMINOPEPTIDASE P

Peptides and Pseudopeptides of 2-Hydroxy-3-Amino Acids

Angela Stöckel, Beate Stiebitz, and Klaus Neubert

Department of Biochemistry and Biotechnology
Institute of Biochemistry
Martin-Luther-University
Halle-Wittenberg, Kurt-Mothes-Str. 3, D-06120 Halle, Germany

1. INTRODUCTION

Aminopeptidase P (APP, EC 3.4.11.9) is a metal-dependent proline-specific peptidase. The enzyme splits N-terminal Xaa-Pro peptide bonds and plays an important role in the regulation of the physiological activity of peptides, e. g. bradykinin.[1, 2] Only recently, the first specific inhibitor of APP, apstatin, was discovered.[3] It is known, however, that peptides containing N-terminal 2-hydroxy-3-amino acids like bestatin and amastatin are inhibitors of the metal-dependent peptidases leucine aminopeptidase (LAP), aminopeptidase B and aminopeptidase M.[4] X-ray crystallografic results demonstrated the bestatin chelation of one of the active Zinc ions of bovine lens LAP.[5] Bestatin seems to be a natural mimetic of the tetrahedral intermediat of LAP-catalyzed substrate hydrolysis.

Both bestatin and amastatin do not show any inhibition on APP, because they do not comply with requirements for APP-catalyzed peptide hydrolysis. Therefore, it should be possible to obtain potent inhibitors of APP by adapting the bestatin structure.

2. RESULTS AND DISCUSSION

Accordingly, we designed and synthesized several bestatin analogous 2-hydroxy-3-aminoacyl-proline peptides (1), 2-hydroxy-3-amino acid pyrrolidides (2) and thiazolidides (3) (figure 1). It is assumed that the side chain of the 2-hydroxy-3-amino acid binds to the S_1-subsite and the proline or related structures binds to the S_1'-subsite.

The synthesis of 2-hydroxy-3-amino acids starting from N-protected R- or S-α-amino aldehydes was performed according to Herranz et al.[6] 2-Hydroxy-3-aminoacyl-proline peptides and 2-hydroxy-3-amino acid pyrrolidides and thiazolidides were obtained using conventional solution methods.

Cellular Peptidases in Immune Functions and Diseases, edited by Ansorge and Langner
Plenum Press, New York, 1997

Table 1. Kinetic parameters of the classical linear mixed-type inhibition of APP

	APP from *E. coli*[a]		APP from rat[b]	
Compound	K_i [μM]	α	K_i [μM]	α
(2S,3R)-HAPB-Pro-OMe·HBr	2.60	1.5	24.0	0.83
(2S,3R)-HAPB-Pro-Phe-OMe·HCl	1.26	1.7	57.0	0.84
(2S,3R)-HAPB-Pyr·HCl	19.8	2.2	37.6	1.1
(2S,3R)-HAPB-Thia·HBr	30.7	0.44	27.4	5.2
(2R,3R)-HAPB-Thia·HBr	60.3	2.3	269	3.8
(2R,3S)-HAPB-Thia·HBr	28.2	0.38	78.9	2.9
(2S,3S)-HAPB-Pro-OMe	158	0.62	854	2.4
(2S,3R)-HAMH-Thia·HBr	22.9	2.1	4.32	6.6

[a]Inhibition $APP_{E. coli}$-catalyzed hydrolysis of Lys(Abz)-Pro-Pro-4NA (S (μM): 4.8, 7.1, 9.5, 14.3) at 30 °C and pH 7.4 in 40 mM Tris/HCl containing 0.75 mM $MnCl_2$ and 24 nM APP.

[b]Inhibition of APP_{rat}-catalyzed hydrolysis of Lys(Abz)-Pro-Pro-4NA (S (μM):4, 6, 8) at 30 °C and pH 7.4 in 40 mM Tricin-Puffer containing 2.4 mM $MnCl_2$, 0.24 nM APP.

The inhibition of the enzyme-catalyzed substrate hydrolysis was investigated in a continuous assay. Lys(Abz)-Pro-Pro-4NA (Abz: 2-amino benzoyl, 4NA: 4-nitroanilid) served as fluorogenic substrate.[7]

Table 1 shows the K_i- and α-values for the inhibition of $APP_{E. coli}$ and APP_{rat} by dipeptide methylesters (HAPB-Pro-OMe, HAPB: 2-hydroxy-3-amino-4-phenyl-butanoic acid), the tripeptide methylester (HAPB-Pro-Phe-OMe), the 2-hydroxy-3-amino acid pyrrolidide (HAPB-Pyr) and thiazolidides (Xaa-Thia, Xaa: HAPB, HAMH - 2-hydroxy-3-amino-5-methylhexanoic acid). All compounds listed exhibit a linear mixed-type kinetics ($a \neq 1$, $\beta = 0$). The double reciprocal plot for the inhibition of APP_{rat} by (2S,3R)-HAMH-Thia and the corresponding replots prove the linear mixed-type inhibition (figure 2).[8]

In contrast, the 2-hydroxy-3-aminoacyl-proline dipeptides having a free C-terminal carboxyl group are slow-binding inhibitors of both enzymes (figure 3). The kinetic parameters are listed in table 2.

Compounds containing the 2-hydroxy-3-amino acid in the (2S,3R)-configuration are the most potent inhibitors.

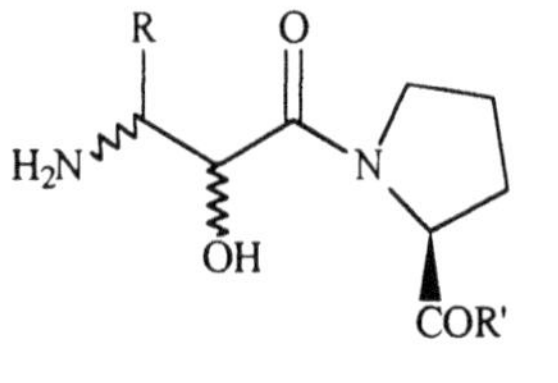

1 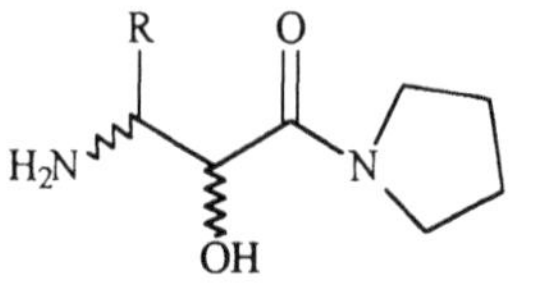 2 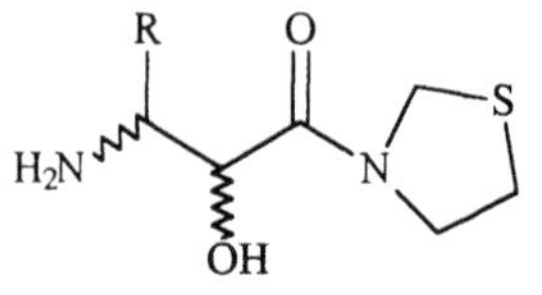 3

R: $CH_2C_6H_5$ (HAPB), $CH_2CH(CH_3)_2$ (HAMH)
R': OH, OCH_3, $PheOCH_3$

Figure 1. Structure of compounds.

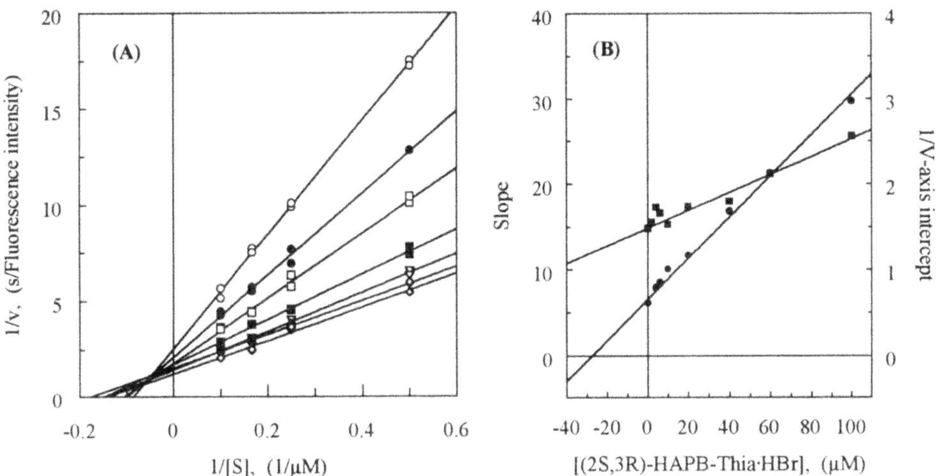

Figure 2. Inhibition of rat intestine APP by (2S,3R)-HAPB-Thia. Initial velocities were measured at pH 7.4 and 30 °C in 40 mM Tricin-buffer containing 2.4 mM $MnCl_2$, different substrate concentrations of (Lys(Abz)-Pro-Pro-4NA) and (2S,3R)-HAPB-Thia. Enzyme concentration was 0.24 nM. (A) Double-reciprocal plot (1/v vs. 1/[S]). The lines represent the following concentrations of (2S,3R)-HAPB-Thia (μM): ◊ = 0, ◆ = 2, ∇ = 10, ■ = 20, ❑ = 40, ● = 60, O = 100. (B) $Slope_{I,S}$ replot (●) and 1/V-axis intercept replot (■) of the double reciprocal plot versus inhibitor.

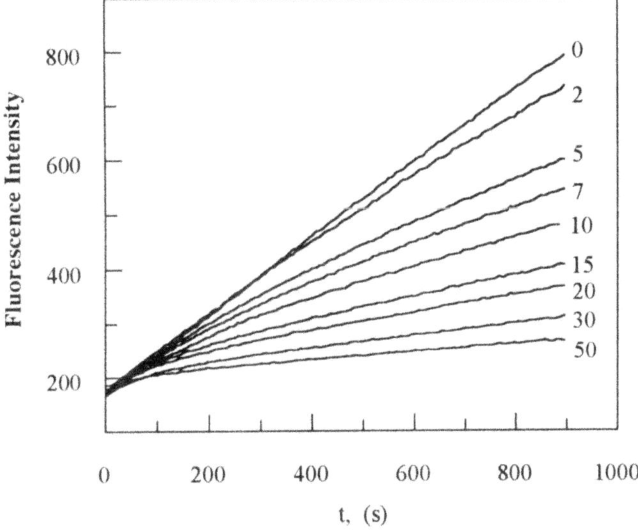

Figure 3. Reaction progress curves for hydrolysis of Lys(Abz)-Pro-Pro-4NA by APP_{rat} in the presence of increasing concentrations of (2S,3R)-HAPB-Pro-OH (μM). Assays were initiated by adding the enzyme. Curves were obtained at pH 7.4 and 30 °C in 40 mM Tricin-buffer containing 2.4 mM $MnCl_2$, 7 μM substrate (Lys(Abz)-Pro-Pro-4NA) and different concentrations of (2S,3R)-HAPB-Thia. Enzyme concentration was 0.12 nM.

Table 2. Kinetic constants of the slow-binding inhibition of APP

Compound	K_i [μM]	K_i^* [μM]	k_{on} [s^{-1}]	k_{on}/K_i^* [s^{-1} M^{-1}]
		APP from *E. coli*[a]		
(2S,3R)-HAPB-Pro-OH	14 ±4	0.17 ±0.04	0.10	588 000
(2R,3R)-HAPB-Pro-OH·TFA	190 ±30	12 ±2	0.015	1 250
(2R,3S)-HAPB-Pro-OH·TFA	16 ±3	0.74 ±0.08	0.023	31 100
(2S,3R)-HAMH-Pro-OH	14 ±5	0.37 ±0.07	0.060	162 000
		Membrane-bound APP from rat intestine[b]		
(2S,3R)-HAPB-Pro-OH	32 ± 5	2.1 ± 0.4	0.026	12 400
(2R,3R)-HAPB-Pro-OH·TFA	2100 ± 400	170 ± 20	0.014	82
(2R,3S)-HAPB-Pro-OH·TFA	120 ± 0	11 ± 1	0.017	155
(2S,3R)-HAMH-Pro-OH	110 ± 3	1.1 ± 0.1	0.064	58 200

[a]Inhibition of $APP_{E. coli}$-catalyzed hydrolysis of Lys(Abz)-Pro-Pro-4NA (S (μM): 4.8, 7.1, 9.5, 14.3) at 30 °C and pH 7.4 in 40 mM Tris/HCl-buffer containing 0.75 mM $MnCl_2$ and 5 nM APP.

[b]Inhibition of APP_{rat}-catalyzed hydrolysis of Lys(Abz)-Pro-Pro-4NA (S (μM): 4, 6, 8) at 30 °C and pH 7.4 in 40 mM Tricin-buffer containing 2.4 mM $MnCl_2$ and 0.12 nM APP.

In general, the kinetic constants for all inhibitors investigated are in the micromolar range. In all cases APP from *E. coli* is inhibited more efficiently than APP from rat. Both enzymes are distinguished by their substrate specificity. The compounds are to short to occupy the S_1, S_1' and S_2'-subsites of the rat APP, which is a requirement for APP_{rat}-catalyzed substrate hydrolysis.[9] Bacterial APP do not have such restrictions.

Nevertheless, the overall inhibition constant (K_i*) of (2S,3R)-HAMH-Pro-OH is 1.1 μM for APP from rat. This compound is the best of all known inhibitors of APP_{rat}. The most effective inhibitor of $APP_{E.coli}$ is (2S,3R)-HAPB-Pro-OH with K_i* = 0,17 μM. Both compounds inhibit APP more efficiently than apstatin (2S,3R)-HAPB-Pro-Pro-Ala-NH_2) which was recently described by Prechel et al.[3] ($APP_{E.coli}$: K_i = 14 μM, $APP_{rat\ lung}$: K_i = 2,6 μM).

In conclusion, the 2-hydroxy-3-aminoacyl-Pro-OH dipeptides are the first slow-binding inhibitors of APP. The configuration of the carbon atoms C_2 and C_3 is crucial for an effective enzyme inhibition.

3. ACKNOWLEDGMENT

This work was supported by the Deutsche Forschungsgemeinschaft, SFB 387, and by a grant to A.S. from the DECHEMA e.V. We would like to thank Prof. J. Lasch and Dr. R. Koelsch for providing APP from rat and Prof. T. Yoshimoto for providing APP from *E. coli*. We thank Dr. C. Mrestani-Klaus for the NMR results and Dr. A. Schierhorn for the mass spectroscopic investigations.

4. REFERENCES

1. Yaron, A. and Naider, F.: Proline-Dependent Structural and Biological Properties of Peptides and Proteins. Crit. Rev. Biochem. Mol. Biol. 28:31–81, 1993.

2. Simmons, W. H. and Orawski, A. T.: Membrane-bound Aminopeptidase P from Bovine Lung. J. Biol. Chem. 267: 4897–4903, 1992.
3. Prechel, M. M., Orawski, A. T., Maggiora, L. L. and Simmons, W. H.: Effect of a New Aminopeptidase P Inhibitor, Apstatin, on Bradykinin Degradation in the Rat Lung. J. Pharmacol. Exp. Ther. 275:1136–1142,1995.
4. Rich, D. H., Moon, B. J. and Harbeson, S.: Inhibition of Aminopeptidases by Amastatin and Bestatin Derivates. Effect of Inhibitor Structure on Slow-Binding Processes. J. Med. Chem. 27: 417–422, 1984.
5. Burley, S. K., David, P. R. and Lipscomb, W. N.: Leucine Aminopeptidase: Bestatin Inhibition and a Model for Enzyme-Catalyzed Peptide Hydrolysis. Proc. Natl. Acad. Sci. USA. 88: 6916–6920, 1991.
6. Herranz, R., Castro-Pichel, J., Vinuesa, S. and Garcia-Lopez, M. T.: An Improved One-Pot Method for the Stereoselective Synthesis of the (2S,3R)-3-Amino-2-hydroxy Acids: Key Intermediates for Bestatin and Amastatin. J. Org. Chem. 55:2232–2234, 1990.
7. Stiebitz. B., Stöckel, A., Lasch, J. and Neubert, K.: Development of a Continuous Assay for Aminopeptidase P: Substrate Design and Substrate Specificity, Symposium: Cellular Peptidases in Immune Functions and Diseases; poster - abstract, Magdeburg, 1996.
8. Segel, I. H.: Enzyme Kinetics. p. 170–178, John Wiley & Sons, Inc., New York, 1993.
9. Yoshimoto, T., Orawski, A. T. and Simmons, W. H.: Substrate Specifity of Aminopeptidase P from *Escherichia coli*: Comparison with Membrane-Bound Forms from Rat and Bovine Lung. Arch. Biochem. Biophys. 311:28–34, 1994.

6

γ-GLUTAMYL TRANSPEPTIDASE, A BLOOD-BRAIN BARRIER ASSOCIATED MEMBRANE PROTEIN

Splitting Peptides to Transport Amino Acids

Sabine Wolf and H. G. Gassen

Institute for Biochemistry
Technical University of Darmstadt
Petersenstr. 22, D-64287 Darmstadt, Germany

1. THE BLOOD-BRAIN BARRIER

In all vertebrates, the blood circulation not only represents the main transport system of nutrients, gases, ions, water, and metabolic products but also of hormones and components of the cellular and humoral immune system. Among the blood vessels, venoles, capillaries, and arteriols (the microcapillary system) have the smallest diameter but they are of greatest significance for the well function of any tissue. In humans, the number of capillaries is estimated to be $40x10^9$, corresponding to an exchange area of about 600 m^2. Whereas the large blood vessels transport the above mentioned compounds to the organs, the microcapillary system penetrates through every tissue and supplies the individual cells with the nutrients necessary. Three types of capillaries can be distinguished: the fenestrated, the discontinuous, and the continuous type. In the continuous type the endothelial cells, which cover the inner surface of the brain capillaries, build up a dense surface. In consequence, there is no possibility for free exchange of blood solutes to the central nervous system.

At the end of the last century, Paul Ehrlich carried out staining experiments indicating that trypan blue, a basic dye, could not penetrate into the central nervous sytem after injection in the peripheral tissue [1]. His coworker Goldmann did the control experiments in 1913: after injection of the dye into the ventricles of the brain, he could demonstrate, that only the spinal cord and the brain became stained [2]. Ultrastructural investigations using electron microscope techniques led in the late sixties to the conclusion, that the endothelial cells of the brain microcapillaries are responsible for this restriction [3,4]. Since the discovery of the morphological peculiarities of brain microcapillaries (for ultrastructure see figure 1), many other properties of biochemical and immunological nature contributed

Cellular Peptidases in Immune Functions and Diseases, edited by Ansorge and Langner
Plenum Press, New York, 1997

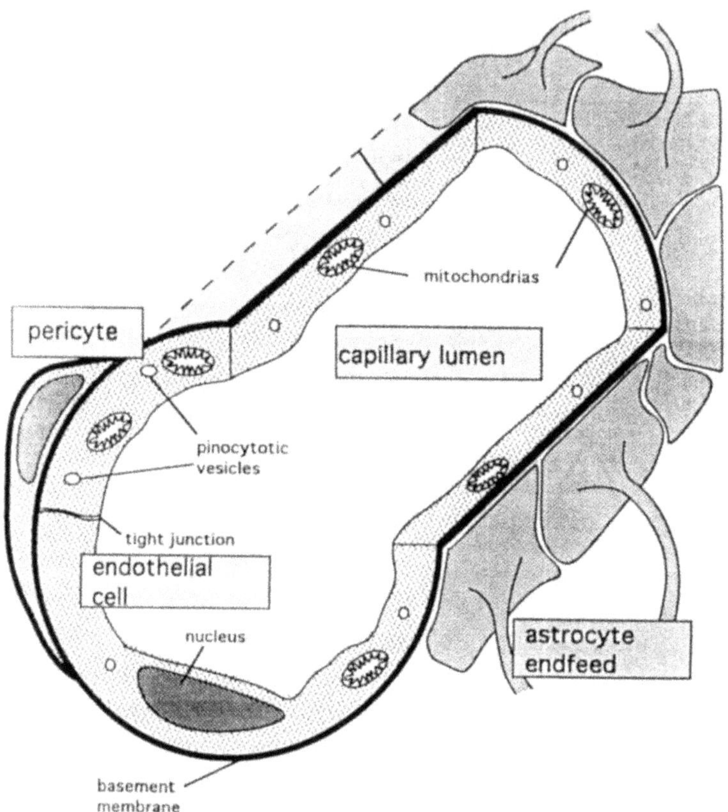

Figure 1. Schematic cross section through a brain capillary. The inner surface is coated with a single layer of endothelial cells of the continuous type, indicated by tight cell-cell contacts (tight junctions). Two membranes can be distinguished: the luminal, which is in contact to the circulating blood, and the abluminal, which faces the vessel towards the extracellular space of the brain. Pinocytotic vesicles are drastically diminished in these cells, indicating the lack of unspecific transcellular transport. In contrast mitochondrias, power stations of the cell, are abundant and obviously responsible for the delivery of ATP, which drives many energy-dependent transport sytems. Another cell type, the pericyte, is located at the abluminal membrane of the endothelial cells, and mainly covers the areas around the tight junctions. Both cell types, endothelial cells and pericytes are embedded in a basement membrane. The outer surface of this vessel "tube" is occupied with the end feet of astrocytes, a glia cell type.

the knowledge about this fascinating phenomenon of controlled exchange between blood and brain and vice versa. The totality of all these properties of brain microcapillaries, which made them distinguishable from any other capillaries of the body, are summarized under the term blood-brain barrier (bbb).

The complex functions of the blood brain barrier can be phrased as follows:

i. During diet or physical activity, the composition of blood solutes fluctuates drastically. The continuous layer of endothelial cells restricts the unspecific transport of these compounds into the brain and thus prevents the penetration of circulating peripheral neurotransmitters or neuroactive substances.
ii. As one of the organs with highest metabolical activities - the brain of an adult human, representing only 2% of the total body weight, consumes 20% of the to-

tal metabolic energy. The central nervous system is provided with nutrients from the blood via specific transport systems.

iii. Many hydrophobic substances (lypophilic xenobiotics) are able to penetrate the brain based upon their lipophilic character. These compounds and further metabolites that are generated within the brain may interfere with neuronal function. For this reason, metabolism and retrograde transport from the brain into the blood is nessecary.

Proteins and structures, that exist only in the endothelial cells of the bbb-type but not in endothelial cells of peripheral capillaries are designated as bbb-markers. Besides morphological features such as tight-junctions a number of transport proteins, such as the glucose transporter GLUT1 [5] or the P-glycoprotein [6], which is responsible for the export of toxic compounds from the brain into the circulating blood, belong to these markers.

2. γ-GLUTAMYL TRANSPEPTIDASE, A bbb MARKER

Besides the transport proteins, a number of enzymes are classified as bbb-markers. In some cases their function within the barrier is known, e.g. DOPA-decarboxylase or carboanhydrase IV. Although the enzymatic properties and the pathways involved i.e. in the kidney are well described, the precise role of γ-glutamyl transpeptidase (GGT) at the bbb is still unkown. In vertebrates, the enzyme is found in the epithelium of secretory tissues such as kidney (epithel of proximal tubules) and in many organ barriers, e.g. the blood-cerebrospinal fluid (CSF) barrier (epithelial cells of the plexus chorioideus) [7]. Since no other capillaries possess GGT in detectable amounts, the activity of this enzyme is used as a biochemical marker for the bbb [8]. Concerning the localization of the enzyme at the barrier conflicting results are reported. Gandhour et al. [9] localized GGT at the luminal membrane of endothelial cells. Other authors, including our own work, found the main fraction of GGT at the abluminal membrane of the capillaries and the highest concentration in pericytes, another microcapillary associated cell type [10,11]. A recent publication supports the luminal localization [12] of the membrane-anchored enzyme.

The GGT catalyzes three reactions: (a) The γ-glutamyl transpeptidation reaction, in which the γ-glutamyl moiety of a donor peptide is transferred to an acceptor molecule; (b) the auto-γ-transpeptidation reaction, where the donor peptide also acts as acceptor; and (c) the hydrolysis reaction, in which the enzyme hydrolyzes γ-glutamyl compounds at the γ-glutamyl amide bond [7]. Glutathione, gluthathione-S-conjugates, γ-glutamyl amino acids as well as glutamine act as donor molecules, whereas acceptors are peptides, amino acids, γ-glutamyl amino acids or water. The transpeptidation reaction is also the main step in an amino acid transport mechanism - the γ-glutamyl cycle - postulated by Orlowski and Meister [13]. The hydrolysis step is found in the leukotriene and mercapturic acid pathway [14] (see section 2.2).

2.1. GGT and Amino Acid Transport at the bbb

The γ-glutamyl cycle, which was postulated by Orlowsky and Meister [13], combines resynthesis and degradation of glutathione within a cyclic pathway (figure 2). In two intracellular reactions, glutathione is synthesized by the action of γ-glutamylcysteine synthetase and glutathione synthetase from cysteine, glutamate and glycine under consumption of of two equivalents ATP. After the translocation of glutathione to the exoplasmatic membrane, GGT transfers the γglutamine amino acid of the molecule to an acceptor amino acid. Both products of the reaction are now transported back into the cell, where further degradation takes place. Cysteinylglycine is hydrolyzed to the corresponding amino acids by a dipeptidase, whereas γ-glutamyl cyclotransferase catalyzes the hydrolysis of the γ-glutamyl-amino acid to the amino acid and 5-oxo-L-proline (pyroglutamate; 5-pyrrolidon-2-carboxylate).

Since the presented pathway results in the transport of an amino acid into a cell, Orlowski and Meister postulated that the cascade of reactions should represent an amino acid transport mechanism. Later on, Griffith and coworkers [15] demonstrated that the γ-glutamyl amino acid is translocated into the cell by a specific transport system, which has no affinity to free amino acids. Additionally, the translocation of glutathione via a specific transporter to the exoplasmatic membrane, where the active site of the exopeptidase GGT is located, was shown [16].

Nevertheless, the amino acid transport system hypothesis has been modified in attempt to incorporate the results from different experiments. Calculations of kinetic data showed that an equal stoichiometric relation between amino acid uptake and glutathione synthesis would lead to a glutathione deficit within the cell since uptake of γ-glutamyl amino acids is much faster than resynthesis of glutathione. Furthermore, amino acid translocation was studied in the mammary gland of lactating rats and in the placenta of pregnant rats [17]. These experiments indicated that the amino acid uptake was stimulated by the administration of γ-glutamyl amino acids. In contrast, transport was decreased after inhibition of the GGT, whereas uptake increased if 5-oxo-L-proline was given following the inhibition of GGT. Additionally, other amino acids were transported into the cells as

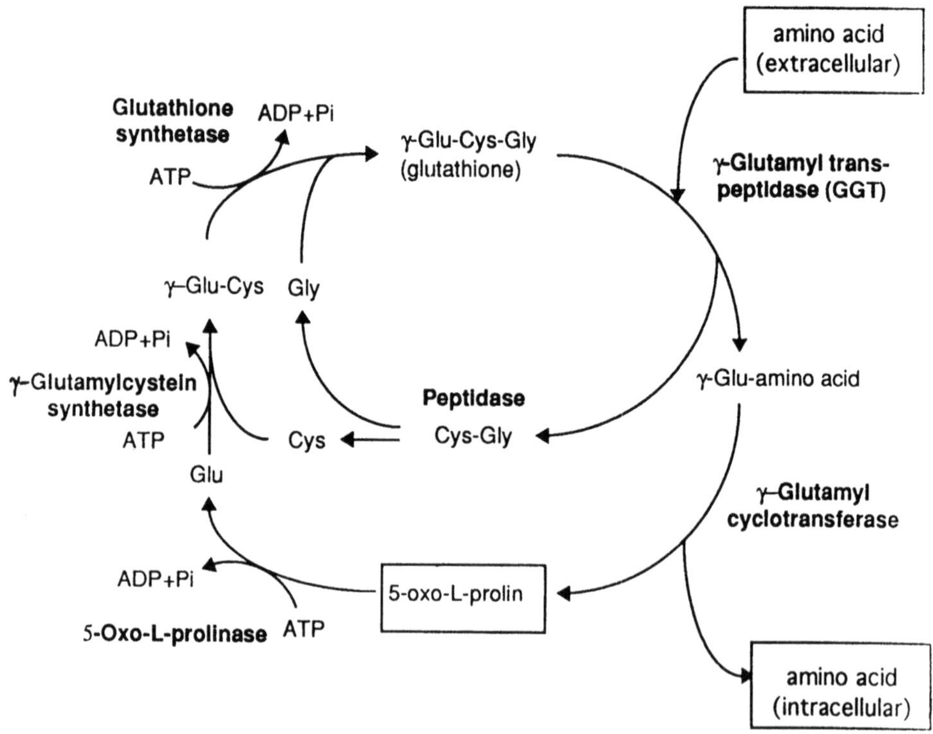

Figure 2. The γ-glutamyl cycle.

well. Based on these data, the authors concluded that the γ-glutamyl cycle acts as a signal transducing pathway: by translocation and gluthathione consumption high, extracellular amino acid levels produce a signal that stimulates other amino acid transport sytems with the same substrate specifity. Presumably this signal is related to 5-oxo-L-proline.

If the function of GGT at the bbb corresponds to this amino acid transport hypothesis, remaining enzymes of the pathway should be present in the brain microvasculature. Since the GGT has a high affinity towards amino acids such as cysteine, glutamine and methionine, the γ-glutamyl cycle would stimulate transport systems of brain capillaries, e.g. the A-, L-, ASC or $B^{+,0}$-system. In fact, Cancilla and Debault [18] could demonstrate, that the induction of GGT in cultured, cerebrovascular endothelial cells is due to an increasing uptake of amino acids via the A-system. Furthermore, the proteolytic degradation of GGT in brain capillaries led to a decrease of amino acid uptake by the A-system.

The γ-glutamyl cycle needs six enzymes, which are all present in the mammalian kidney. Within this tissue, the synthesis of renal glutathione and the resorption of amino acids from the primary urine are the prime functions of this pathway. If the GGT is involved in amino acid transport at the bbb, at least one of the other 5 enzymes- besides GGT- must be detectable in the brain microcirculating system. Regarding the enzymes involved, we selected the 5-oxo-L-prolinase (5-OP) as a candidate to prove our hypothesis. This cytosolic enzyme has been detected in different mammalian tissues [18] such as kidney, liver and brain. Furthermore, Tate and coworkers [19] could localize high activities in plexus chorioideus, the place of the blood-CSF barrier. 5-OP catalyzes the hydrolysis of 5-oxo-L-proline- the postulated second messenger molecule for amino acid transport- to glutamate under consumption of one equivalent ATP [20]. 5-OP is the only known enzyme which accepts 5-oxo-L-proline as a substrate.

Our primary interest was to investigate the specific activities of 5-OP in the brain and especially in brain microvessels and brain microvessel endothelial cells (BMECs). Therefore, we developed a sensitive enzyme test which is based on the fluorometric evaluation of the reaction product glutamate, by amino acid analysis and precolumn derivatisation with o-phthaldialdehyde (OPA) [21,22]. Different brain homogenates including isolated microvessels and BMECs as well as aorta endothelial cells were investigated with regard to their 5-OP and GGT [23] activities. The data are given in table 1.

For both enzymes, the specific activities increased from total homogenate to brain capillaries and further to BMECs. These data, as well as the lack of activity in aorta endothelial cells were taken as evidence, that 5-OP may be considered to be a bbb marker enzyme and thus should be involved in amino acid transport at the bbb. The lower 5-OP activity in comparison to GGT in brain microvessels and BMECs may be due to the prepa-

Table 1. Comparison of specific activities of γ-glutamyl transpeptidase and 5-oxo-L-prolinase in different regions of the brain and aorta endothelial cells

Region of the brain	GGT [mU/mg]	5-OP [mU/mg]
Total homogenate	2	4.3
Cerebrum	3.5	3.8
Cerebellum	4.6	3.5
Brain capillaries	19	6.4
Brain capillary endothelial cells	114	13.4
Plexus chorioideus	10	3.8
Aorta endothelial cells	0	0

ration procedure. 5-OP, which exists as a dimer, contains 26 cysteine thiol groups per subunit, and activity is lost in absence of reducing agents if the measurements take longer than a few hours.

A model pathway, that explains the function of GGT coupled amino acid transport at the bbb including 5-OP is shown in figure 3.

The A-system, an amino acid transport protein present in the abluminal membrane of BMECs [24], is stimulated by a second messenger (5-oxo-L-proline), which is an intermediate product of the γ-glutamyl cycle. This proposal, which is consistent with our findings that GGT is mainly present in the abluminal membrane of the brain capillaries [10], postulates the participation of GGT in the retrograde transport mechanisms of amino acids at the bbb. For example, glycine as a substrate of the A-sytem is permanently produced in the nervous system is acting as neurotransmitter. Possibly, the γ-glutamyl cycle supports the retrograde transport of this neuroactive compound and therefore maintains the accurate function of the central nervous system by preventing an accumulation of glycine.

Our recent work concentrates upon the localization of 5-OP in brain capillaries by immunostaining. For the preparation of 5-OP as an antigene we selected porcine kidney as a starting material since this tissue showed the highest activity (11 U/mg) in previous tests. Several attempts to get an immune serume against 5-OP failed or resulted only in a

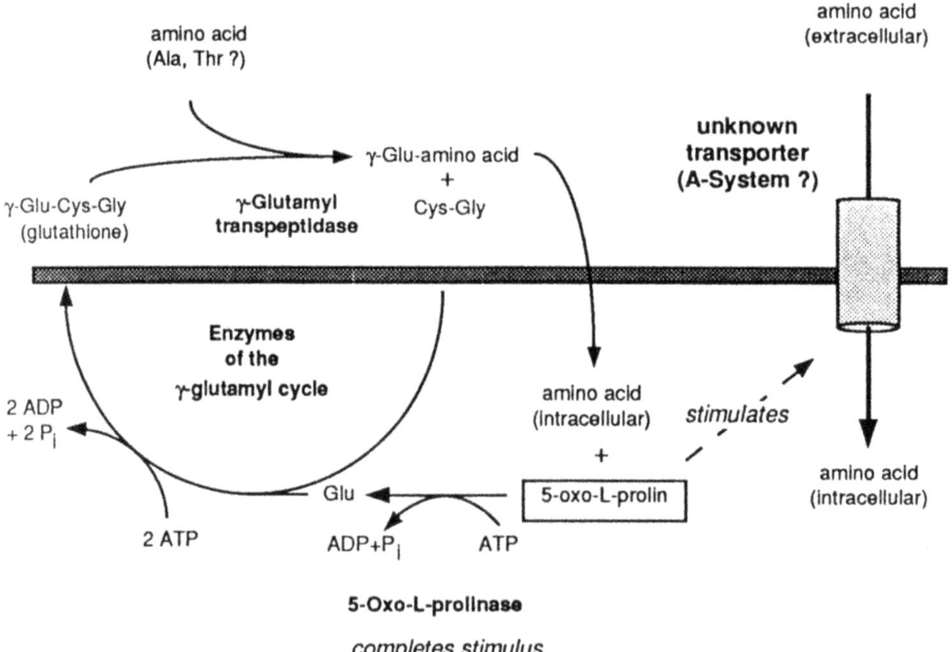

Figure 3. Function of GGT, a bbb-marker: participation via the γ-gutamyl cycle in amino acid transport at the bbb - a model.

weak immune response. Our data from amino acid sequencing (400 amino acids out of 1300 of one subunit) indicate that the enzyme is related to microbial enzymes that metabolize hydantoin derivatives. Furthermore, part of the sequence data point out similarities within conserved regions of ATP binding sites.

Techniques such as PCR with specific oligonucleotides deduced from peptide sequences additionally prove the presence of 5-OP in BMECs. In future, our main interest will focus on the determination of the complete sequence, the localization of the enzyme at the bbb and futhermore on the influence of the γ-glutamyl cycle on amino acid transport at the barrier.

2.2. GGT and Detoxication

In a physiologically different pathway GGT participates in the mercapturic acid pathway. Mercapturic acids are S-substituted N-acetyl-L-cysteine conjugates, which orginate from their respective S-substituted glutathione compounds. Since leukotriene C_4 (LTC_4) can be regarded as a S-substituted glutathione conjugate, the metabolism of peptidoleukotrienes can be considered as a special case of the mercapturic acid pathway. In general, this pathway is a scavenger route devoted to detoxify potentially harmful, electrophilic xenobiotics and endogenous toxins. The initial event is the glutathione-S-transferase (GST) catalyzed conjugation of such an electrophilic compound with glutathione. Subsequently, its γ-glutamyl residue is removed by GGT before a dipeptidase liberates the glycine moiety of the conjugate. The resulting S-substituted cysteine is then N-acetylated by the action of the cysteine-S-conjugate specific N-acetyltransferase (NAcT) to form a mercapturic acid. In contrast to the cysteine derivative, this compound is a good substrate for the organic anion transporter of the kidney by which the mercapturate is excreted into the urine. For more details in detoxicatin pathways at the bbb see Aigner et al.[25]

Since GGT and GST are present in the cerebrovascular endothelium, these cells are capable of performing at least the primary two steps of the mercaturic pathway or the leukotriene metabolism, respectively.

In accordance to our research interest - to identify the function of GGT at the bbb - we selected a further enzyme, the microsomal NAcT, as a representative for the above outlined pathways. Quantification of specific activities of NAcT by a newly established sensitive test [26] led to the conclusion, that GGT also plays a role in the detoxication via the mercapturic acid pathway. The details of the role of the leukotriene pathway at the bbb remain to be elucidated. We purified NAcT from porcine kidney [27] and obtained antibodies against the highly purified enzyme. Our present work centers in characterization of the antibodies and the localization of the NAcT at sites of the bbb.

3. CONCLUSIONS

Detailed knowledge about the function of GGT at the bbb will help to clarify different kinds of pharmacological aspects. Baba et al. [28] showed, that GGT activity was dramatically deminished as a result of brain ischema. Since GGT acts as a key enzyme in the γ-glutamyl cycle, the retrograde transport of neurotransmitter amino acids into the blood is also affected, resulting in an additional misfunction: namely accumulation of these compounds within the ischemic areas of the brain. Possibly, by administration of 5-oxo-L-proline as a drug, these secondary effects of brain ischema might be reduced.

Another aspect of GGT at the bbb is the participation in detoxication reactions via the mercapturic acid pathway. Since lipophilic xenobiotics are removed from brain, in consequence many brain affecting drugs are a target for these detoxication reactions, too. A better knowledge about the drug metabolizing routes at the bbb will lead to new conceptions for specific drug delivery into the brain.

4. ACKNOWLEDGMENTS

We greatfully acknowledge the support by the Deutsche Forschungsgemeinschaft (Wo 569-1/1).

5. REFERENCES

1. Ehrlich, P.(1885) "Das Sauerstoffbedürfnis des Oraganismus" in: "Eine farbanlytische Studie", p.69–72, A. Hirschwald, Berlin.
2. Goldmann. E. E. (1919) "Vitalfärbung am Zentralnervensytem", Abh. Preuss. Akad. Wiss. Phys. Math. 1, 1–60.
3. Reese, T. S. and Karnowsky, M. J. (1967) "Fine structural localization of a blood-brain barrier to exogenous peroxidase", J. Cell. Biol. 34, 207–217.
4. Brightman, M. W. and Reese, T. S. (1969) "Junctions between intimated apposed cell membranes in the vertebrate brain", J. Cell. Biol. 40, 648–677.
5. Weiler-Güttler, H., Zinke, H. , Möckel, B., Frey, A. and Gassen, H. G. (1989) "cDNA Cloning and Sequence Analysis of the Glucose Transporter from Porcine Blood-Brain Barrier". Biol. Chem. Hoppe-Seyler 370, 467–473.
6. Greenwood, J. (1992) "Characterization of a rat retinal endothelial cell culture and the expression of P-glycoprotein in brain and retinal endothelium in vitro", J. Neuroimmunol. 39, 123–132.
7. Frey, A. (1993) "Gamma-Glutamyl Transpeptidase: Molecular Cloning and Structural and Functional Features of a Blood-Brain Barrier Marker Protein" in: The Blood-Brain Barrier: Cellular and Molecular Biology (ed. Pardridge, W. M.), Raven Press Ltd., New York, p. 339–368.
8. Goldstein, G. W., Wolinsky, J. S. Csejtey, J. and Diamond, I (1975) "Isolation of metabolically active capillaries from rat brain", J. Neurochem. 25, 715–717.
9. Ghandour, M. S., Langley, O. K. and Varga, V. (1980) "Immunological localization of γ-glutamyl transpeptidase in cerebellum at light and electron microscope levels", Neurosci. Lett. 20, 125–129.
10. Frey, A., Meckelein, B., Weiler,-Güttler, H., Möckel, B. Flach, R. and Gassen H.G. (1991) "Pericytes of the brain microvasculature express γ-glutamyl transpeptidase", Eur. J. Biochem. 202, 421–429.
11. Risau, W., Dingler, A., Albrecht, U., Dehouck, M. P. and Cecchelli, R. (1992) "Blood-Brain Barrier Pericytes Are the Main Source of γ-Glutamyltranspeptidase Activity in Brain Capillaries", J. Neurochem. 58, 667–672.
12. Sánchez del Pino, M. M., Hawkins, R. A. and Peterson, D. R. (1995) "Biochemical Discrimination between Luminal and Abluminal Enzyme and Transport Activities of the Blood-Brain Barrier", J. Biol. Chem. 270, 14907–14912.
13. Orlowsky, M. and Meister, A. (1970) "The γ-Glutamyl Cycle: A Possible Transport System for Amino Acids", Proc. Natl. Acad. Sci. USA 67, 1248–1255.
14. Meister, A., Tate, S. S. and Griffith, O. W. (1981) "The γ-Glutamyl-Transpeptidase", Methods. Enzymol. 77, 237–253.
15. Griffith, O., Bridges, R. and Meister, A. (1979) "Transport of γ-glutamyl amino acids: Role of glutathione and γ-glutamyl transpeptidase", Proc. Natl. Acad. Sci. USA 76, 6319–6322.
16. Griffith, O. and Meister, A. (1979) "Translocation of intracellular glutathione to membrane bound γ-glutamyl transpeptidase as a discrete step in the γ-glutamyl cycle: Glutathionuria after inhibition of the transpeptidase", Proc. Natl. Acad. Sci. USA 76, 268–272.
17. Vina, J. R., Puertes, I. R., Hernandez, R. and Vina, J. (1989) "Role of γ-glutamyl cycle in the regulation of amino acid translocation", Am. J. Physiol. 257, 916–922.
18. Meister A. and Anderson, M. E. (1983) "Glutathione", Annu. Rev. Biochem. 52, 711–760.

19. Tate, S.S., Ross, L.R. and Meister, A. (1973) "The γ-Glutamyl Cycle in the Choroid Plexus: Its possible Function in Amino Acid Transport", Proc. Natl. Acad. Sci. USA 70, 1447–1493.
20. Van der Werf, P., Orlowski, M. and Meister, A. (1971) "Enzymatic Conversion of 5-Oxo-L-Proline to L-glutamate Coupled with Cleavage of Adenosine Triphosphate, a Reaction in the γ-Glutamyl Cycle", Proc. Natl. Acad. Sci. USA 68, 2982–2985.
21. Hill, D. W., Walters, F. H., Wilson, T. D. and Stuart, J. D. (1979) "High Performance Liquid Chromatographic Determination of Amino Acids in the Picomole Range", Anal. Chem. 51, 1338–1341.
22. Lindroth, P. and Mopper, K. (1979) "High Performance Liquid Chromatographic Determination of Subpicomole Amounts of Amino Acids by Precolumn Fluorescence Derivatisation with o-Phthaldialdehyde", Anal. Chem. 51, 1667–1674.
23. Orlowski, M. and Meister, A. (1963) "The γ-Glutamyl-p-Nitroanilid: A new convenient substrate for determination and study of L-and D-γ-glutamyl transpeptidase activities", Biochim. Biophys. Acta 73, 679–683.
24. Kannan, R., Kuhlenkamp, J. F., Ookhtens, M. and Kaplowitz, N. (1992) "Transport of Glutathione at the Blood-brain Barrier of the Rat: Inhibition by Glutathione Analogs and Age-Dependence", J. Exp. Pharm. Ther. 263(3), 964–970.
25. Aigner, A. , Wolf, S. and Gassen, H.G. (1997) "Transport und Entgiftung: Grundlagen, Ansätze und Perspektiven für die Blut-Hirn-Schranke-Forschung", Angewandte, in press.
26. Aigner, A., Jäger, M., Weber, P. and Wolf, S. (1994) "A Nonradioactive Assay for Microsomal Cysteine-S-Conjugate N-Acetyltransferase Activity by High-Pressure Liuid Chromatography", Anal. Biochem. 223, 227–231.
27. Aigner, A., Jäger, M., Pasternack, R., Weber, P., Wienke, D. und Wolf, S. (1996), "Purification and characterization of cysteine-S-conjugate N-acetyltransferase from pig kidney", Biochem. J. 317, 213–218.
28. Baba, T., Black, K. L. Ikezaki, K., Chen, K. and Becker, D. P. (1991) "Intracaroid Infusion of Leukotriene C_4 Selectively Increases Blood-Brain-Barrier Permeability after Focal Ischemia in Rats", J. Cereb Blood Flow Metab. 11 (4), 638–643.

7

STRUCTURE AND EXPRESSION OF AMINOPEPTIDASE N

Jørgen Olsen, Klaus Kokholm, Ove Norén, and Hans Sjöström

Department of Medical Biochemistry and Genetics
Biochemistry Laboratory C
The Panum Institute
University of Copenhagen
Blegdamsvej 3. DK-2200 Copenhagen N. Denmark

1. STRUCTURE OF AMINOPEPTIDASE N

Aminopeptidase N (APN) is a very abundant membrane protein in the microvillar membrane of the small intestinal absorptive epithelial cell the enterocyte [1,2]. APN is an ectopeptidase and from its position in the brush border membrane it faces the small intestinal lumen. The enzyme is therefore readily supplied with substrate molecules in the form of oligopeptides derived from nutritional proteins following the actions of gastric and pancreatic proteases. It is generally accepted that the physiological function of APN (and other brush border peptidases) is to convert oligopeptides in the small intestinal lumen into amino acids which can subsequently be absorbed by amino acid carriers. Besides the small intestine APN is also found in a wide variety of other tissues such as the endometrium [3], the kidney, the spleen and the brain [2]. The physiological role of APN in these alternative locations is unknown but it has been suggested that APN might be involved in the degradation of regulatory peptides [2]. In the recent years the realisation that APN is expressed in cells of the immune system has lent support to the idea that APN might be involved in the processing of antigens [4].

Knowledge obtained from classical biochemical studies of APN [5,6], from electron-miscroscopic studies of liposomes containing reconstituted APN [7] and from the cloning of the human APN cDNA [8,9] permit a relatively detailed picture of APN to be drawn (fig. 1).

In the membrane APN is found as a dimer of two non covalently associated subunits with a relative molecular mass of 160 kDa. The APN monomer is a type II membrane protein and is anchored to the membrane via an uncleaved signal sequence. The membrane anchor is preceded by a short cytoplasmic tail of eight amino acids. A junctional segment of approximately 40 amino acids rich in serine and threonine residues separates the largest and catalytically active C-terminal part of the enzyme from the membrane by approximately 5 nm. Computer analysis of the amino acid sequence of the junctional segment show that it has a high probability of being glycosylated at one or more serine and threonine residues and recent experimental data using radioactively labelled sugars have confirmed that this part of the enzyme is indeed O-linked glycosylated (Kristina Norén et al., submitted). The cDNA sequence revealed 10 putative N glycosylation sites and previous analysis suggest that all of these sites are utilized. Two potential tyrosine sulfation sites were also predicted which is in agreement with the reported modification of tyrosine residues by sulphation in APN [10].

The cDNA sequence revealed the presence of the amino acid sequence His Glu x x His which is a Zn^{++} binding motif found in one class of metallo peptidases [11]. The APN HEXXH motif is located within a region displaying a relatively high homology to the E.coli aminopeptidase N (the product of the PEPN gene[12,13]). The bacterial enzyme has a cytosolic location which is in agreement with the enzyme lacking the region corresponding to the membrane anchor and the junctional segment. Thus these entities have been added or preserved during the evolution of the mammalian enzyme.

APN hydrolyses oligopeptides by successive cleavage of the N-terminal peptide bond. APN display highest activity towards oligopeptides with a aliphatic branched N-terminal amino acid, however, apart from oligopeptides with the structure NH_2 aminoacyl prolyl which can not be hydrolysed the enzyme is able to cleave a broad spectrum of oligopeptides [2].

The determination of the primary structure of APN led to two additional important discoveries. First it was shown that APN is identical to the myeloid surface marker CD13[9] which is a marker of haematopoietic cells committed to the myeloid lineage. Later it was discovered

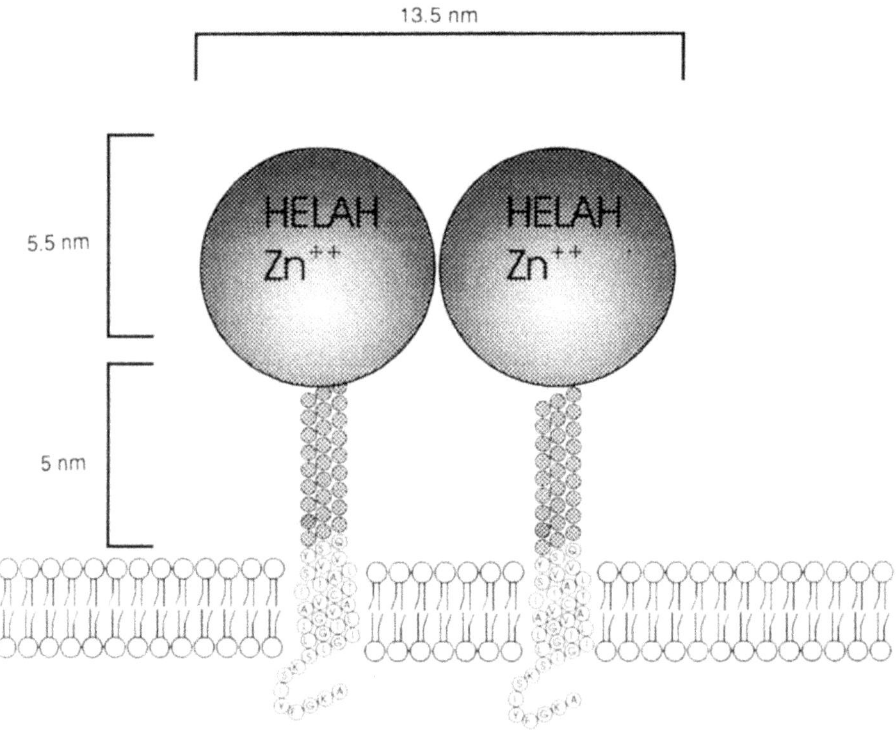

Figure 1. Structure of aminopeptidase N. APN is a type II membrane protein anchored to the membrane via an uncleaved signal sequence. The enzyme is found as a dimer of two non covalently associated monomers. The largest part of the enzyme has a globular shape and carries the active site which includes a HEXXH motif for Zn^{++} binding. The globular headgroup is separated from the membrane by a stalk which is O-linked glycosylated. In addition the APN amino acid sequence revealed 10 N linked glycosylation sites and 2 tyrosine sulphation sites.

that APN is used as a receptor for corona viridae. In pigs the transmissible gasteroenteridis virus (TGEV) which causes a severe gastroenteritis in piglets utilizes APN to enter the enterocytes [14] and likewise the pig respiratory corona virus (PRCV) make use of APN as receptor. In humans the 229E corona virus uses APN to enter alveolar cells and establish an upper respiratory tract infection [15].

2. AMINOPEPTIDASE N IN THE SMALL INTESTINE

Most of the knowledge obtained about APN is derived from studies on the enzyme in the small intestinal mucosa. The functional unit in the small intestinal epithelium is the crypt / villus entity. Immature stem cells located near the bottom of the crypts give rise to daughter cells which migrate out of the crypts and towards the tip of the villus where they undergo apoptosis and are eventually extruded to the intestinal lumen. The migration last from 2 to 3 days depending on the species and the epithelium is thus constantly renewed[16,17]. During the migration the enterocytes differentiate and start to express brush border enzymes. The expression of APN starts at the crypt / villus junction and this increase in APN expression is due to an increased level of APN mRNA [18,19]. Following its synthesis in the rough endoplasmic reticulum of the enterocyte APN is targeted to the apical microvillar membrane [20]. In the midvillar enterocyte which express the highest levels of APN the enzyme appears to be transported directly from the Golgi apparatus to the apical membrane [5,18] whereas in the more immature enterocytes in the upper crypt region some of the APN molecules may pass the basolateral membrane on their route to the apical membrane [18]. These findings have been extended by cell culture models. In the Madine Darby Canine Kidney (MDCK) cell line APN has been shown to be mainly directly transported to the apical membrane [21,22] whereas in the human colon carcinoma cell line Caco 2 which has the charasteristicts of an undifferentiated fetal enterocyte a substantial fraction of APN passes the basolateral membrane on the route to the apical microvillar membrane[23].

3. THE AMINOPEPTIDASE N GENE

The human and porcine APN genes have been cloned [24,25] (fig. 2) and mapped to chromosome 15 in man [26] and 7 in pigs [27]. The part of the human gene which encodes the

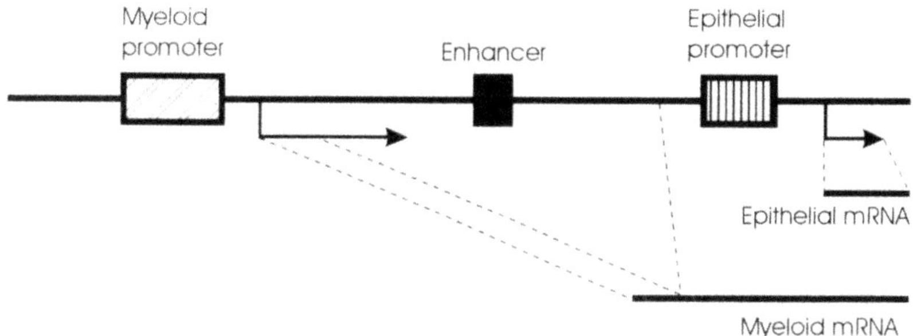

Figure 2. The APN gene is controlled by two promoters. An epithelial promoter close to the coding part of the gene drives transcription in epithelial cells. In cells of the hematopoetic system transcription occurs from an upstream myeloid promoter. The two promoters are separated by an intron which in the human gene measures approximately 8 kb. In this intron an enhancer with activity in both myeloid and epithelial cells is located. The mRNAs generated from the two promoters differ in their 5′ untranslated regions but encode the same enzyme as they share the same initiator methionine codon.

protein is divided by 20 exons [28]. Analysis of the transcriptional initiation sites in different tissues revealed the presence of two different APN promoters [29,30]. The promoters are arranged in tandem and in the human gene the distance between the two promoters have been mapped to approximately 8 kb. The epithelial promoter is closest to the coding part of the gene and is active in the enterocyte and other epithelial cells such as hepatocytes and endometrial cells [29,31]. The other myeloid promoter is placed upstream of the epithelial promoter and is active mainly in myeloid cells [30]. The transcripts from both promoters encode the same APN protein as they only differ in their 5′ untranslated regions. Sequence analysis shows that the epithelial promoter contains a TATA box whereas the myeloid promoter is GC rich and lacks a TATA box. In the large intron which separates the two promoters an enhancer with activity in both epithelial and myeloid cells is located [32].

4. THE EPITHELIAL AMINOPEPTIDASE N PROMOTER

When various deletions of the APN epithelial promoter were placed in front of a reportergene and transfected into epithelial cell lines (Caco-2 and HepG2) it was found that the first 123 bp upstream of the transcriptional initiation site were sufficient to secure full activity of the promoter [29,33]. DNase I footprinting of the epithelial promoter using nuclear extract from the Caco 2 colon carcinoma cell line reveals three protected regions from position +44 to -307 (fig. 3 and [29]).

The region from position -53 to -30 binds the Sp1 transcription factor [29]. Sp1 is a ubiquitously expressed transcription factor which binds to its recognition element via a DNA binding region containing three Zn fingers [34–38]. Sp1 has been shown to be important for the transcription of a wide variety of genes including both the so-called housekeeping genes which have promoters without a TATA box but usually contains multiple Sp1 sites as well as genes which contains a TATA box and are expressed in a more restricted tissue specific manner. Sp1 is one of the members in a gene family with at least two other members [39,40] which all share the ability to bind to a GC rich sequence with the core consensus sequence 5′ GGCGGG 3′. When an electrophoretic mobility shift assay is carried out with the APN Sp1 element and nuclear extract from Caco 2 cells three retarded bands are observed (fig. 4). By the use of specific antisera it was demonstrated that the most slowly migrating band is due to the binding of Sp1 whereas the two faster migrating bands are due to the binding of another Sp1 family member Sp3 (fig. 4). Sp3 lacks a transactivation domain and its binding therefore leads to a competitive inhibition of Sp1 mediated transcription [41].

The region from position -85 to -58 binds members of the hepatocyte nuclear factor 1 family [29,31,33]. HNF1α is a homedomain protein [42–45] which binds to DNA as a dimer. Dimerization occurs prior to DNA binding and is mediated through an N terminally located dimerization domain [46]. HNF1β is the other HNF1 family member encoded by a seperate gene [47–49]. At least in humans additional forms of the HNF1 proteins are generated through alternative splicing [50]. The HNF1 proteins are expressed in a variety of epithelial cell types. High amounts are found in hepatocytes, kidney proximal tubular cells and enterocytes [48,49,51,52]. In some cell types such eg. endometrial cells [31] only HNF1β is expressed. Otherwise a mixture of HNF1α and HNF1β is expressed and both homo and heterodimers can be formed. Both HNF1α and HNF1β are able to activate the APN epithelial promoter (fig. 5).

The region from position -112 to -90 has been designated the UF region (the upstream footprint region). A heterogeneous population of proteins bind to this element in

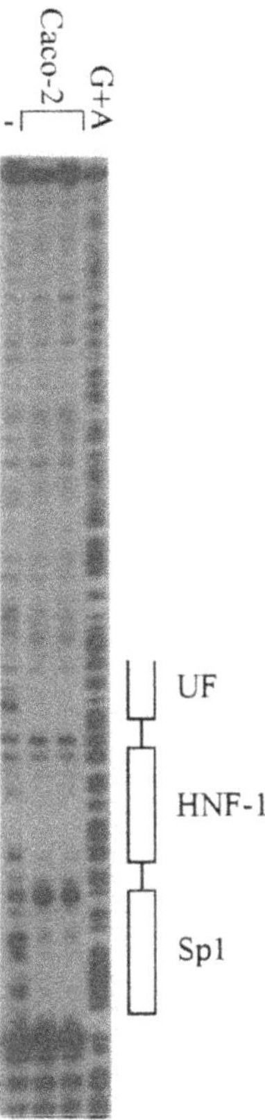

Figure 3. DNase I footprinting of the APN epithelial promoter. A DNA fragment (positions 307 to +44 of the pocine APN gene) was labelled on the upper strand, incubated without () or with 40μg nuclear proteins extracted from differentiated Caco 2 cells and digested with DNAse I. The resulting digestion products were eletrophoresed on a sequencing gel followed by autoradiography. The UF, the HNF1 and the Sp1 regions are indicated. A Maxam Gilbert G+A sequencing reaction is used as a marker.

different tissues. The sequence of the pig and the human UF element is an almost perfect inverted repeat of the sequence 5′ AGGTCA 3′ with a one bp spacing. This suggest that members of the steroid hormone nuclear receptor family bind to this element. Indeed competitive EMSA suggests that the COUP-tf factor which is a member of the steroid hormone nuclear receptor family [53] binds to the UF element [33]. Despite intensive investigations no function of the UF element could be demonstrated in neither Caco 2 nor in HepG2 cells [29,33]. However, it cannot be ruled out that the UF element under certain circumstances is important for stimulating transcription from the APN epithelial promoter.

5. THE MYELOID AMINOPEPTIDASE N PROMOTER

Little is known about functional elements present within the first 200 bp upstream of the major initiation site of the myeloid promoter. However, using nuclear extract prepared from the HUT 78 T cell line a footprint of potential importance have been identified (Kehlen et al., in preparation). The fact that the myeloid promoter is active in the HUT 78 cell line and that APN mRNA can be detected in these cells as well as in some other T cell

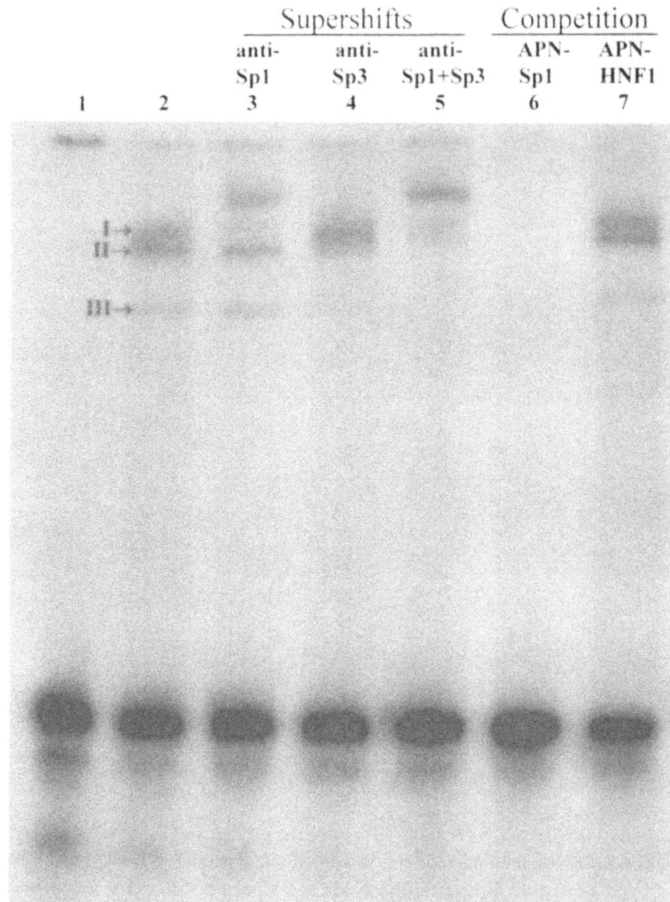

Figure 4. Sp1 and Sp3 bind to the APN promoter. Electrophoretic mobility shift assays (EMSA) with Caco-2 nuclear extracts (3 μg) was carried out with a ^{32}P labelled double-stranded oligonucleotide covering the Sp1 site of the APN promoter (APNSp1). Lane 3 shows a supershift with an antibody (1μl diluted 1/10) against Sp1 (Santa Cruz Biotechnologi): complex I is retarded. Addition of an antibody (1μl diluted 1/10) against Sp3 (Santa Cruz Biotechnologi) in lane 4 results in the disappearance of complexes II and III. The addition of both antibodies (lane 5) results in a supershift of complex I and the disappearance of complex II and III. Lanes 6 and 7 show the specificity of the three complexes. In lane 6 it is seen that all three complexes can be competed away with a 100 fold excess of unlabelled APNSp1 whereas a 100 fold excess of an unrelated probe (APNHNF1) does not affect any of the complexes.

lines [54–56] demonstrates that the CD13 marker is not entirely restricted to cells commited to the myeloid lineage although the highest levels of APN mRNA is found in myeloid cell types. With respect to putative enhancer elements present further upstream of the initiation sites utilized by the myeloid promoter, it has been elegantly shown that the Myb and Ets transcription factors cooperate to transactivate the myeloid APN promoter from sites present in the region from positions -291 to -411 [57]. Myb is a transcription factor which is known to be very important for hematopoesis as demonstrated by the severe disturbancies in the normal hematopoetic development observed in mice homozygous for a mutation in the myb gene [58]. Myb often cooperate with the Ets transcription factor [59–61] and this transcription factor was initially discovered as a fusion protein with a truncated Myb protein

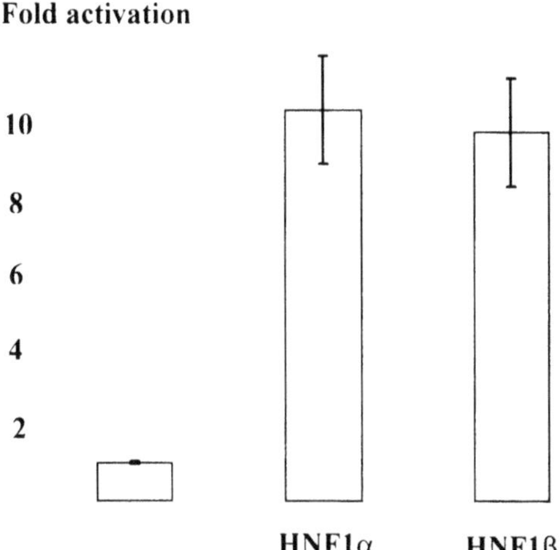

Figure 5. HNF1α and HNF1β activate the APN epithelial promoter. Transient transfections were carried out in C33-A cells which do not express HNF1 proteins. A plasmid (0.82 pmol) containing the CAT-gene under the control of a partial APN promoter (including the TATA-box, Sp1 site and the HNF1 site) was used together with a reference plasmid (0.25 pmol of pGL2 promoter; Promega Biotech, USA) containing the luciferase gene under the control of the constitutive SV40 promoter in all transfections. Co-transfections were performed as indicated with ekspression plasmids (0.67 pmol) for HNF1α (RSVHNF1α) and HNF1β (CMVHNF1β). This led to a ten fold activation from the APN promoter in either case.

in the E26 chicken leukemia virus [62,63]. Ets is the prototype in the large Ets transcription factor family which includes more than 20 members all sharing a common DNA binding motif the Ets domain [64]. Although Ets family members are also found in eg. epithelial cells many of the Ets domain factors are thought to be involved in the regulation of cell proliferation and differentiation in hematopoetic cells [64].

6. THE AMINOPEPTIDASE N ENHANCER

The screening of 5 kb upstream of the porcine epithelial APN promoter led to the identification of an enhancer fragment of 300 bp [32]. The enhancer is very well conserved between pig and man (Olsen et al., unpublished) and it is located between the two APN promoters 2.7 kb upstream of the epithelial promoter. In Caco 2 cells the addition of the enhancer increases transcription from the basal epithelial promoter almost 4 fold. The enhancer is a *bona fide* enhancer as it retains activity when it is transferred to a heterologous promoter (the SV40 early promoter) and the stimulation by the APN enhancer fragment is independent of its orientation and position relative to the heterologous promoter. Analysis of the promoter in 6 different cell lines showed that the activity of the enhancer is cell type dependent. Relatively high enhancer activity was found in both epithelial cell lines (Caco 2 and HepG2) and in the leukemia cell line K562 indicating that the enhancer might be able to cooperate with both APN promoters. The K562 cell line was established from a patient with chronic myelogeneous leukemia in blast chrisis [65] and has subsequently been shown to be able to differentiate into cells of both the erythroid and myeloid lineages [66]. Interestingly the enhancer was without effect in the adult T cell leukemia cell line Jurkat which expresses very little APN mRNA suggesting that the enhancer might be involved in

Figure 6. Structure of the APN enhancer. A schematical view of the 300 bp APN enhancer fragment from the porcine APN gene. The relative positions of regions protected by nuclear extract from Caco 2 cells are indcated as are the potential interacting transcription factors.

driving the transctription from the myeliod APN promoter preferentially in cells of the myeloid lienage. DNase I footprinting of the enhancer revealed 5 protected regions (fig. 6, I to V) using Caco 2 nuclear extracts. The footprints I to IV are all required for activity in Caco 2 cells as demonstrated by deletion analysis and by individual point mutations. The footprint V contains the consensus sequence for the serum response factor [67,68] and although the region containing this element was not required for activity in Caco 2 cells the possibility that the footprint V can confer responsiveness to growth factors still needs to be investigated. This is of particular importance as renal carcinoma cells have been shown to upregulate the expression of APN following stimulation with interleukin 4 [69]. Moreover epidermal growth factor (EGF) has also been shown to increase the intestinal expression of APN *in vivo* [70].

The regions covered by footprint I and II each contain a putative binding site for the PEA3 transcription factor. PEA3 is a member of the Ets transcription factor family and it is expressed in some epithelial cells [71]. As the Ets transcription factors recognise very similar DNA sequences [64] it is very probable that an Ets family member might bind to footprint I and II in myeloid cells and therefore have the posibility of cooperating with Myb bound upstream of the myeloid promoter. The footprint III contains a putative C/EBP binding site. C/EBP is a transcription factor which is mainly expressed in hepatocytes, enterocytes and adipocytes and it is involved in regulating transcription from several genes expressed in the liver [45]. Footprint IV contains a Sp1 element. From the mutational analysis it is clear that the most severe mutations were located to the middle of footprint I and in the 3′ end of footprint II. None of these mutations were directly in the Ets elements mentioned above and it is possible that other and yet unidentified factors bind to these regions.

7. TISSUE SPECIFIC TRANSCRIPTION FROM THE APN GENE

In a careful study carried out 30 years ago Rehfeld [72] could list 22 organs that expressed APN activity. The activity found in these organs varied by more than two orders of magnitude. We are now closer to understanding the molecular basis for the part of this remarkable regulation that occurs at the level of transcription. First of all two different promoters divide the task of mediating transcription of the common transcription unit. The epithelial promoter is a target for a ubiquitously expressed and a tissue specific transcription factor. The net result of this combination is a promoter with its activity limited to a subset of epithelial cells. The myeloid promoter responds to two well established haematopoietic transcription factors (Myb and Ets) which probably cooperates with other haematopoietic transcription factors binding to more proximal elements in the myeloid promoter. An enhancer with cell type dependent activity is located between the two promoters. The enhancer is active both in cell types which expresses APN as well as in cell types that do not. However, the enhancer will only affect APN transcription in cell types where either the upstream or the downstream promoter is already active as the function of an enhancer is to modulate the activity of a corresponding promoter.

One aspect of APN expression that is still poorly understood is its regulation by cytokines and other growth factors. The enhancer display some properties suggesting that it may be involved in this process and the role of the enhancer as target for intracellular signal transduction molecules is a topic that will be important to illuminate in future experiments.

REFERENCES

1. Norén O, Sjöström H, Danielsen EM, et al. Desnuelle P, Sjöström H, Norén O, editors.Molecular and cellular basis of digestion. Elsevier Science Publishers B.V. 1986; 19, The enzymes of the enterocyte plasma membrane. p. 335–65.
2. Kenny AJ, Stephenson LS, Turner AJ. ; Kenny AJ, Turner AJ, editors.Mammalian ectoenzymes. Amsterdam: Elsevier Science Publishers B.V. 1987; 7, Cell surface peptidases. p. 169–210.
3. Classen-Linke I, Denker H-W, Winterhager E. Apical plasma membrane-bound enzymes of rabbit uterine epithelium. Pattern changes during the periimplantation phase. Histochemistry 1987;87:517–29.
4. Larsen SL, Pedersen LO, Buus S, Stryhn A. T cell responses affected by aminopeptidase N (CD13)-mediated trimming of major histocompatibility complex class II-bound peptides. J Exp Med 1996;184:183–9.
5. Danielsen EM, Cowell GM, Norén O, et al. Kenny AJ, Turner AJ, editors.Mammalian ectoenzymes. Amsterdam: Elsevier Science Pbulishers B.V. 1987; 3, Biosynthesis. p. 47–85.
6. Danielsen EM, Cowell GM, Norén O, Sjöström H. Biosynthesis of microvillar proteins. Biochem J 1984;221:1–14.
7. Hussain MM, Tranum-Jensen J, Norén O, Sjöström H, Christiansen K. Reconstitution of purified amphiphilic pig intestinal microvillus aminopeptidase. Mode of membrane insertion and morphology. Biochem J 1981;199:179–86.
8. Olsen J, Cowel GM, K nigsh fer E, Danielsen EM, M ller J, Laustsen L, Hansen OC, Welinder KG, Engberg J, Hunziker W, Spies M, Sjöström H, Norén O. Complete amino acid sequence of human intestinal aminopeptidase N as deduced from cloned cDNA. FEBS Lett 1988;238:307–14.
9. Look AT, Ashmun RA, Shapiro LH, Peiper SC. Human myeloid plasma membrane glycoprotein CD13 (gp150) is identical to aminopeptidase N. J Clin Invest 1989;83:1299–307.
10. Danielsen EM. Tyrosine sulfation, a post-translational modification of microvillar enzymes in the small intestinal enterocyte. EMBO J 1987;6:2891–6.
11. Hooper NM. Families of zinc metalloproteases. [Review]. FEBS Letters 1994;354:1–6.
12. McCaman MT, Gabe JD. The nucleotide sequence of the pepN gene and its over-expression in Escherichia coli. Gene 1986;48:145–53.
13. Foglino M, Gharbi S, Lazdunski A. Nucleotide sequence of the pepN gene encoding aminopeptidase N of Escherichia coli. Gene 1986;49:303–9.
14. Delmas B, Gelfi J, L'Haridon R, Vogel LK, Sjöström H, Norén O, Laude, H. Aminopeptidase N is a major receptor for the entero-pathogenic coronavirus TGEV. Nature 1992;357:417–20.
15. Yeager CL, Ashmun RA, Williams RK, Cardellichio CB, Shapiro LH. Look, AT, Holmes KV. Human aminopeptidase N is a receptor for human coronavirus 229E. Nature 1992;357:420–2.
16. Gordon JI. Intestinal epithelial differentiation: new insights from chimeric and transgenic mice. J Cell Biol 1989;108:1187–94.
17. Simon TC, Gordon JI. Intestinal epithelial cell differentiation: new insights from mice, flies and nematodes. Curr Opin Genet Develop 1995;5:557–86.
18. Hansen GH, Niels-Christiansen LL, Poulsen MD, Norén O, Sjöström H. Distribution of three microvillar enzymes along the small intestinal crypt-villus axis. J Submicrosc Cytol Pathol 1994;26:453–60.
19. Norén O, Dabelsteen E, Høyer PE, Olsen J, Sjöström H, Hansen GH. Onset of transcription of the aminopeptidase N (leukemia antigen CD13) gene at the crypt/villus transition zone during rabbit enterocyte differentiation. FEBS Lett 1989;259:107–12.
20. Hansen GH, Sjöström H, Norén O, Dabelsteen E. Immunomiscroscopic localization of aminopeptidase N in pig enterocyte. Implications for the route of intercellular transport. Eur J Cell Biol 1987;43:253–9.
21. Wessels HP, Hansen GH, Fuhrer C, Look AT, Sjöström H, Norén O, Spiess M. Aminopeptidase N is directly sorted to the apical domain in MDCK cells. J Cell Biol 1990;111:2923–30.

22. Vogel LK, Spiess M, Sjöström H, Norén O. Evidence for an apical sorting signal on the ectodomain of human aminopeptidase N. J Biol Chem 1992;267:2794–7.
23. Vogel LK, Norén O, Sjöström H. Transcytosis of aminopeptidase N in Caco-2 cells is mediated by a non-cytoplasmic signal. J Biol Chem 1995;270:22933–8.
24. Look AT, Peiper SC, Rebentisch MB, Ashmun RA, Roussel MF, Lemons RS, Le Beau MM, Rubin CM, Sherr CJ. Molecular cloning, expression, and chromosomal localization of the gene encoding a human myeloid membrane antigen (gp150). J Clin Invest 1986;78:914–21.
25. Olsen J, Sjöström H, Norén O. Cloning of the pig aminopeptidase N gene. Identification of possible regulatory elements and the exon distribution in relation to the membrane-spanning region. FEBS Letters 1989;251:275–81.
26. Kruse TA, Bolund L, Grzeschik KH, Ropers HH, Olsen J, Sjöström H, Norén O. Assignment of the human aminopeptidase N (peptidase E) gene to chromosome 15q13-qter. FEBS Letters 1988;239:305–8.
27. Poulsen PH, Thomsen PD, Olsen J. Assignment of the porcine aminopeptidase N (PEPN) gene to chromosome 7cenq21. Cytogenet Cell Genet 1991;57:44–6.
28. Lerche C, Vogel LK, Shapiro LH, Norén O, Sjöström H. Human aminopeptidase N is encoded by 20 exons. Mamm Genome 1996;7 In Press.
29. Olsen J, Laustsen L, Kärnström U, Sjöström H, Norén O. Tissue-specific interactions between nuclear proteins and the aminopeptidase N promoter. J Biol Chem 1991;266:18089–96.
30. Shapiro LH, Ashmun RA, Roberts WM, Look T. Separate promoters control transcription of the human aminopeptidase N gene in myeloid and intestinal epithelial cells. J Biol Chem 1991;266:11999–2007.
31. Olsen J, Classen-Linke I, Sjöström H, Norén O. Pseudopregnancy induces the expression of hepatocyte nuclear factor-1 beta and its target gene aminopeptidase N in rabbit endometrium via the epithelial promoter. Biochem J 1995;312:31–7.
32. Olsen J, Kokholm K, Troelsen JT, Laustsen L. An enhancer with cell-type specific activity is located between the myeloid and epithelial aminipeptidase N (CD13) promoters. Manuscript 1996.
33. Olsen J, Laustsen L, Troelsen J. HNF1a activates the aminopeptidase N promoter in intestinal (Caco-2) cells. FEBS Lett 1994;342:325–8.
34. Courey AJ, Tjian R. Analysis of Sp1 in vivo reveals multiple transcriptional domains, including a novel glutamine-rich activation motif. Cell 1988;55:887–98.
35. Kadonaga JT, Courey AJ, Ladika J, Tjian R. Distinct regions of Sp1 modulate DNA binding and transcriptional activation. Science 1988;242:1566–70.
36. Kadonaga JT, Carner KR, Masiarz FR, Tjian R. Isolation of cDNA encoding transcription factor Sp1 and functional analysis of the DNA binding domain. Cell 1987;51:1079–90.
37. Briggs MR, Kadonaga JT, Bell SP, Tjian R. Purification and biochemical characterization of the promoter-specific transcription factor, Sp1. Science 1986;234:47–52.
38. Kadonaga JT, Katherine A, Tjian R. Promoter-specific activation of RNA polymerase II transcription by Sp1. Trends Biochem Sci 1986;11:20–3.
39. Hagen G, Dennig J, Preiss A, Beato M, Suske G. Functional analyses of the transcription factor Sp4 reveal properties distinct from Sp1 and Sp3. J Biol Chem 1995;270:24989–94.
40. Hagen G, Muller S, Beato M, Suske G. Cloning by recognition site screening of two novel GT box binding proteins: a family of Sp1 related genes. Nucleic Acids Res 1992;20:5519–25.
41. Hagen G, Muller S, Beato M, Suske G. Sp1-mediated transcriptional activation is repressed by Sp3. EMBO J 1994;13:3843–51.
42. Tronche F, Yaniv M. HNF1, a homeoprotein member of the hepatic transcription regulatory network. BioEssays 1992;14:579–87.
43. Mendel DB, Crabtree GR. HNF-1, a member of a novel class of dimerizing homeodomain proteins. J Biol Chem 1991;266:677–80.
44. Lai E, Darnell JEJ. Transcriptional control in hepatocytes: a window on development. Trends Biochem Sci 1991;16:427–9.
45. De Simone V, Cortese R. Transcription factors and liver-specific genes. Biochim Biophys Acta 1992;1132:119–26.
46. Nicosia A, Monaci P, Tomei L, De Francesco R, Nuzzo M, Stunnenberg H, Cortese R. A myosin-like dimerization helix and an extra-large homeodomain are essential elements of the tripartite DNA binding structure of LFB1. Cell 1990;61:1225–36.
47. Mendel DB, Hansen LP, Graves MK, Conley PB, Crabtree GR. HNF-1α and HNF1 (vHNF-1) share dimerization and homeo domains, but not activation domains, and form heterodimers in vitro. Genes & Dev 1991;5:1042–56.
48. Rey-Campos J, Chouard T, Yaniv M, Cereghini S. vHNF1 is a homeoprotein that activates transcription and forms heterodimers with HNF1. EMBO J 1991;10:1445–57.

49. De Simone V, De Magistris L, Lazzaro D, Gerstner J, Monaci P, Nicossia A, Cortese R. LFB3, a heterodimer-forming homeoprotein of the LFB1 family, is expressed in specialized epithelia. EMBO J 1991;10:1435–43.
50. Bach I, Yaniv M. More potent transcriptional activators or a transdominant inhibitor of the HNF1 homeoprotein family are generated by alternative RNA processing. EMBO J 1993;12:4229–42.
51. Cereghini S, Ott M-O, Power S, Maury M. Expression patterns of vHNF1 and HNF1 homeoproteins in early postimplantation embryos suggest distinct and sequential developmental roles. Development 1992;116:783–97.
52. Blumenfeld M, Maury M, Chouard T, Yaniv M, Condamine H. Hepatic nuclear factor 1 (HNF1) shows a wider distribution than products of its known target genes in developing mouse. Development 1991;113:589–99.
53. Wang LH, Tsai SY, Cook RG, Beattie WG, Tsai MJ, O'Malley BW. COUP transcription factor is a member of the steroid receptor superfamily. Nature 1989;340:163–6.
54. Riemann D, Gohring B, Langner J. Expression of aminopeptidase N/CD13 in tumour-infiltrating lymphocytes from human renal cell carcinoma. Immunol Lett 1994;42:19–23.
55. Wex T, Lendeckel U, Wex H, Frank K, Ansorge S. Quantification of aminopeptidase N mRNA in T cells by competitive PCR. FEBS Lett 1995;374:341–4.
56. Lendeckel U, Wex T, Kahne T, Frank K, Reinhold D, Ansorge S. Expression of the aminopeptidase N (CD13) gene in the human T cell lines HuT78 and H9. Cell Immunol 1994;153:214–26.
57. Shapiro LH. Myb and Ets proteins cooperate to transactivate an early myeloid gene. J Biol Chem 1995;270:8763–71.
58. Mucenski ML, McLain K, Kier AB, Swerdlow SH, Schreiner CM, Miller TA, Pietryga DW, Scott WJ, Jr., Potter SS. A functional c-myb gene is required for normal murine fetal hepatic hematopoiesis. Cell 1991;65:677–89.
59. Metz T, Graf T. v-myb and v-ets transform chicken erythroid cells and cooperate both in trans and in cis to induce distinct differentiation phenotypes. Genes & Dev 1991;5:369–80.
60. Melotti P, Calabretta B. Ets-2 and c-Myb act independently in regulating expression of the hematopoietic stem cell antigen CD34. J Biol Chem 1994;269:25303–9.
61. Dudek H, Tantravahi RV, Rao VN, Reddy ES, Reddy EP. Myb and Ets proteins cooperate in transcriptional activation of the mim-1 promoter. Proc Natl Acad Sci USA 1992;89:1291–5.
62. Nunn MF, Seeburg PH, Moscovici C, Duesberg PH. Tripartite structure of the avian erythroblastosis virus E26 transforming gene. Nature 1983;306:391–5.
63. Leprince D, Gegonne A, Coll J, de Taisne C, Schneeberger A, Lagrou C, Stehelin D. A putative second cell-derived oncogene of the avian leukaemia retrovirus E26. Nature 1983;306:395–7.
64. Wasylyk B, Hahn SL, Giovane A. The Ets family of transcription factors [published erratum appears in Eur J Biochem 1993 Aug 1;215(3):907]. [Review]. Eur J Biochem 1993;211:7–18.
65. Lozzio CB, Lozzio BB. Human chronic myelogenous leukemia cell-line with positive Philadelphia chromosome. Blood 1975;45:321–34.
66. Hay R, Caputo J, Chen TR, et al.American Type Culture Collection. Catalogue of Cell Lines and Hybridomas. 7th ed. Rockville, Maryland: American Type Culture Collection; 1992.
67. Treisman R. Ternary complex factors: growth factor regulated transcriptional activators. Curr Opin Genet Develop 1994;4:96–101.
68. Treisman R. The serum response element. Trends Biochem Sci 1992;17:423–6.
69. Riemann D, Kehlen A, Langner J. Stimulation of the expression and the enzyme activity of aminopeptidase N/CD13 and dipeptidylpeptidase IV/CD26 on human renal cell carcinoma cells and renal tubular epithelial cells by T cell-derived cytokines, such as IL-4 and IL-13. Clin Exp Immunol 1995;100:277–83.
70. Foltzer-Jourdainne C, Raul F. Effect of epidermal growth factor on the expression of digestive hydrolases in the jejunum and colon of newborn rats. Endocrinology 1990;127:1763–9.
71. Xin JH, Cowie A, Lachance P, Hassell JA. Molecular cloning and characterization of PEA3, a new member of the Ets oncogene family that is differentially expressed in mouse embryonic cells. Genes & Dev 1992;6:481–96.
72. Rehfeld N, Peters JE, Giesecke H, Haschen RJ. Untersuchungen über aminosaüre-arylamidasen. I Verteilung und isoenzyme der aryl-amidase im menschlichen organismus. Acta Biologica et Medica Germanica 1967;19:819–30.

8

ACTIVATION-DEPENDENT INDUCTION OF T CELL ALANYL AMINOPEPTIDASE AND ITS POSSIBLE INVOLVEMENT IN T CELL GROWTH

U. Lendeckel, T. Wex, D. Reinhold, M. Arndt, A. Ittenson, K. Frank, and S. Ansorge

Institute of Experimental Internal Medicine
Center of Internal Medicine
Otto-von-Guericke University Magdeburg
Leipziger Str. 44, D-39120 Magdeburg, Germany

1. INTRODUCTION

Membrane alanyl aminopeptidase (APN, EC 3.4.11.2) is a 150-kDa metalloprotease which has been identified as the leukocyte surface differentiation antigen CD13[1]. In humans the APN gene is located on the long arm of chromosome 15 (q11-qter) [2] with the coding part of the gene encoded by 20 exons[3]. Within the haematopoetic system APN is dominantely expressed in cells of the myelo-monocytic lineage and is used, therefore, as a standard marker in the diagnosis of leukaemia. Aminopeptidase N of leukocytes is supposed to be involved in the degradation of neuropeptides[4–7] and cytokines[8,9], but its function remains to be fully elucidated. APN may function as a corona virus receptor[10–13] and seems to contribute in tumor invasion and matrix degradation[14–15]. APN is also implicated in antigen processing[16]. Furthermore, anti-CD13 monoclonal antibodies have been shown to neutralize CMV[17]. In recent years evidence accumulated showing that malignant B and T cells[18–23] as well as activated T cells[24–26] are capable of expressing APN on the cell surface. A similar mechanism of induction may underlie the CD13 surface expression of tumor infiltrating T cells[27], or T cells[28] and NK cells[29] derived from local sites of inflammation.

T cell lines H9 and HuT78 have been shown to contain both neutral aminopeptidase activity and high amounts of APN-mRNA[30]. Determination of the exact copy number of APN transcripts by competitive PCR[31] indicated that APN-mRNA represents an abundance class II mRNA-species in these cell lines.

Here we describe the induction of aminopeptidase activity, CD13 surface expression, and APN-mRNA contents in response to T cell activation. Furthermore, we show that inhibition of either aminopeptidase activity or APN-gene expression decreases T cell proliferation and alters their cytokine production.

Cellular Peptidases in Immune Functions and Diseases, edited by Ansorge and Langner
Plenum Press, New York, 1997

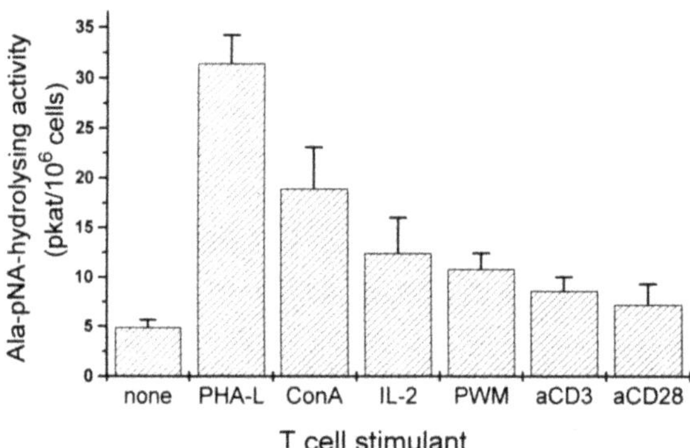

Figure 1. Capability of a panel of T cell activating agents to induce Ala-pNA-hydrolysing activity. T cells enriched by nylon wool adherence were cultured in RPMI1640 / 10% FCS in the presence of the stimulants indicated together with 10 nM PMA for 3 days. Cells were harvested and Ala-pNA-hydrolysing activity was measured in the cell lysate as described previously[30]. Concentrations of the stimulants were as follows: phytohaemagglutinine (PHA-L) and concanavalin A (ConA) 1μg/ml each, 2 μg/ml pokeweed mitogen (PWM), 50 units/ml IL-2, 100 ng/ml aCD3 or aCD28.

2. INDUCTION OF Ala-pNA-HYDROLYSING ACTIVITY

Resting peripheral T cells contain only minute amounts of Ala-pNA-hydrolysing activity (4.8 pkat/10^6 cells). However, in response to activation T cells gradually increase their neutral aminopeptidase activity reaching peak levels of up to 31 pkat/10^6 cells 3 to 4 days after activation. As shown in figure 1, different stimulating agents all induce T cell aminopeptidase activity, although with different efficiency. Ala-pNA-hydrolysing activity of activated T cells could not simply attributed to aminopeptidase N. However, support for the assumption that the increase in activity is due to an increase of APN expression comes from the observed induction of CD13 surface expression and APN-mRNA levels in response to T cell activation (see below).

3. INDUCTION OF CD13 SURFACE EXPRESSION

Freshly isolated peripheral T cells are CD13-negative by cytofluorimetric anlysis or other techniques of similar sensitivity, however, when activated by PHA-L/PMA or ConA/PMA T cells rapidly become CD13-positive. The increase in CD13 surface expression is detectable as early as 20 hours after activation and is reaching peak levels after 48 to 72 hours. Although the CD13-density per cell is rather weak, about half of the activated cells are positive by cytofluorimetric analysis and nearly 100 % of cells are CD13-positive when analyzed by fluorescence microscopy as shown in figure 2. Obviously, there is an induction of both intracellular and surface-associated CD13 in response to T cell activation.

4. ACTIVATION-DEPENDENT INCREASE OF APN-mRNA CONTENTS

Activation of T cells goes along with a dramatic increase of APN-mRNA contents, as shown recently by both RNase protection assay and dot-blot hybridization[25]. Studying the time course of APN-mRNA induction by RT-PCR we found that trace amounts only of

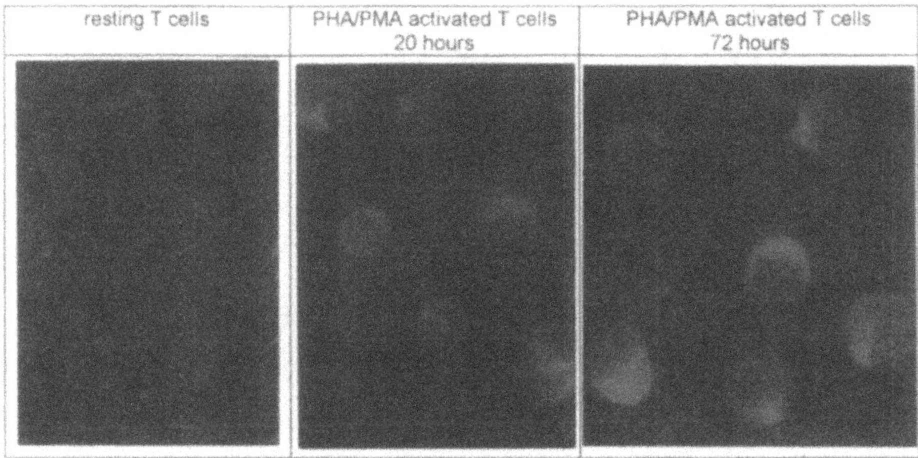

Figure 2. CD13-immunostaining of resting and activated T cells. T cells were cultured as described in legend to figure 1. Resting cells were analyzed after one day of culture in medium only. PHA-L/PMA-activated T cells were analyzed 20 or 72 hours after stimulation. Fifty microliters of cell suspension ($5x10^6$ cells/ml) were used for immunostaining of internal or membrane associated CD13 by means of the antiCD13 mab (clone WM15, Dianova) and the cell permeabilization kit (Dianova), essentially as recommended by the supplier. Normal goat serum was included to block unspecific binding of the secondary antibody, goat-anti-mouse conjugated to texas red). Controls lacking the primary antibody showed no fluorescence (not shown). Resting T cells (left) showed low CD13-immuno-reactivity, which increased gradually in response to T cell activation. Middle: 20 hours, right: 72 hours after activation. Cells were analyzed by video fluorescence microscopy (Axiovert 135 TV (Zeiss); Fluar x100 oil immersion, filter set 00 (Zeiss), 3CCD video camera (Sony), contron frame grabber) Integration: 250 fields for resting cells, 150 fields for cells activated 20 hours, 50 fields for cells activated 72 hours.

APN-mRNA are detectable until 24 hours after stimulation by PHA-L/PMA, but maximum APN-mRNA levels occur 3 days after activation and represent an 8-fold increase, approximately (figure 3).

5. INHIBITION OF APN ENZYMATIC ACTIVITY OR APN-TRANSCRIPTION

5.1. Aminopeptidase Inhibitors

The potent aminopeptidase inhibitor bestatin has been reported to influence the proliferation of a wide variety of cells[32–35]. Bestatin strongly inhibits both leucyl-aminopeptidase and soluble arginyl-aminopeptidase at sub-micromolar concentrations and partially inhibits APN and probably a number of other aminopeptidases at higher concentrations as well. The aminopeptidase inhibitors probestin and actinonin appear to be more specific, at concentrations up to 10 μM both effectors predominantely inhibit APN[36–38]. Therefore, the effects of bestatin on the DNA synthesis of activated T cells were compared to those observed with probestin and actinonin. As it is shown in figure 4, all inhibitors provoked a significant reduction of DNA synthesis at micromolar concentrations in a dose-dependent manner. At 5 μM probestin caused a 50% reduction of DNA synthesis, whereas bestatin and actinonin reduced DNA synthesis to less than 35% of control. Actinonin and bestatin were equally effective in inhibiting DNA synthesis suggesting that this effect is indeed due to the inhibition of APN.

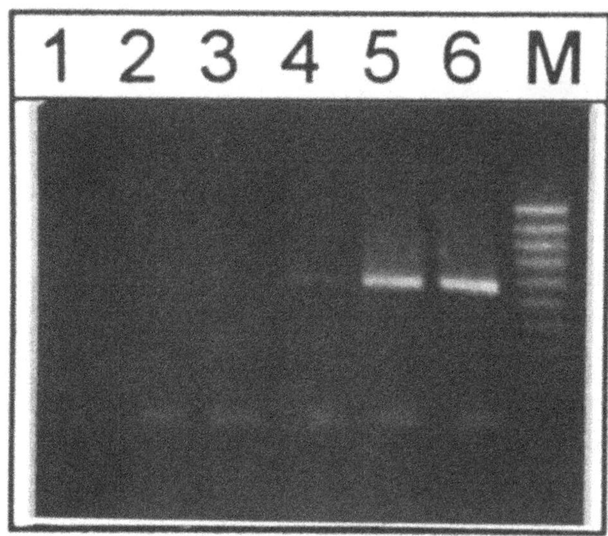

Figure 3. Induction of APN-mRNA in the course of T cell activation. RT-PCR. was performed using 500 ng total RNA from T cells harvested after 1, 3, 5, 24, 48, or 72 hours (lanes 1 to 6) of activation by PHA-L/PMA. In a 20 µl -volume RNA was transcribed by 20 units of AMV reverse transcriptase (Boehringer Ingelheim) in the supplied buffer with the addition of 0.5 mmol/l dNTP, 10 mmol/l DTT, 50 mmol/l random hexanucleotides (Boehringer Mannheim) and 50 units of placenta RNase inhibitor (Ambion) during a 1 hr incubation at 37°C. The enzyme was inactivated by a 10 min incubation at 65°C and then one tenth of the reverse transcription reaction was used as the template for the amplification reaction. Twenty five cycles were performed in an Autogene II (CLF) in 50 µl reaction buffer containing 0.5 units Goldstar Taq-polymerase (Eurogentec), 0.5 mmol/l dNTP, and 100 ng of APN-specific primers (forward: gtctactgcaacgctatcg; reverse: gatggacacatgtgggcaccttg; yielding a 573 bp product). The initial denaturing step was for 1.5 min at 95°C. Each cycle consisted of annealing for 0.7 min at 60°C, elongation for 72°C for 1.0 min, and denaturing at 96°C for 0.3 min. The final elongation step was extended to 3.0 min. Ten microliters of each reaction mixture was loaded on a 1.9% agarose gel and electrophoresed at 5V/cm in 1xTBE buffer and then stained with ethidium bromide.

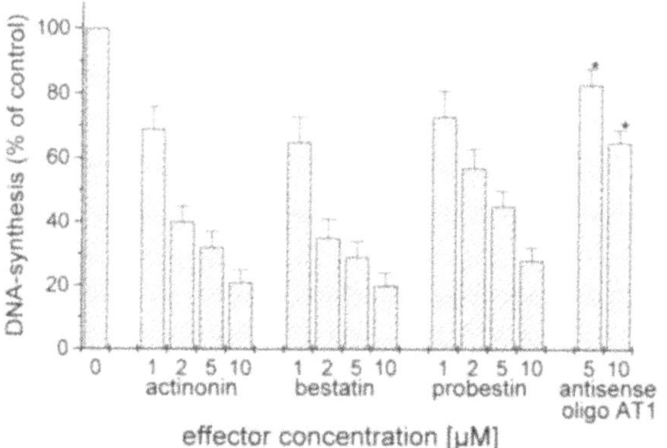

Figure 4. Inhibition of DNA-synthesis of ConA/PMA-activated peripheral T cells by amino-peptidase inhibitors and APN-specific antisense oligodesoxyribonucleotides. T cells were grown as described in legend to fig.2. Aminopeptidase inhibitors were applied simultaneously with the activation and oligodesoxyribonucleotides were given one day prior to the activation. The final concentration of the effectors was as indicated at the bottom axis. After 3 days cells were pulsed with ^{3}H-methyl-thymidine (0.2µCi per well; Amersham) for six hours. Cells were harvested onto glass fiber filters, lysed, washed with TCA and the TCA-unsoluble fraction of the incorporated radioactivity was determined by scintillation spectrometry (betaplate, LKB Pharmacia, Sweden). Given is the mean of 6 experiments ± sd. * In the case of the oligonucleotides the antisense data were compared to the sense-control.

5.2. APN-Specific Antisense Desoxyribonucleotides

Further support for the assumption that APN contributes in the cell proliferation machinery comes from the observation, that APN-specific antisense desoxyribonucleotides are capable of significantly reducing the DNA-synthesis of activated T cells. Pre-incubation of T cells for one day in the presence of 5 or 10 μM of the antisense desoxyribonucleotide AT1 (5′-thio-CACCACTGACAGTGCGATG) followed by activation with ConA/PMA reduced DNA-synthesis measured 3 days after activation to 83% or 65% of the corresponding sense control AT2 (5′-thio-CATCGCACTGTCAGTGGTG) (figure 4). Similar results were obtained using the T cell line H9 instead of peripheral T cells (not shown).

5.3. APN-Specific Antisense-Vector

As outlined above, both aminopeptidase inhibitors and APN-specific antisense desoxyribonucleotides interfer with T cell DNA synthesis. This work has been extended by showing that H9 cells transfected with an APN-antisense vector also grow at a markedly reduced rate, compared to H9 cells transfected with the corresponding sense-vector.

The vector was constructed by inserting a 2785 bp HindIII-fragment of the APN-cDNA (kindly provided by T.A.Look[1]) into pREP4 (Invitrogene). The cDNA-fragment was inserted adjacent to the RSV-promoter in either sense or antisense orientation.

Transfected H9 cells were grown in RPMI 1640 medium containing 10% fetal calf serum for two days and then DNA-synthesis was measured as described earlier[29].

Figure 5 shows a 75% decrease of DNA-synthesis in the H9 cells transfected with the APN-antisense vector compared to the sense control.

Inhibition of APN expression in H9 cells by an APN-antisense vector not only changed the proliferation rate of H9 cells, but was also found to alter the production of IL1-RA. Whereas in the supernatant of wild type H9 cells or sense transfectants less than 2 pg/ml of IL1-RA was detectable, that of H9 cells transfected with APN-antisense plasmid contained about 10 pg/ml IL1-RA, representing an fivefold increase in secreted IL -

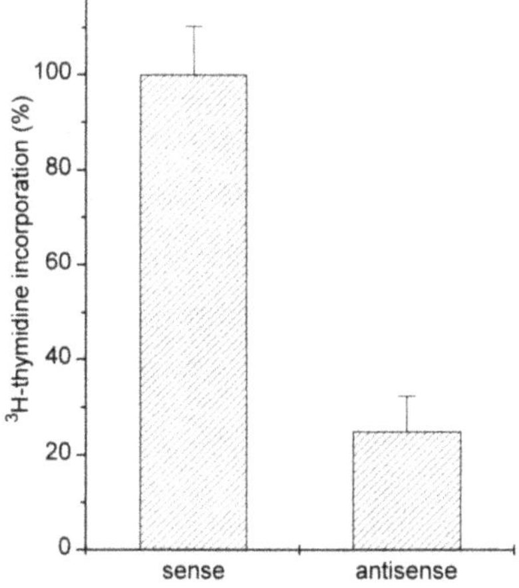

Figure 5. Inhibition of DNA-synthesis in H9-transfectants lacking APN expression. H9 cells and the derived transfectants were seeded in 96well plates at a density of 10.000 to 50.000 cells/ 200μl medium per well. After 24h cells were pulsed with ^{3}H-methyl-thymidine and DNA-synthesis was measured as described in legend to figure 4. Activities of antisense transfectants and sense transfectants were in the range of 120 - 1100 or 1100 - 3580 cpm, respectively. Given is the mean ± sd of three independent experiments, with each value determined in triplicate.

RA. IL1-RA concentrations in the supernatant were measured by ELISA (R&D Systems) according to the protocol recommended by the supplier.

6. DISCUSSION

The data presented here prove the induction of APN gene and surface expression in response to T cell activation. This induction has been shown on the level of APN-mRNA as well as on the protein and enzymatic activity levels.

With respect to CD13 surface expression and neutral aminopeptidase activity, an increase could be observed as early as 24 hours after stimulation, whereas a significant increase in APN-mRNA levels has been detected after 2 or 3 days.

The combination of PHA-L/PMA appeared to be the most efficient one capable of inducing APN/CD13 expression, but other stimulants were also effective. This suggests that there is a link between T cell activation and the expression of APN/CD13.

As it is demonstrated by fluorescence microscopy (figure 2), nearly all T cells acquire CD13 surface expression in response to T cell activation. Therefore, activation-dependent CD13 surface expression cannot be assigned to a certain T cell subset, but should be regarded as a general T cell activation marker instead.

Actinonin, bestatin and probestin have been shown recently to suppress DNA synthesis of PHA-stimulated peripheral blood mononuclear cells and of IL-1-, IL-2- or PHA/PMA-stimulated T cells[29,30]. Extending this work, we show that APN-specific antisense-oligodesoxyribonucleotides also significantly decrease DNA-synthesis of activated T cells. This led us to propose APN playing a role in regulation of cell proliferation. Further support for this assumption comes from the work on H9 cells transfected with APN antisense or sense plasmids. As seen with peripheral T cells, the T cell line H9 showed a markedly reduced DNA-synthesis when APN gene expression was inhibited by transfection of an APN-antisense plasmid. The context provided by the data on aminopeptidase inhibitors, APN-specific antisense oligodesoxyribonucleotides and the APN-antisense vector strongly implies that APN indeed contributes in the cell proliferation process. This gives a first clue on the function of APN in/on human T cells and might explain the observed activation-dependent induction of APN gene and surface expression.

Consistent with this view of APN as a regulator/modulator of cell growth is the abnormal expression on malignant lymphocytes of CD13 in cases of acute lymphocytic leukaemia (B-ALL) or chronic T cell lymphoma. It may well be, that CD13 contributes to the malignant phenotype by promoting cell proliferation. In this context it is notable that there are a few CD13-positive T cell lines existing, e.g. KARPAS (ACC31) or P12/Ichikawa (ACC34, German Collection of Microorganisms and Tissue Cultures, Braunschweig, Germany).

Unfortunately, at present the answers are largely speculative, as to how APN might affect the proliferation process. We propose, that APN proteolytically modifies peptides (and/or their precursors) involved in growth stimulation or retardation. This might also explain the changes in the level of secreted IL-1RA in the culture supernatant of H9 cells transfected with the APN-antisense plasmid.

In summary, our data demonstrate the induction of APN gene and surface expression in response to T cell activation. In addition the data are suggestive of APN playing a role in the regulation or modulation of T cell proliferation. Although the mechanisms underlying this postulated action of APN remain to be elucidated fully, one can expect APN proteolytically activating or inactivating peptides involved in T cell growth.

7. ACKNOWLEDGMENTS

This work has been supported by the Deutsche Forschungsgemeinschaft, SFB 387. Our thanks go to Christine Wolf, Ruth Hilde Hädicke and Helga Ossyra for expert technical assistance.

8. REFERENCES

1. Look, A.T., Ashmun, R.A., Shapiro, L.H., and Peiper, S.C. (1989). Human myeloid plasma membrane glycoprotein CD13 (gp150) is identical to aminopeptidase N. J. Clin. Invest. *83*, 1299–1307.
2. Watt, V.M. and Willard, H.F. (1990). The human aminopeptidase N gene: isolation, chromosome localization, and DNA polymorphism analysis. Hum. Genet. *85*, 651–654.
3. Lerche, C., Vogel, L.K., Shapiro, L.H., Noren, O., and Sjöström, H. (1996). Human aminopeptidase N is encoded by 20 exons. Mammalian Genome *7*, 712–713.
4. Furuhashi, M., Mizutani, S., Kurauchi, O., Kasugai, M., Narita, O., and Tomoda, Y. (1988). In vitro degradation of opioid peptides by human placental aminopeptidase M. Exp. Clin. Endocrinol. *92*, 235–237.
5. Giros, B., Gros, C., Solhonne, B., and Schwartz, J.C. (1986). Characterization of amino- peptidases responsible for inactivating endogenous (Met5)enkephalin in brain slices using peptidase inhibitors and anti-aminopeptidase M antibodies. Mol. Pharmacol. *29*, 281–287.
6. Ahmad, S., Wang, L., and Ward, P.E. (1992). Dipeptidyl(amino)peptidase IV and aminopeptidase M metabolize circulating substance P in vivo. J. Pharmacol. Exp. Ther. *260*, 1257–1261.
7. Shimamura, M., Hazato, T., and Iwaguchi, T. (1991). Enkephalin-degrading aminopeptidase in the longitudinal muscle layer of guinea pig small intestine: its properties and action on neuropeptides. J. Biochem. Tokyo. *109*, 492–497.
8. Kanayama, N., Kajiwara, Y., Goto, J., el Maradny, E., Maehara, K., Andou, K., and Terao, T. (1995). Inactivation of interleukin-8 by aminopeptidase N (CD13). J. Leukoc. Biol. *57*, 129–134.
9. Hoffmann, T., Faust, J., Neubert, K., and Ansorge, S. (1993). Dipeptidyl peptidase IV (CD 26) and aminopeptidase N (CD 13) catalyzed hydrolysis of cytokines and peptides with N-terminal cytokine sequences. FEBS Lett. *336*, 61–64.
10. Delmas, B., Gelfi, J., L'Haridon, R., Vogel, L.K., Sjöström, H., Noren, O., and Laude, H. (1992). Aminopeptidase N is a major receptor for the entero-pathogenic coronavirus TGEV. Nature *357*, 417–420.
11. Delmas, B., Gelfi, J., Sjöström, H., Noren, O., and Laude, H. (1993). Further characterization of aminopeptidase-N as a receptor for coronaviruses. Adv. Exp. Med. Biol. *342*, 293–298.
12. Delmas, B., Gelfi, J., Kut, E., Sjöström, H., Noren, O., and Laude, H. (1994). Determinants essential for the transmissible gastroenteritis virus-receptor interaction reside within a domain of aminopeptidase-N that is distinct from the enzymatic site. J. Virol. *68*, 5216–5224.
13. Yeager, C.L., Ashmun, R.A., Williams, R.K., Cardellichio, C.B., Shapiro, L.H., Look, A.T., and Holmes, K.V. (1992). Human aminopeptidase N is a receptor for human coronavirus 229E. Nature *357*, 420–422.
14. Saiki, I., Fujii, H., Yoneda, J., Abe, F., Nakajima, M., Tsuruo, T., and Azuma, I. (1993). Role of aminopeptidase N (CD13) in tumor-cell invasion and extracellular matrix degradation. Int. J. Cancer *54*, 137–143.
15. Fujii, H., Nakajima, M., Saiki, I., Yoneda, J., Azuma, I., and Tsuruo, T. (1995). Human melanoma invasion and metastasis enhancement by high expression of aminopeptidase N/CD13. Clin. Exp. Metastasis *13*, 337–344.
16. Hansen, A.S., Noren, O., Sjöström, H., and Werdelin, O. (1993). A mouse aminopeptidase N is a marker for antigen-presenting cells and appears to be co-expressed with major histocompatibility complex class II molecules. Eur. J. Immunol. *23*, 2358–2364.
17. Giugni, T.D., Söderberg, C., Ham, D.J., Bautista, R.M., Hedlund, K.-O., Möller, E., and Zaia, J.A. (1996). Neutralization of human cytomegalovirus by human CD13-specific antibodies. The Journal of Infectious Diseases *173*, 1062–1071.
18. Hsu, P.N., Tien, H.F., Wang, C.H., Chen, Y.C., Shen, M.C., Lin, D.T., Lin, K.H., Liang, D.C., and Lin, K.S. (1991). A subset of acute lymphoblastic leukemia with co-expression of myeloid antigens: prevalence and clinical significance. J. Formos. Med. Assoc. *90*, 225–231.
19. Hara, J., Kawa Ha, K., Yumura Yagi, K., Kurahashi, H., Tawa, A., Ishihara, S., Inoue, M., Murayama, N., and Okada, S. (1991). In vivo and in vitro expression of myeloid antigens on B-lineage acute lymphoblastic leukemia cells. Leukemia *5*, 19–25.

20. Ferrara, F., De Rosa, C., Fasanaro, A., Mele, G., Finizio, O., Schiavone, E.M., Spada, O.A., Rametta, V., and Del Vecchio, L. (1990). Myeloid antigen expression in adult acute lymphoblastic leukemia: clinicohematological correlations and prognostic relevance. Hematol. Pathol. *4*, 93–98.
21. Drexler, H.G., Thiel, E., and Ludwig, W.-D. (1991). Review of the incidence and clinical relevance of myeloid antigen-positive acute lymphoblastic leukemia. Leukemia *5*, 637–645.
22. Dreno, B., Bureau, B., Stalder, J.F., and Litoux, P. (1990). MY7 monoclonal antibody for diagnosis of cutaneous T-cell lymphoma. Arch. Dermatol. *126*, 1454–1456.
23. Dixon, J., Kaklamanis, L., Turley, H., Hickson, I.D., Leek, R.D., Harris, A.L., and Gatter, K.C. (1994). Expression of aminopeptidase-n (CD 13) in normal tissues and malignant neoplasms of epithelial and lymphoid origin. J. Clin. Pathol. *47*, 43–47.
24. Kunz, D., Bühling, F., Hütter, H.J., Aoyagi, T., and Ansorge, S. (1993). Aminopeptidase N (CD13, EC 3.3.4.11.2) occurs on the surface of resting and concanavalin A-stimulated lymphocytes. Biol. Chem. Hoppe Seyler *374*, 291–296.
25. Lendeckel, U., Wex, T., Reinhold, D., Kähne, T., Frank, K., Faust, J., Neubert, K., and Ansorge, S., (1996) Biochem. J. *319*, 817–821.
26. Ansorge, S., Schön, E., and Kunz, D. (1991). Membrane-bound peptidases of lymphocytes: functional implications. Biomed. Biochim. Acta *50*, 799–807.
27. Riemann, D., Göhring, B., and Langner, J. (1994). Expression of aminopeptidase N/CD13 in tumour-infiltrating lymphocytes from human renal cell carcinoma. Immunol. Lett. *42*, 19–23.
28. Riemann, D., Schwachula, A., Hentschel, M., and Langner, J. (1993). Demonstration of CD13/aminopeptidase N on synovial fluid T cells from patients with different forms of joint effusions. Immunobiology *187*, 24–35.
29. Bühling, F., Kunz, D., Lendeckel, U., Reinhold, D., and Ansorge, S. (1992). Aminopeptidase N (APN; CD13) on natural killer cells from peripheral blood and synovial fluid. Immunobiology *186*, 157–158.
30. Lendeckel, U., Wex, T., Kähne, T., Frank, K., Reinhold, D., and Ansorge, S. (1994). Expression of the aminopeptidase N (CD13) gene in the human T cell lines HuT78 and H9. Cell. Immunol. *153*, 214–226.
31. Wex, T., Lendeckel, U., Wex, H., Frank, K., and Ansorge, S. (1995). Quantification of aminopeptidase N mRNA in T cells by competitive PCR. FEBS Lett. *374*, 341–344.
32. Ino K., Goto S., Kosaki A., Nomura S., Asada E., Misawa T., Furuhashi Y., Mizutani S., and Tomoda Y. (1991). Groth inhibitory effect of bestatin on choriocarcinoma cell lines in vitro. Biotherapy *3*, 351–357.
33. Ino, K., Isobe, K., Goto, S., Nakashima, I., and Tomoda, Y. (1992). Inhibitory effect of bestatin on the growth of human lymphocytes. Immunopharmacology *23*, 163–171.
34. Ino, K., Goto, S., Okamoto, T., Nomura, S., Nawa, A., Isobe, K., Mizutani, S., and Tomoda, Y. (1994). Expression of aminopeptidase N on human choriocarcinoma cells and cell growth suppression by the inhibition of aminopeptidase N activity. Jpn. J. Cancer Res. *85*, 927–933.
35. Sakurada K., Imamura M., Kobayashi M., Tachibana N., Abe K., Tanaka M., Okabe M., Morioka M., Kasai M., and Sugiura T. (1990). Inhibitory effect of bestatin on the growth of human leukemic cells. Acta Oncologica *29*, 799–802.
36. Tieku, S. and Hooper, N.M. (1992). Inhibition of aminopeptidases N, A, and W. Biochemical Pharmacology *44*, 1725–1730.
37. Aoyagi T., Yoshida S., Nakamura Y., Shigihara Y., Hamada M., and Takeuchi T. (1990). Probestin, a new inhibitor of aminopeptidase M, produced by streptomyces azureus MH663- 2F6. The Journal of Antibiotics *XLIII*, 143–148. 38. Yoshida S., Nakamura Y., Naganawa H., Aoyagi T., and Takeuchi T. (1990). Probestin, a new inhibitor of aminopeptidase M, produced by streptomyces azureus MH663–2F6. The Journal of Antibiotics *XLIII*, 149–156.

9

ANTISENSE-MEDIATED INHIBITION OF AMINOPEPTIDASE N (CD13) MARKEDLY DECREASES GROWTH RATES OF HEMATOPOIETIC TUMOUR CELLS

T. Wex, U. Lendeckel, D. Reinhold, T. Kähne, M. Arndt, K. Frank, and S. Ansorge

Institute of Experimental Internal Medicine
Center of Internal Medicine
Otto-von-Guericke University Magdeburg
Leipziger Str. 44, D-39120 Magdeburg, Germany

1. INTRODUCTION

There are several reports describing neutral aminopeptidase activities expressed in immune cells [1-7]. One of the best studied members of these enzymes is the Zn-dependent aminopeptidase N (CD13, E.C.3.4.11.2, APN) Within the hematopoietic system the expression of this leukocyte surface antigen was thought to be restricted to myelomonocytic cells [1, 8, 9]. During recent years, it has been proven that lymphoid cells contain APN-mRNA and express corresponding enzymatic activity, too [2-4, 7, 10, 11]. These cells contain very low amounts of CD13 only, and therefore, are predominantly CD13-negative [1, 9, 12, 13]. Recently, different groups have been provided conclusive evidence that CD13 is strongly induced during processes such as T cell activation or inflammation, suggesting that this antigen may well be involved in the regulation of proliferation and differentiation processes [3, 12, 14-15].

In this study, we established transfectants lacking APN expression of two hematopoietic cell lines, namely the T cell line H9 and the histiocytic cell line U937. We show that the antisense-mediated inhibition of APN expression of these cells causes a strong decrease of their growth rates.

2. ANTISENSE-MEDIATED INHIBITION OF APN EXPRESSION

APN antisense and sense plasmids based on the expression vector pREP4. In this vector transcription of the inserted APN-cDNA fragments is controlled by the Rous Sarkom Virus LTR in these plasmids (figure 1).

Cellular Peptidases in Immune Functions and Diseases, edited by Ansorge and Langner
Plenum Press, New York, 1997

U937 and H9 were transfected by electroporation with the plasmids pREP4, pREP4AA (antisense) and pREP4AS (sense). The hygromycin B phosphotransferase encoded by pREP4 allowed the selection of successfully transfected cells within 4 - 6 weeks.

Due to their strong CD13 expression U937 cells represent a model for proving the chosen antisense approach. The CD13-immunoreactivity of U937 cells and the derived transfectants was investigated by immunofluorescence analysis. The range of CD13-positive cells

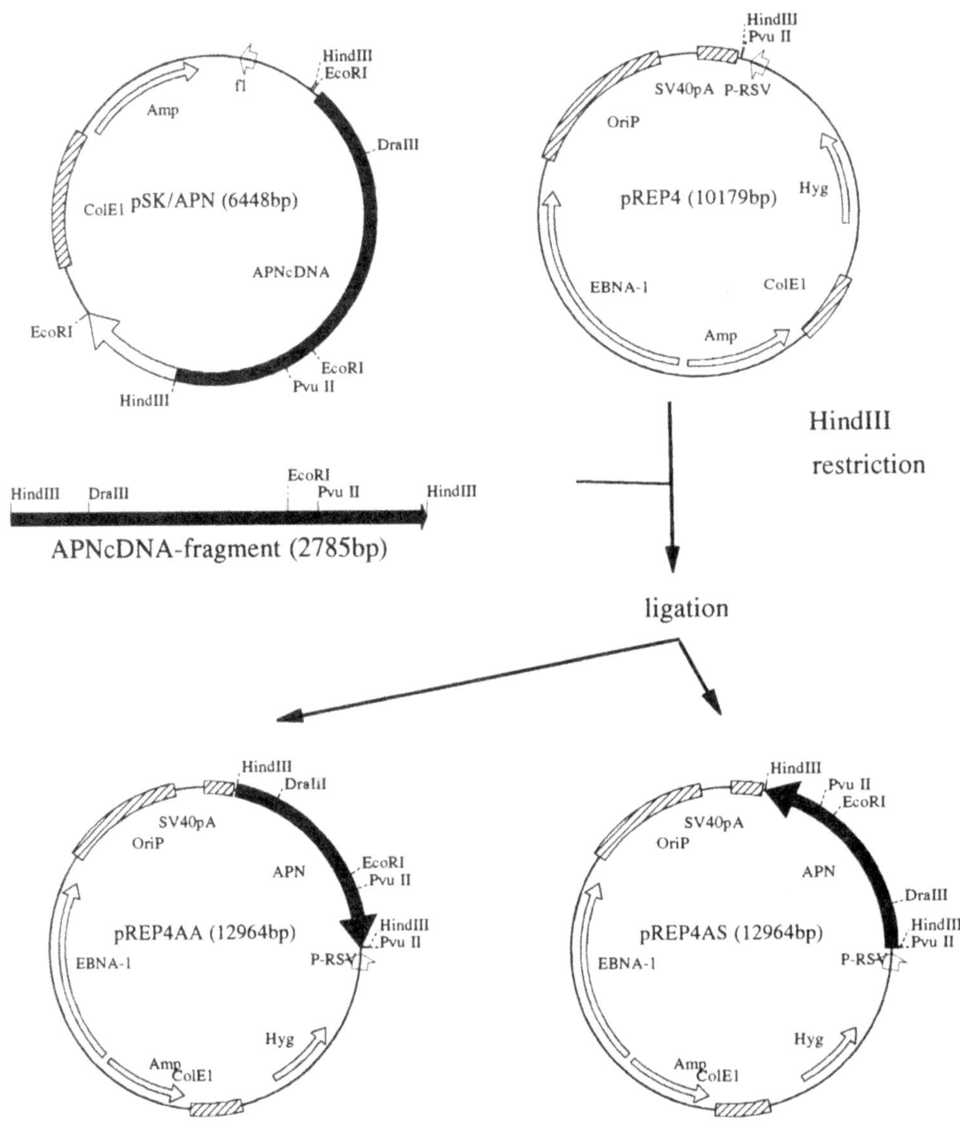

Figure 1. Construction of the plasmids pREP 4AA /AS. The 2785bp HindIII-fragment of the APNcDNA (kind gift from A.T. Look[1]) was ligated into the appropriate site of the plasmid pREP4 (ITC Biotechnology, Germany). Cells were transfected using these plasmids by electroporation (4 μg DNA per 10^6 cells; H9: 960μF, 300V; U937: 250μF, 300V; cuvette 0.4cm, Gene Pulser, Bio-Rad) and selected in medium containing 250μg/ml hygromycin (Boehringer Mannheim, Germany).

Table 1. Neutral aminopeptidase activities of U937 cells and the derived transfectants

U937	Ala-pNA hydrolysis (pkat/10^6 cells)[a], residual activity after incubation with the inhibitory anti-CD13 mab WM15 (%)[b] Lysate[a]	WM15[b]	Particulate fraction[a]	WM15[b]
Wild type	63 ± 10	58 ± 14	40 ± 2	56 ± 15
Sense transfectants	58 ± 16	n.d.	37 ± 8	n.d.
Antisense transfectants	137 ± 69	99 ± 1	11 ± 4	92 ± 10

Neutral aminopeptidase activities of cells or subcellular fractions were determined in triplicate by measuring the hydrolysis of the chromogenic substrate Ala-pNA (2.5mM) as described previously.[4] The WM15-mediated inhibition was quantified after the incubation of samples with this antibody for 30 min.

was determined to be 0–5%, 95–100%, and 85–98% for antisense, sense or wildtype cells, respectively. The particulate fraction of antisense transfectants contained 8% of aminopeptidase activity only, compared to 63% of wildtype cells (table 1). Furthermore, the anti-CD13 monoclonal antibody (mab) WM15 completely failed to inhibit Ala-pNA hydrolysis in the lysate of antisense transfectants (table 1). Interestingly, the Ala-pNA hydrolysis determined in the lysate of antisense transfectants was increased up to 200% (table 1). This opposite effect was clarified by studying the corresponding activity of isoelectrically focused (IEF) cell lysates. Using this technique, we showed that beside APN at least one other aminopeptidase activity is expressed in U937 cells, and that this activity is strongly increased in the antisense transfectants (figure 2A). The complete lack of APN-derived activity with isoelectric points (IP) between 3.5 and 4.8 gave further evidence that the APN expression was inhibited (figure 2A).

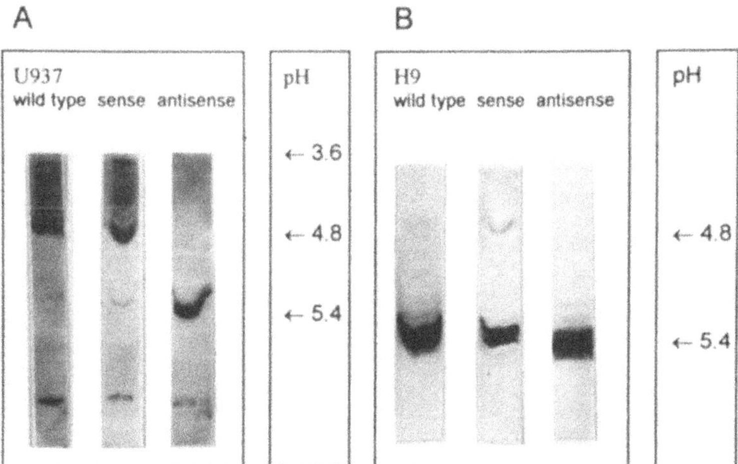

Figure 2. Neutral aminopeptidase activities (Ala-βNA-hydrolysis) of cell lines and the derived transfectants after isoelectric focusing of their cell lysates (A: U937; B: H9). Cells were solubilised in 1% n-octyl-β-D-glucopyranoside in PBS (pH7.4) and centrifuged at 100.000 x g for 30 min. The resulting supernatants were used for IEF which was performed as previously described[4]. After capillary blotting using nitro-cellulose membrane (0.45µm, Schleicher & Schüll), the blots were stained enzymatically for neutral aminopeptidase activity using Ala-β-methoxy naphthylamide and Fast Blue B.

CD13-negative H9 cells express very small amounts of APN as shown previously[4,13]. The successful inhibition of this low APN level (IP of 4.8) was demonstrated by IEF showing the lack of this activity (figure 2B). The enzymatic activities in the particulate fractions of H9 and the derived transfectants were similar (below 10 pkat/10^6 cells). Furthermore, the anti-CD13 mab WM15 failed to inhibit Ala-pNA hydrolysis of H9 cells determined in the lysates, the particulate and the cytosolic fractions as well as of vital cells (data not shown).

3. INHIBITION OF THE GROWTH RATES OF ANTISENSE TRANSFECTANTS

A strong reduction in the growth rates of both the U937 and H9 derived antisense transfectants was observed. The proliferation rates of these cells were decreased to 25–50% in comparison with sense transfectants as quantified by ^{3}H-thymidine incorporation (figure 3). The observed growth inhibition was also confirmed by the determination of cell numbers (data not shown).

Further studies were performed to investigate changes in the cytokine pattern of the H9 and the derived transfectants. The level of IL-1RA in the supernatant of antisense transfectants was increased up to fivefold compared to sense transfectants (figure 4), whereas the levels of other cytokines such as interleukin-2, -6, -10, -1beta as well as of the soluble IL-2 receptor were not changed as shown for TNF-alpha (figure 4).

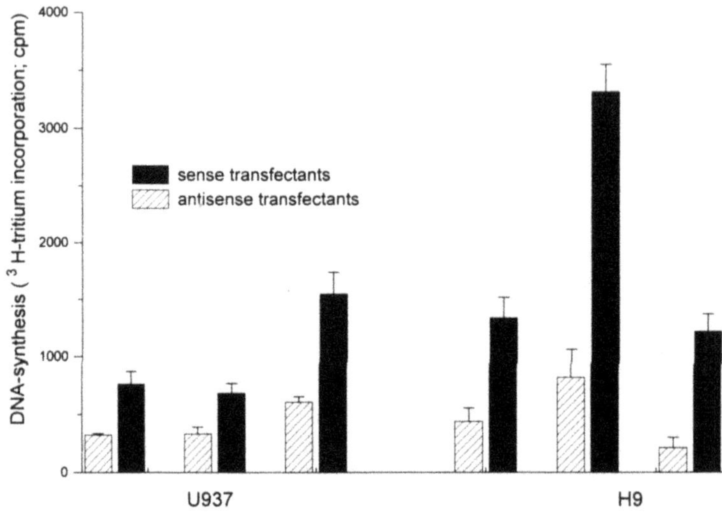

Figure 3. Inhibition of DNA synthesis in transfectants lacking APN expression. Cells were seeded in a density of 10.000 to 50.000 cells/ 200µl medium per well of a 96-well plate. After different time (24h-72h), the cultures were pulsed with ^{3}H-methyl-thymidine (0.2µCi per well; Amersham) for additional six hours. Cells were harvested onto glass fiber filters, lysed, washed with TCA and the TCA-insoluble part of the incorporated radioactivity was determined by scintillation spectrometry (betaplate, LKB Pharmacia, Sweden). Data represent three independent experiments of transfectants derived from U937 as well as H9 cells, each values was determined in triplicate. The values of wildtype cells were in the range of 8 - 18 x 10^3 cpm.

4. DISCUSSION

We investigated effects of the antisense-mediated inhibition of the APN in two hematopoietic cell lines. The complete loss of CD13-immunoreactivity in the antisense U937 cells demonstrates the capability of the constructed antisense-vector to inhibit high cellular APN expression. This was also shown by the biochemical studies mentioned above (part 2). Therefore, the resulting antisense populations could be regarded as CD13-negative cells and represent an appropriate model for functional investigations concerning the APN of hematopoietic cells. For the first time, the existence of different APN enzyme forms with respect to the isoelectric points (IP) was shown. The variability in the IPs of APN-derived activity in U937 cells (between 3.5 and 4.8, figure 2A) implies the occurrence of such enzyme forms which may result from different extent of glycosylation as shown by O'Connell and co-workers [16].

The antisense-mediated inhibition of APN expression in CD13-negative H9 cells gave conclusive evidence for an APN expression of these T cells, as suggested previously [4, 11]. However, the APN represents a small fraction of the neutral aminopeptidase activity in these cells only. In addition to the APN lymphocytes express at least one other neutral aminopeptidase activity (IP of 5.4). Since this expression is not decreased in antisense transfectants, it is not related to the APN-gene and therefore, it does not represent a modified APN as it has been suggested earlier [4]. Neutral aminopeptidase activities in lymphoid cells which differ from APN were described by other groups [17–19]. Preliminary studies aimed on the identification of the aminopeptidase expressed in H9 cells revealed that it is not identical to other known cytosolic aminopeptidase such as leucine-aminopeptidase, alanine-aminopeptidase or puromycin-sensitive aminopeptidase.

The strong reduction in the growth rates of APN-antisense transfectants supports the hypothesis that APN is involved in the regulation of immune cell's growth and proliferation. This hypothesis is underlined by the observation that the monoclonal anti-CD13 antibody WM15 inhibits the cell invasion of tumour cell lines [20] as well as the proliferation of

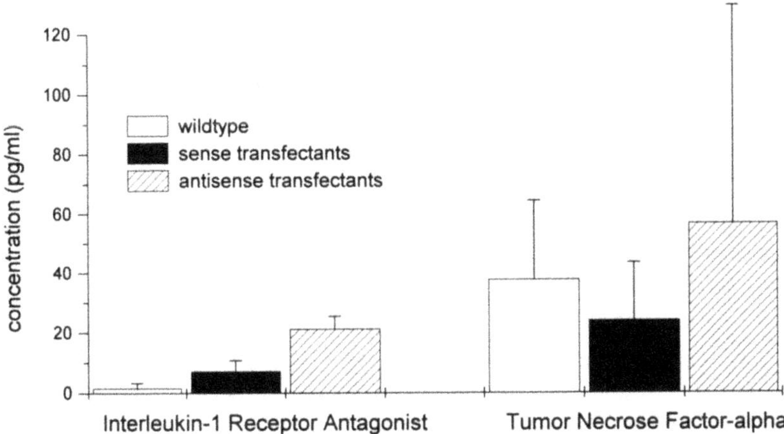

Figure 4. Levels of IL-1RA and TNF-α in the supernatant of antisense transfectants derived from H9. Cells were seeded in 24-well plates (200.000 per 2ml medium) and supernatants were harvested after 72 hours time of cultivation. Concentrations of both cytokines the IL1-RA and TNF-α were determined in double using commercially available sandwich immunoassays (R&D Systems, UK). Given are the means and standard deviations of at least 6 experiments.

choriocarcinoma cells [21]. Furthermore, it was shown that the DNA synthesis of mitogenic activated T cells could be prevented by the APN-specific inhibtor probestin and other aminopeptidase inhibitors such as bestatin, puromycin and actinonin [3, 12].

Investigations aimed on studying cytokine pattern of the transfectants revealed that IL1-RA level is increased in the supernatant of the antisense transfectants derived from H9. Interestingly, Reinhold and co-workers showed that the proliferation rate of hematopoietic cells is strongly inhibited in the presence of IL1-RA [22]. Similar inhibitory effects of this cytokine on cellular proliferation and differentiation were demonstrated by others [23–24]. Furthermore, it was shown that both the IL1-RA and the anti CD13 mab SY12 inhibits the IL-6 production of monocytes cocultured with synoviocytes [25].

Taking together, data of this study and literature imply that the growth inhibition of the antisense transfectants is due to the increased IL1-RA level. But, the question whether this increase is directly caused by inhibiting APN derived activity or APN-gene expression has not been answered. Possibly, APN is involved in the inactivation process of this cytokine and therefore, inhibition of APN leads to the increase of IL-1RA. However, there is also the possibility that other biologically active proteins are firstly influenced by inhibiting APN and changes of their levels cause the increase of IL-1RA.

To identify the process which is directly linked to the APN function, further studies will be necessary for which the established transfectants seem to be an appropriate model.

5. ACKNOWLEDGMENTS

This work was supported by the "Sonderforschungsbereich 387" of the Deutsche Forschungsgemeinschaft, Germany. We are very grateful to Christine Wolf, Ruth Hilde Hädicke, Helga Ossyra and Silke Möhring for their excellent technical assistance.

6. REFERENCES

1. Look, A.T., Ashmun, R.A., Shapiro, L.H., Peiper, S.C. (1989) Human Myeloid Plasma Membrane Glycoprotein CD13 (gp150) is identical to Aminopeptidase N. J. Clin.Invest. *83*, 1299–1307.
2. Amoscato, A.A., Stramkowski, R.M., Babcock, G.F., Alexander, J.W. (1990) Neutral surface aminopeptidase activity of human tumour cell lines. Biochim. Biophys. Acta *1041*(3), 317–319.
3. Ansorge, S., Schön, E., Kunz, D. (1991) Membrane-bound peptidases of lymphocytes: Functional implications. Biomed. Biochim. Acta *50*(4–6), 799–807.
4. Lendeckel, U., Wex, T., Kähne, T., Reinhold, D., Frank, K., Ansorge, S. (1994) Expression of the Aminopeptidase N (CD13) Gene in the Human T Cell Lines HuT78 and H9. Cell. Immunol. *153*, 214–226.
5. Rautenberg, W. and Tschesche, H. (1984) Aminopeptidases from human leukocytes. Hoppe Seylers Z Physiol Chem. *365*, 49–58.
6. Razak, K. and Newland, A.C. (1992) The Significance of Aminopeptidases and Hematopoietic Cell Differentation. Blood Rev. *6*, 243–250.
7. Riemann, D., Schwachula, A., Hentschel, M., Langner, J. (1993) Demonstration of CD13/ aminopeptidase N on synovial fluid T cells from patients with different forms of joint effusions. Immunobiol. *187*, 24–35.
8. Shipp, M.A. and Look, A.T. (1993) Hematopoietic Differentiation Antigens That Are Membrane-Associated Enzymes: Cutting Is the Key? Blood *82*, 1052–1070.
9. Olsen, J., Laustsen, L., Kärnström, U., Sjöström, H., Noren, O. (1991) Tissue-specific Interactions between Nuclear Proteins and Aminopeptidase N Promoter. J. Biol. Chem. *266* (27), 18089–96.
10. Riemann, D., Göhring, B., Langner, J. (1994) Expression of the aminopeptidase N / CD13 in tumour-infiltrating lymphocytes from human renal cell carcinoma. Immunol. Lett. *42* (1–2), 19–23.
11. Wex, T., Lendeckel, U., Wex, H., Frank, K., Ansorge, S. (1995) Quantification of aminopeptidase N mRNA in T cells by competitive PCR. FEBS Lett. *374*, 341–344.

12. Lendeckel, U., Wex, Th., Reinhold, D., Kähne, T., Frank, K., Faust, J., Neubert, K., Ansorge , S. (1996) Induction of Membrane Alanyl Aminopeptidase Gene and Surface Expression in Human T Cells by Mitogenic Activation. Biochem. J. *319*, 817–821.
13. Wex, Th., Lendeckel, U., Kähne, T., Ittenson, A., Frank, K., Ansorge, S. (1996) The Main Neutral Aminopeptidase Activity of Human Lymphoid Tumour Cell Lines Does Not Originate From The Aminopeptidase N- (APN; CD13) Gene. Biochim. Biophys. Acta (in press).
14. Kunz, D., Bühling, F., Hütter, H.-J., Aoyagi, T., Ansorge, S. (1993) Aminopeptidase N (CD13, E.C. 3.4.11.2) Occurs on the Surface of Resting and ConcanavalinA-stimulated Lymphocytes. Biol. Chem. Hoppe-Seyler *374*, 291–296.
15. Riemann, D., Kehlen, A., Langner, J. (1995) Stimulation of the expression and the enzyme activity of aminopeptidase N/CD13 and dipeptidylpeptidase IV/CD26 on human renal cell carcinoma cells and renal tubular epithelial cells by T cell-derived cytokines, such as IL-4 and IL-13. Clin. Exp. Immunol. *100* (2), 277–283.
16. O'Connell, P.J., Gerkis, V., d'Apice, A.J.F. (1991) Variable O-Glycosylation of CD13 (Aminopeptidase N). J. Biol. Chem *260*, 4593–4597
17. Murray, H., Turner, A.J., Kenny, A.J. (1994) The aminopeptidase activity in the human T- cell lymphoma line (Jurkat) is not at the cell surface and is not aminopeptidase N (CD13). Biochem. J. *298*, 353–360.
18. Belhacene, N., Mari, B., Rossi, B., Aubèrger, P. (1993) Characterisation and purification of T lymphocyte aminopeptidase B: a putative marker of T cell activation. Eur. J. Immunol. *23*, 1948–1955.
19. Favaloro, E. J., Browning, T., Nandurkar, H., Sartor, M., Bradstock, K.F., Koutts, J. (1995) Aminopeptidase-N (CD13; GP 150): Contrasting patterns of enzymatic activity in blood from patients with myeloid or lymphoid leukemia. Leuk. Res. *19* (9), 659–666.
20. Saiki, I., Fujii, H., Yoneda, J., Abe, F., Nakajima, M., Tsuruo, T., Azuma, I. (1993) Role of aminopeptidase N (CD13) in tumor-cell invasion and extracellular matrix degradation. Int. J. Cancer *54*, 137–143.
21. Ino, K., Goto, S., Okamoto, T., Nomura, S., Nawa, A., Isobe , K., Mizutani, S., Tomoda, Y. (1994) Expression of aminopeptidase N on human choriocarcinoma cells and cell growth suppression by the inhibition of aminopeptidase N activity. Jpn. J. Cancer Res. *85*, 927–933.
22. Reinhold, D., Bank, U., Bühling, F., Kähne, T., Kunz, D., Faust, J., Neubert, K., Ansorge, S. (1994) Inhibitors of dipeptidyl peptidase IV (DP IV, CD26) specifically suppress proliferation and modulate cytokine production of strongly CD26 expressing U937 cells. Immunobiol. *19*, 121–136.
23. McKenzie, R.C., Oran, A., Dinarello, C.A., Sauder, D.N. (1996) Interleukin-1 receptor antagonist inhibits subcutaneous B16 melanoma growth *in vivo*. Anticanc. Res. *16* (1), 437–441.
24. Jovcic, G., Ivanovic, Z., Biljanovic-Paunovic, L., Bugarski, D., Stosic-Grujicic, S., Milenkovic, P. (1996) *In vivo* effects of interleukin-1 receptor antagonist on hematopoietic bone marrow progenitor cells in normal mice. Leukemia *10* (3), 564–569.
25. Chomarat, P., Rissoan, M.C., Pin, J.J., Banchereau, J., Miossec, P. (1995) Contribution of IL-1, CD14, and CD13 in the increased IL-6 production induced by in vitro monocyte-synoviocyte interactions. J Immunol. *155* (7), 3645–3652.

10

CO-INCUBATION OF LYMPHOCYTES WITH FIBROBLAST-LIKE SYNOVIOCYTES AND OTHER CELL TYPES CAN INDUCE LYMPHOCYTIC SURFACE EXPRESSION OF AMINOPEPTIDASE N/CD13

Dagmar Riemann,* Astrid Kehlen, Katja Thiele, Matthias Löhn, and Jürgen Langner

Institute of Medical Immunology
Martin Luther University Halle
Straße der OdF 6, D-06097 Halle, Germany

1. INTRODUCTION

Aminopeptidase N (APN, EC 3.4.11.2) is an ubiquitously occurring transmembrane ectoenzyme. Sequence comparisons of the cloned cDNA showed that APN is identical to CD13 antigen[1]. The coding part of the CD13 gene is encoded by 20 exons[2]. CD13 has been considered to be a myeloid-specific marker because lymphocytes of peripheral blood and tonsills always are CD13 negative. However, we have previously demonstrated the expression of CD13 on synovial T cells from patients with different forms of arthritis[3] or on tumor-infiltrating lymphocytes from renal cell carcinoma[4]. To learn more about conditions in tissues resulting in the expression of CD13 on lymphocytes, we incubated tonsillar T and B cells as well as peripheral blood lympho-cytes (PBL) with various mediators in the absence and presence of different stromal cells.

2. CD13 INDUCTION ON LYMPHOCYTES

Mononuclear cells were separated from human tonsills by the standard Ficoll-Hypaque gradient method. Tonsillar cell suspensions were separated from plastic-adherent cells, T and B cells were then separated by rosetting with SRBC and were >> 96 % pure. Similarly, PBL were prepared on Ficoll-Hypaque, and T cells were isolated by rosetting

* E-mail address: dagmar.riemann@medizin.uni-halle.de. Fax: 0049-345-5574055.

Cellular Peptidases in Immune Functions and Diseases, edited by Ansorge and Langner
Plenum Press, New York, 1997

with SRBC. Monocytes were depleted by adherence to plastic. Tonsillar B cells were cultivated in Iscove medium containing 10 % FCS on adherent $CD32^+$ L cells (kindly provided by Dr. Banchereau, Schering-Plough, France) with CD40 mAb (Laboserv, 100 ng/ml)[5]. Incubation of tonsillar B cells in this system with IL-2, SAC (0.001 %), IL-10, IgM (Dianova, 10 μg/ml), IL-4, PMA (5 ng/ml) and combinations of these mediators for up to 5 days could not induce CD13 surface expression as measured by direct immunofluorescence and flow cytometry (Leu-M7, PE labeled, Becton Dickinson). Tonsillar T cells were cultured in AIMV medium with cytokines, such as IL-1, IL-2, IL-4, IL-6, IL-8 or with mitogens, such as Con A (5 μg/ml), PMA, PHA-L (1 - 5 μg/ml) or combinations of these mediators. Cytokines did not induce lymphocytic CD13 expression, and with mitogens, CD13 expression never exceeded 10 % of T cells. Similar results we observed with purified T cells from peripheral blood, e. g. with an incubation of purified T cells on $CD32^+$ L cells with CD3 mAb (Ortho, 50 μg/ml) and IL-2 (100 U/ml). CD25 expression increased to 50 - 80 % of T cells whereas CD13 expression occured in <<5 % of T cells. Contrasting with these results, in the presence of monocytes (mononuclear cell preparation), CD13 expression could be induced by mitogenic activation after 24 hours to 2 days as already described[3].

Since B and T cells have been described to survive in coculture with fibroblast-like synoviocytes (SFC)[6,7] we cocultured tonsillar lymphocytes with SFC (native or fixed with paraformaldehyde) between 1 hour and 7 days. We observed a rapid increase in lymphocytic CD13 expression (up to 90 % of cells; 1 hour values varied between 10 and 45 % of T cells and 16 and 38 % of B cells, resp.) in the coculture system[8]. SFC were obtained from patients with rheumatoid arthritis undergoing surgical synovectomy as described[9]. The coincubation-mediated induction of CD13 on lymphocytes required direct intercellular contact because cultures of SFC and lymphocytes separated by a microporous membrane insert induced no expression of CD13 on lymphocytes. The cocontact-induced CD13 expression was accompanied by an increasing Ala-pNA cleaving activity. Coculture of tonsillar lymphocytes in the presence of human macrophages from pericardial fluid[10], HUVEC, kidney tubular epithelial cells[11] or tumor cell lines, such as Caki-1 (renal cell carcinoma) and PC3 (prostate adenocarcinoma) similarly resulted in the expression of CD13 in tonsillar T or B cells[8]. Lymphocytic expression of CD13 after coculture on SFC or other cells could be further enhanced by mitogens as shown for T cells in Fig. 1. T cells preincubated with SFC for 1 hour and stimulated therafter with mitogens in the absence of SFC could be induced to express CD13 as shown in Fig. 2. Additionally, T cells prepared

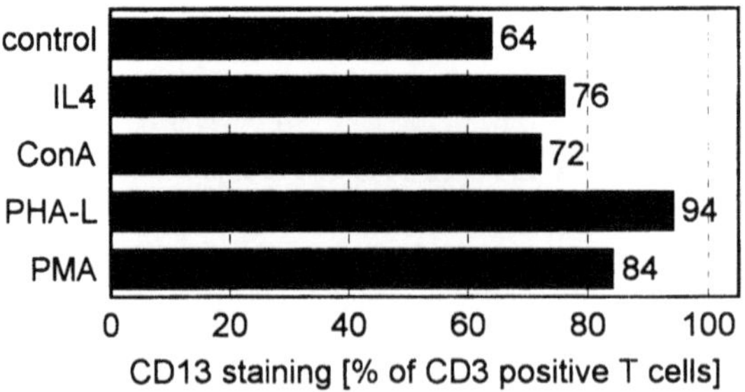

Figure 1. Effects of different mediators on the expression of CD13 on tonsillar T cells after 24 hours. T cells were cultivated on renal tubular epithelial cells in the presence of IL-2 (100 U/ml), or IL-2 in combination with IL-4 (200 U/ml), Con A (5 μg/ml) PHA-L (1 μg/ml) or PMA (5 ng/ml). Results are given as a percentage of CD13 staining of $CD3^+$ cells using two-color flow cytometry.

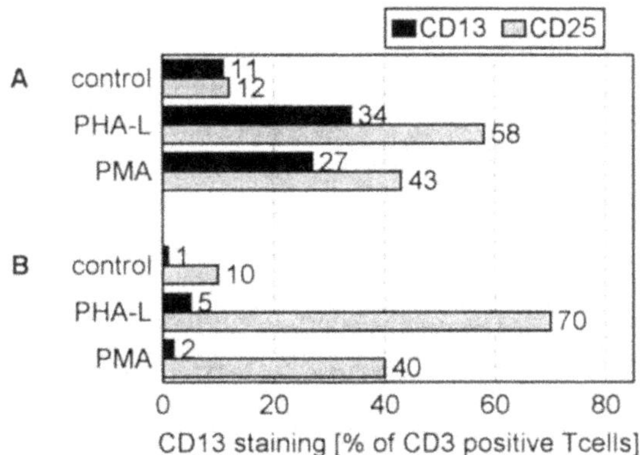

Figure 2. Mitogenic activation (1 d) of tonsillar T cells after a 1-hour incubation with (A) or without (B) SFC. T cells were cultured in serum-free AIMV with IL-2 (100 U/ml). Data are expressed as percent of $CD3^{+}$ cells.

from PBL became $CD13^{+}$ after coincubation with SFC, although with stronger interindividual scattering and after a prolonged coincubation period[8].

3. INDUCTION OF CD13 mRNA

Expression of CD13 mRNA has already been described for the leukemic T cell lines H9 and HUT78[12]. We also used quantitative RT-PCR to investigate induction of CD13 mRNA in tonsillar T or B cells after mitogenic activation or after coincubation with membranes prepared from SFC[7]. Sense (5'GTG ATG GCA GTG GAT GCA CT 3') and antisense (5' CGT CAC ATT GAG GTT CAG CAG 3') primers were synthesized according to Look[1], for details of standard construction see[11]. Tonsillar T lymphocytes in the basal state expressed 0.2 - 6.9 pg CD13 mRNA/µg total RNA. Mitogens, such as Con A, PHA-L or PMA enhanced CD13 mRNA expression of T cells after as short as 1 hour to 120 to 600 % of the control. Culture of lymphocytes in the presence of membrane preparations of SFC resulted in a marked induction of CD13 mRNA to up to 1000 % of the control. CD13 mRNA already increased after 30 min and maximal induction was seen after 2 - 16 hours of incubation with membranes[8]. A 1-hour incubation with actinomycin (1 - 5 µg/ml) decreased basal expression of tonsillar B and T cells to amounts between 10 and 20 % of the control. Furthermore, preincubation with actinomycin nearly prevented the increase in CD13 mRNA induced by SFC membranes (Fig. 3).

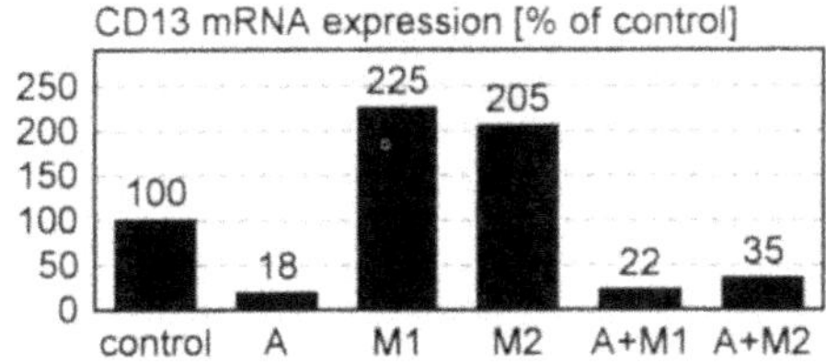

Figure 3. Expression of CD13 specific mRNA in tonsillar B and T cells cultured with two different membrane preparations in the absence or presence of actinomycin. Data are given as percent of the control.

4. DISCUSSION

We describe the capacity of cell-cell contact to induce expression of the membrane ectopeptidase CD13 on lymphocytes. Our findings can aid in the explanation of the occurrence of CD13$^+$ T cells in vivo, e. g. in the inflamed joint where direct interactions of SFC and lymphocytes have been observed. Furthermore, we describe that mitogens such as PMA or PHA increase CD13 mRNA in tonsillar lymphocytes whereas membrane expression of CD13 protein is scarcely induced. However, mitogens are able to increase the co-contact-induced lymphocytic CD13 expression. Furthermore, a short contact with SFC is sufficient to induce afterwards a surface expression of CD13 with mitogens. The time course of CD13 protein expression differs from that of the CD25 molecule induction.

Additionally, we found that membranes prepared from SFC as well as mitogens can induce CD13 mRNA in a very short time. Our results are in contrast with those of Lendeckel et al[13]. These authors, however, used peripheral blood lymphocytes, whereas we use purified tonsillar lymphocytes. Tonsillar T cells express higher percentages of CD45RO$^+$, CD69 and HLA-DR$^+$ cells than lymphocytes within the peripheral blood. A prerequisite for the contact mediated induction of CD13 on lymphocytes could be their state of maturation or activation.

Cell surface peptidases play an important role in signal transduction and communication between cells. CD13 catalyses the removal from small peptides of N-terminal, preferentially neutral, amino acid residues[14]. Expression of CD13 on B and T cells represents a potentially increased cellular ability to inactivate inflammatory mediators, e. g. the degradation of enke-phalins shown already for rat splenocytes[15] or the inactivation of the chemotactic cytokine IL-8[16]. Furthermore, CD13 has been shown to play a role in adhesion and migration processes[17,18] as well as in antigen processing by the trimming of peptides that protrude out of MHC class II molecules[19]. A thorough understandig of the interaction of lymphocytes with synoviocytes leading to the induction of CD13 surface expression on lymphocytes may prove useful in devising new therapeutic strategies for the treatment of rheumatoid arthritis.

REFERENCES

1. Look AT. Ashmun RA, Shapiro LH, Peiper SC (1989) Human myeloid plasma membrane glycoprotein CD13 (gp 150) is identical to aminopeptidase N. J Clin Invest 83:1299–1307.
2. Lerche C. Vogel LK, Shapiro LH, Noren O, Sjöström H (1996) Mamm Genome 712–713.
3. Riemann D, Schwachula A, Hentschel M, Langner J (1993) Demonstration of CD13/aminopeptidase N on synovial fluid T cells from patients with different forms of joint effusions. Immunobiol 187:24–35.
4. Riemann D, Göhring B, Langner J (1994) Expression of aminopeptidase N/CD13 in tumour-infiltrating lymphocytes from human renal cell carcinoma. Immunol Lett 42:19–23.
5. Banchereau J, de Paoli P, Vallé A, Garcia E, Rousset F (1991) Long term human B cell lines dependent on interleukin 4 and anti-CD40. Science 251:70–72.
6. Scott S, Pandolfi F, Kurnick T (1990) Fibroblasts mediate T cell survival: A proposed mechanism for retention of primed cells. J Exp Med 172:1873–1876.
7. Dechanet J, Merville P, Durand I, Banchereau J, Miossec P (1995) The ability of synoviocytes to support terminal differentiation of activated B cells may explain plasma cell accumulaltion in rheumatoid synovium. J Clin Invest 95:456–463.
8. Riemann D, Kehlen A, Thiele K, Löhn M, Langner J (1997) Induction of aminopeptidase N/CD13 on human lymphocytes upon adhesion to fibroblast-like synoviocytes, endothelial cells, epithelial cells and monocytes/macrophages. J Immunol in press.

9. Schwachula A, Riemann D, Kehlen A, Langner J (1994) Characterization of the immunophenotype and functional properties of fibroblast-like synoviocytes in comparison to skin fibroblasts and umbilical vein endothelial cells. Immunobiol 190:67–92.
10. Riemann D, Wollert H-G, Menschikowski J, Mittenzwei S, Langner J (1994) Immunophenotype of lymphocytes in pericardial fluid from patients with different forms of heart disease. Int Arch Allerg Immunol 104:48–56.
11. Riemann D, Kehlen A, Langner J (1995) Stimulation of the expression and the enzyme activity of aminopeptidase N/CD13 and dipeptidylpeptidase IV/CD26 on human renal cell carcinoma cells and renal tubular epithelial cells by T cell-derived cytokines, such as IL-4 and IL-13. Clin Exp. Immun 100:277–283.
12. Wex T, Lendeckel U, Wex H, Frank K, Ansorge S (1995) Quantification of aminopeptidase N mRNA in T cells by competitive PCR. FEBS Lett 374:341–344.
13. Wex T, Lendeckel U, Reinhold D, Kähne T, Arndt M, Frank K, Ansorge S (1996) Antisense-mediated inhibition of aminopeptidase N (CD13) markedly decreases growth rates of hematopoietic tumour cells. Chapter in this book.
14. Sanderink GJ, Artur Y, Siest G (1988) Human aminopeptidases. A review of the literature. J Clin Chem Clin Biochem 26:795–807.
15. Amoscato AA, Spiess RR, Sansoni SB, Herberman RB, Chambers WH (1993) Degradation of enkephalins by rat lymphocyte and purified rat natural killer cell surface aminopeptidases. Brain Behav Immunity 7:176–179.
16. Kanayama N, Kajiwara Y, Goto E, El Maradny E, Macharak, Andou K, Tereo T (1995) Inactivation of interleukin-8 by aminopeptidase N (CD13). J Leukocyte Biol 57:129–134.
17. Menrad A, Speicher D, Walker J, Herlyn M (1993) Biochemical and functional characterization of aminopeptidase N expressed by human melanoma cells. Cancer Res 53:1450–1455.
18. Saiki I, Fujii H, Yoneda H, Abe F, Nakajima M, Tsuruo T, Azuma I (1993) Role of aminopeptidase N (CD13) in tumor-cell in vasion and extracellular matrix degradation. Int J Cancer 54:137–143.
19. Larsen SL, Ostergaard L, Buus S, Stryhn A (1996) T cells responses affected by aminopeptidase N (CD13)-mediated trimming of major histocompatibility complex class II - bound peptides. J Exp Med 184:183–189.

11

TWO TRANSFECTED ENDOTHELIAL CELL LINES EXPRESSING HIGH LEVELS OF MEMBRANE BOUND OR SOLUBLE AMINOPEPTIDASE N

Katja Thiele,[1] Dagmar Riemann,[1] Astrid Kehlen,[1] Matthias Löhn,[1]
Lotte K. Vogel,[2] and Jürgen Langner[1]

[1]Institute of Medical Immunology
Martin Luther University Halle
Straße der OdF 6, D-06097 Halle, Germany
[2]Department of Biochemistry
Panum Institute Copenhagen
Blegdamsvej 3, DK-2200 Copenhagen N, Denmark

1. INTRODUCTION

The CD13 molecule, found to be identical with aminopeptidase N (APN/EC 3.4.11.2)[1] has been considered as a specific marker for the myeloid lineage within the hematopoietic system, however the occurrence of APN on lymphoid cells was recently described e.g. on tumor-infiltrating lymphocytes[2], on T cells in the synovial fluid from patients with rheumatoid arthritis[3] and some kinds of leukemia or lymphoma[4,5,6]. Our group very recently obtained evidence that the coculture of lymphocytic cells with various adherent APN positive cells such as fibroblast-like synoviocytes, HUVEC, kidney tubular epithelial cells or some tumor cell lines can induce the APN molecule on lymphocytes[7]. In need of a coculture system to further investigate the influence of the CD13/APN molecule we focused our interest on transfecting a APN negative adherent cell line with an expression vector for APN. To this end we used the spontaneously transformed immortal endothelial cell line ECV304 (ATCC CRL-1998) established from the vein of an apparently normal human umbilical cord[8]. Some reports describe the existence of a solubilized form of aminopeptidase N in serum, originating from liver and increased in hepatobiliary disease[9,10,11]. Therefore it seemed to us of further interest, wether the soluble form of this molecule is also able to induce the lymphocytic expression and if direct cell contact is necessary for induction. For that reason we additionally transformed the ECV304 cell line with an expression vector for the secretory form of aminopeptidase N.

2. CHARACTERIZATION OF TWO ENDOTHELIAL CELL LINES TRANSFECTED WITH EXPRESSION VECTORS FOR MEMBRANE BOUND AND SOLUBLE APN

ECV304 cells were electroporated with either the expression vector pTEJ4 carrying the full length wild type cDNA for APN or pTEJ8 containing an APN cDNA, in which the coding region for the first 33 amino acids is replaced by the coding region for the 17 amino acids of hemagglutinin signal peptide [12,13]. While pTEJ8 is coding for neomycin resistance as a selection marker, pTEJ4 had to be cotransfected with the pSV2neo vector for neomycin resistance [14]. The APN cDNA is controlled by the human ubiquitin promotor, which allows a high level constitutive expression of the protein. Stable clones were established by neomycin selection (1 mg/ml) and additionally fluorescent staining and cell sorting (FACS Vantage, Becton Dickinson) in the case of membrane expression of APN and the limited dilution method and enzyme activity determination in the case of soluble APN. The two clones ECV1d-s (membrane expression) and ECVp8-28 (soluble form), which are representative for 6 and 32, respectively, different clones obtained from the selection and cloning procedures were further characterized with regard to their ability to express APN. The expression of APN mRNA was determined by competitive RT-PCR. The basic level of APN mRNA expression in the original cell line ECV304 is 13 pg/µg total RNA. Clone ECV1d-s contains 6.6 ng/µg and clone ECVp8-28 contains the maximum level of 20–120 ng/µg total RNA. The expression of membrane protein by the clone ECV1d-s was shown by immunofluorescent staining with anti-CD13 mAb (Leu-M7, Becton Dickinson) and flow cytometric measurement (FACScan, Becton Dickinson). Fig. 1 shows histograms of ECV304 and of the transfected cell clone ECV1d-s, which expresses high levels of APN on all cells. For the determination of the enzymatic activity of APN the substrate

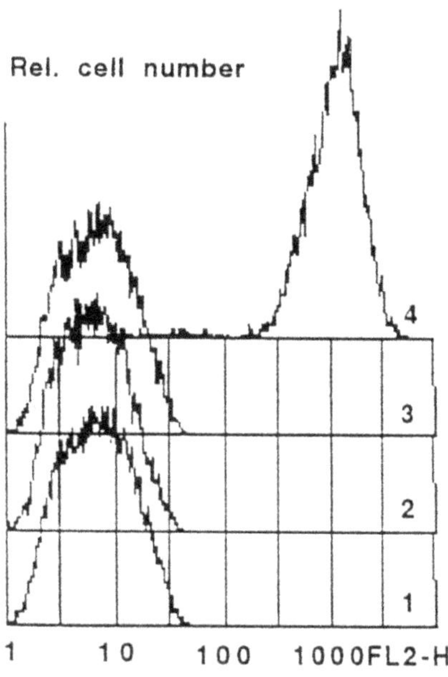

Figure 1. Expression of APN on ECV304 cells and ECV1d-s cells. Cells were stained with an IgG2a PE control antibody (1- ECV304; 3- ECV1d-s) or the Leu-M7 PE antibody (2-ECV304; 4- ECV1d-s).

alanine p-nitroanilide (Ala-pNA) was used. On intact cultured cells activities of 87–166 pkat/10^6 cells for ECV304 and 359–400 pkat/10^6 cells for ECV1d-s were found. The activity of the soluble form was measured in the medium harvested after 2 or 3 days of cultivation of ECVp8-28. An Ala-pNA cleaving activity of 0–5 pkat/ml for ECV304 and ECV1d-s and of 500–1300 pkat/ml for ECVp8-28 was found.

3. COCULTURE OF TRANSFECTED ENDOTHELIAL CELLS WITH JURKAT T CELLS

Jurkat T cells were incubated on confluent ECV304, ECV1d-s or ECVp8-28 cells for one day. After collecting the T cells from the culture, cells were stained with an anti-CD13 mAb and analysed by flow cytometry. As a result of coculture with ECV304 and ECVp8-28, 1±1% of cells express APN. After coculture with ECV1d-s, 34% of Jurkat cells express APN.

4. DISCUSSION

We developed a model to study the influence of the APN molecule expressed on endothelial cells in a coculture system with lymphocytes. The results indicate that the APN molecule seems to be deeply involved in the induction of lymphocytic expression of APN. The cell lines ECV304, which cannot induce APN on lymphocytes and ECV1d-s, which is able to induce this molecule, differ only in the presence of APN on their membranes. The cell line ECVp8-28 can produce and secrete high amounts of catalytically activ soluble APN but it cannot induce the lymphocytic expression in coculture experiments. Our results therefore suggest that direct contact with cells expressing APN on their surface is necessary for induction of the expression of the APN molecule on lymphocytic cells whereas the enzymatic activity of APN alone is not sufficient

REFERENCES

1. Look AT, Ashmun RA, Shapiro LH, Peiper SC (1989) Human myeloid plasma membrane glycoprotein CD13 (gp150) is identical to aminopeptidase N. J Clin Invest 83:1299–1307
2. Riemann D, Göhring B, Langner J (1994) Expression of aminopeptidase N/CD13 in tumour-infiltrating lymphocytes from human renal cell carcinoma. Immunol Lett 42:19–23
3. Riemann D, Schwachula A, Hentschel M, Langner J (1993) Demonstration of CD13/ aminopeptidase N on synovial fluid T cells from patients with different forms of joint effusions. Immunobiol 187:24–35
4. Dreno B, Bureau B, Stalder JF, Litoux P (1990) My7 monoclonal antibody for diagnosis of cutaneous T cell lymphoma. Arch Dermatol 126:1454–1456
5. Drexler HG, Thiel E, Ludwig WD (1991) Reviews of the incidence and clinical relevance of myeloid antigen-positive acute lymphoblastic leukemia. Leukemia 5:637–645
6. Ohsaka A, Sato N, Imai Y, Hirai S, Oka Y, Kikuchi M, Takahashi A (1996) Multiple gastric involvement by myeloid antigen CD13-positive non-secretory plasma cell leukemia. Brit J Haematol 92:134–136
7. Riemann D, Kehlen A, Thiele K, Löhn M, Langner J (1997) Induction of aminopeptidase N/CD13 on human lymphocytes upon adhesion to fibroblast-like synoviocytes, endothelial cells, epithelial cells and monocytes/macrophages. J Immunol in press
8. Hughes SE (1996) Functional characterization of the spontaneously transformed human umbilical vein endothelial cell line ECV304: Use in an in vitro model of angiogenesis. Exp Cell Res 225:171–185
9. Peters JE, Nilius R, Otto L (1973) Elektrophoretische Varianten der Alanin-Aminopeptidase in Serum bei hepatobiliären Erkrankungen. Clin Chim Acta 45:177–187

10. De Broe JE, Roels F, Nouwen EJ, Claes L, Wieme RJ (1985) Liver plasma membrane: the source of high molecular weight alkaline phosphatase in human serum. Hepatology 5:118–128
11. Sanderink GJ, Artur Y, Schiele F, Gueguen R, Siest G (1988) Alanine aminopeptidase in serum: Biological variations and reference limits. Clin Chem 34:1422–1426
12. Vogel LK, Noren O, Sjöström H (1992) The apical sorting signal of human aminopeptidase N is not located in the stalk but in the catalytic head group. FEBS Lett 308:14–17
13. Vogel LK, Spiess M, Sjöström H, Noren O (1992) Evidence for an apical sorting signal on the ectodomain of human aminopeptidase N. J Biol Chem 267:2794–2797
14. Johansen TE, Schøller MS, Tolstoy S, Schwartz TW (1990) Biosynthesis of peptide precursors and protease inhibitors using new constitutive and inducible eucaryotic expression vectors. FEBS Lett 267:289–294

AMINOPEPTIDASE N-MEDIATED SIGNAL TRANSDUCTION AND INHIBITION OF PROLIFERATION OF HUMAN MYELOID CELLS

M. Löhn,[1] C. Mueller,[1] K. Thiele,[1] T. Kähne,[2] D. Riemann,[1] and J. Langner[1]

[1]Institute of Medical Immunology
Martin Luther University Halle/Saale
OdF Str. 06, D-06097 Halle/Saale, Germany
[2]Institute of Experimental Internal Medicine
Otto von Guericke University
Leipziger Str. 44, D-39120 Magdeburg, Germany

1. INTRODUCTION

Cell surface peptidases are involved in degradative processes of various peptides, in cell-cell communcation, in lymphocyte activation and in cell growth[1, 2, 3, 4]. Aminopeptidase N (APN, CD13), a ubiquitously occurring surface molecule of many tissues, among them human liver, kidneys, intestine, placenta, bone marrow or brain microvessels, acts with a relatively broad substrate specifity on peptides with N-terminal neutral amino acids [5]. Within the haematopoietic system CD13 is localized on monocytes, granulocytes and activated T lymphocytes[5, 6, 7, 8].

Similar to dipeptidyl peptidase IV (CD26), CD13 could be involved in signal transduction, early activation steps and proliferation processes of cells [9, 10]. In this study the influence of CD13 antibodies on intracellular concentrations of free calcium ions ($[Ca^{2+}]_i$) was analysed by flow cytometry. Furthermore, we show that treatment of the myeloid cell line U937 with monoclonal antibodies (mAb) to CD13 results in a decrease of cell number in a dose-dependent manner.

2. CD13 MEDIATED SIGNAL TRANSDUCTION

The measurements of $[Ca^{2+}]_i$ were performed by flow-cytometry on a FACS Vantage (Becton Dickinson, USA) using the membrane permeable Ca^{2+}-sensitive dye INDO-1/AM. $[Ca^{2+}]_i$ was analysed by calculating the ratio of emitted wavelenghts of the dye.

The analyses of $[Ca^{2+}]_i$ were done with myeloid U937 cells which constitutively express the CD13 molecule and bulk cultures of T cells expressing CD13. As a model for li-

Cellular Peptidases in Immune Functions and Diseases, edited by Ansorge and Langner
Plenum Press, New York, 1997

gand-receptor binding, we used mAb specific to CD13 of various clones and subclasses. To avoid errors due to mechanical disturbations during the mAb application, an application of bath solution was performed prior to the mAb application.

After application of CD13 mAb to the cells a transient increase in $[Ca^{2+}]_i$ was measured. In these experiments mAb to CD13 of different clones were used. The rise of $[Ca^{2+}]_i$ depended on the clone of mAb used in the experiment. The maximum increase of $[Ca^{2+}]_i$ (~ 300 nmol/l Ca^{2+}) was reached after application of the mAb Leu-M7 (subclass IgG_1, Becton Dickinson). MAb of the subclass IgG_{2a}, like CD13 mAb 22A5 did also evoke Ca^{2+} influx through the cell membrane (Fig. 1). Repeated applications of CD13 mAb did not induce further rises of $[Ca^{2+}]_i$ (Fig. 2).

The rise in $[Ca^{2+}]_i$ evoked by CD13 mAb was blocked by chelating the extracellular Ca^{2+} (Fig. 3) and by application of actinonin (100 μmol/l) (Fig. 4). Actinonin at high concentrations does not reduce the binding of CD13 mAb to CD13 on the cell membrane (Fig. 5). These approaches demonstrate that the mAb-mediated increase of $[Ca^{2+}]_i$ probably requires the enzyme activity.

3. CD13 MEDIATED INHIBITION OF PROLIFERATION

Several authors have shown that inhibition of CD13 with bestatin has modulatory effects on the proliferation on blood progenitors and U937 cells [11, 12].

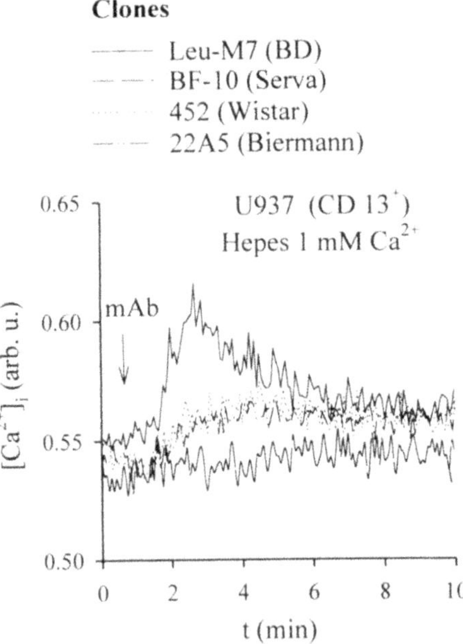

Figure 1. Registration of the intracellular concentration of free calcium ions ($[Ca^{2+}]_i$) using the Ca^{2+}-sensitive dye INDO-1/AM in U937 cells. The mean value of the cells at each sample unit is proportional to the $[Ca^{2+}]_i$. The slope of the mAb induced rise of $[Ca^{2+}]_i$ depends on the origin of mAb used in this experiment. Maximum response was evoked using the CD13mAb of clone Leu-M7. Bath solution (mmol/l): 140 NaCl, 5 KCl; 1 $CaCl_2$, 10 Glucose, 10 Hepes; CD13mAb (μg/l): 25.

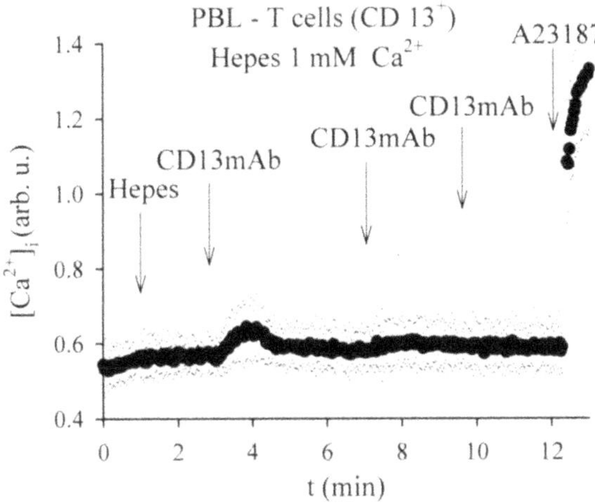

Figure 2. Registration of $[Ca^{2+}]_i$ in peripheral blood CD13-positive ($CD13^+$) T-lymphocytes. Repeated application of CD13mAb does not evoke a subsequent rise of $[Ca^{2+}]_i$. The application of the Ca^{2+}-ionophore A23187 was done for the determination of the fluorescence maximum. Bath solution (mmol/l): 140 NaCl, 5 KCl; 1 $CaCl_2$, 10 Glucose, 10 Hepes; CD13mAb (μg/l): 25; A23187 (μmol/l): 30.

To test the influence of CD13 on proliferation, U937 cells were treated with different concentrations of CD13 mAb (Fig. 6).

The number of cells treated with mAb was calculated relative to the control of cells cultivated in normal culture medium. After a 24-h application of CD13 mAb (clone 452, kindly provided by Dr. Herlyn, Wistar Institute, Philadelphia, PA).), the cell number decreased in a dose-dependent manner. At a concentration of 25 μg/ml the mAb induced decrease of the cell number was almost as intense as the decrease induced by 12-o-tetradecanoyl-phorbol-13-acetate (TPA), a potent inducer of apoptosis in U937 cells [13].

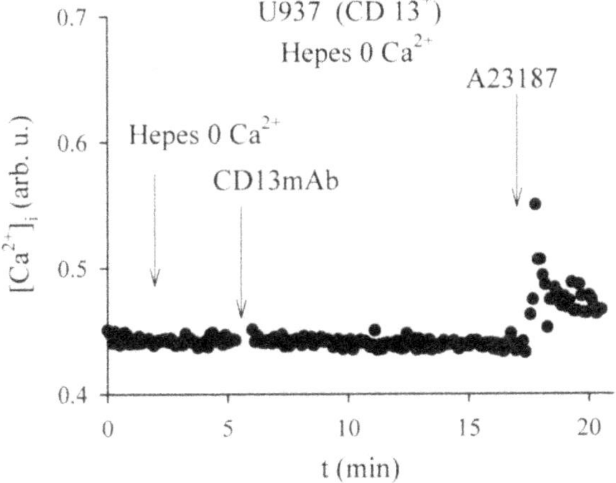

Figure 3. Attempt to induce Ca^{2+}-release from intracellular stores after antibody application to U937 cells. The application of the Ca^{2+}-ionophore A23187 shows that the intracellular stores are intact and were emptied through this approach. Bath solution (mmol/l): 140 NaCl, 5 KCl; 0 $CaCl_2$, 0.5 HEDTA, 10 Glucose, 10 Hepes; CD13mAb (μg/l): 25; A23187 (μmol/l): 30.

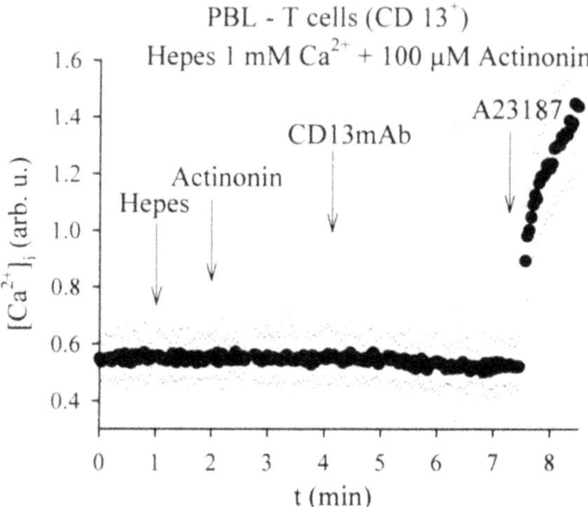

Figure 4. Registration of $[Ca^{2+}]_i$ in peripheral blood CD13-positive ($CD13^+$) T-lymphocytes. Inhibition of the CD13mAb evoked rise of $[Ca^{2+}]_i$ with actinonin. Bath solution (mmol/l): 140 NaCl, 5 KCl; 1 $CaCl_2$, 10 Glucose, 10 Hepes; CD13mAb (µg/l): 25; Actinonin (µmol/l): 100; A23187 (µmol/l): 30.

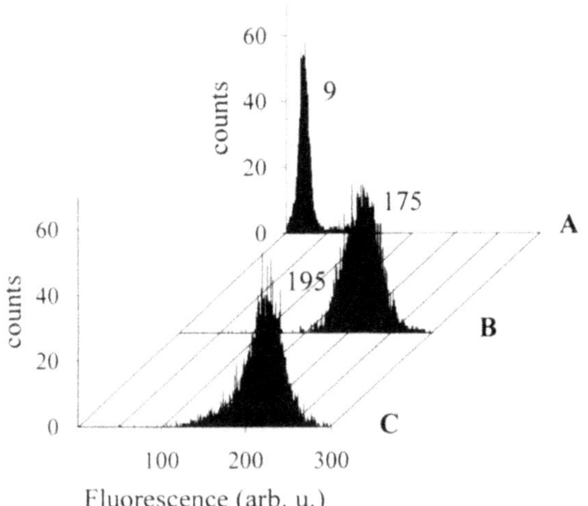

Figure 5. Registration of CD13 density on the cell surface. Test for competetive binding of CD13 mAb and actinonin at high concentration. Beneath the histogram given values represent the median of the fluorescence. (A) Base fluorescence level of U937 with secondary goat-anti mouse mAb (smAb). (B) Cell surface density of CD13 detected with CD13 mAb Leu-M7 and smAb. (C) Test for competitive binding: Cell surface density of CD13 detected with CD13 mAb Leu-M7, addition of actinonin and smAb. Bath solution (mmol/l): 140 NaCl, 5 KCl: 1 $CaCl_2$, 10 Glucose, 10 Hepes; CD13mAb (µg/l): 25; Actinonin (µmol/l): 100.

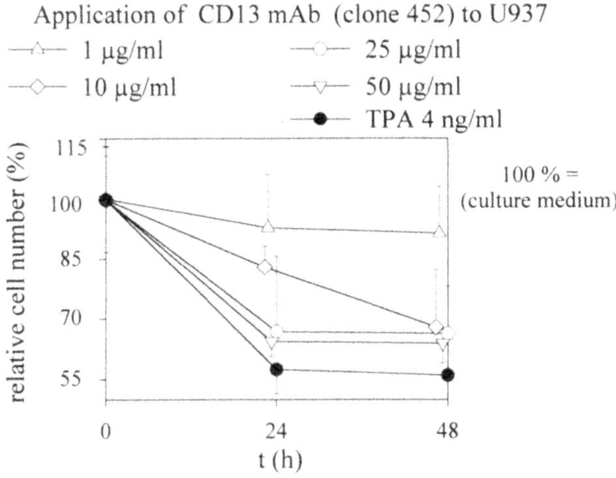

Figure 6. Dose-response curves of the effect of 24 and 48 hours application of mAb to CD13 (Wistar) on relative changes in cell number (compared to culture medium application for 24 and 48 hours) of U937 cells. CD13mAb (µg/l): 1, 10, 25, 50; TPA (ng/l): 4.

4. DISCUSSION

The aim of the present study was to analyse possible physiological roles of CD13 with regard to signal transduction and proliferation of blood cells.

CD13 shows a highly significant homology with bacterial CD13 and this suggests that this molecule received evolutionary conservation with well defined function [5]. On the one hand CD13 could be a molecule with restricted functions, either to degrade active peptides to inactivate them or to cleave inactive peptides in order to produce activated peptides or ligands for ion channel coupled receptors. On the other hand CD13 could represent a multifunctional protein with separated autonome actions of cleavage processes (enzymatic activity) and of signal transduction (receptor activity).

It is well known that aminopeptidases are involved in signal transduction e.g. signal transduction via the thrombin receptor activated by thrombin or via insulin receptor activated by trypsin [17]. We show in the present study that an aminopeptidase must have a direct influence on ion currents through the cell membrane which, to the best of our knownledge, seems to be a novel property of aminopeptidases.

Additionally, we show that binding of CD13 mAb to U937 cells and bulk cultures of T cells expressing CD13 can induce an increase of $[Ca^{2+}]_i$. This may suggest that CD13 directly or indirectly could be involved in control of ion channels and by this way could influence and regulate the influx of extracellular Ca^{2+}. Increase of $[Ca^{2+}]_i$ was inhibited by the removal of extracellular Ca^{2+}, so the increase of $[Ca^{2+}]_i$ could be the result of an influx of extracellular Ca^{2+} and not due to a Ca^{2+}-release from intracellular stores. However, chelating the extracellular Ca^{2+} also results in a block of CD13 activity due to chelating zinc ions [14]. Furthermore, the CD13 mAb evoked increase of $[Ca^{2+}]_i$ is also blocked by actinonin, an inhibitor of enzyme activity.

Our results are in contradiction to the data of MacIntyre et al [15, 16]. These authors have discussed that the CD13 mAb-evoked increase of $[Ca^{2+}]_i$ could be a result of Ca^{2+}-mobilization after formation of Ag-mAb-Fc-receptor complexes, because $F(ab')_2$ fragments of CD13 mAb did not induce a Ca^{2+}-mobilization and the pattern of response varied with the isotype of the mAb tested [15, 16]. However, the inhibition of CD13 enzyme activity by actinonin or by chelating the extracellular Ca^{2+} should not influence the formation of Ag-mAb-Fc-receptor complexes. Actinonin did not act as a competitor for binding sites. The binding of CD13 mAb was almost unaffected by actinonin. In our experiments we tested CD13 mAb from different clones but all of them were from the same isoform (IgG_1) except the CD13 mAb 22A5 (IgG_{2a}). Moreover, if the mAb evoked response should be mediated via the Fc-receptor, then the amplitude of the responses must be roughly the same. In our experiment the contrary case was detected. MAb of IgG_{2a} subclass also evoke a strong influx of extracellular Ca^{2+}. The different increase of $[Ca^{2+}]_i$ evoked by different CD13 mAb may be due to binding to different parts of the "CD13 receptor site".

Inducing changes in $[Ca^{2+}]_i$, CD13 may serve as a regulatory protein in global cell processes such as proliferation and cell death. Similar to experiments with CD26 [9], the treatment of U937 cells with CD13 mAb results in a decrease of the cell number in a dose-dependent manner, suggesting that CD13 is involved in regulating proliferation or death of cells.

At high mAb concentrations, the mAb-induced decrease of cell number is similar to the TPA-induced decrease indicating that at this mAb concentration the CD13 mediated response results in specific effects to the cells. Since TPA is a potent inducer of apoptosis in U937 cells this could suggest that the CD13 mediated increase of $[Ca^{2+}]_i$ might result in apoptosis due to activating Ca^{2+}-dependent endonucleases.

Further studies will be necessary to clearify the physiological role of CD13 with regard to signal transduction and cell growth.

5. REFERENCES

1. Brownlees, J., Williams, C.H. (1993) Peptidases, peptides, and the mammalian blood-brain barrier. J. Neurochem. 60, 793–803
2. Kohno, H., Kanno, T. (1985) Properties and activities of aminopeptidases in normal and mitogen stimulated human lymphocytes. J.Biochem. 226, 59–65
3. Kohno, H., Kanno, T. (1985) Intracellular localization and properties of aminopeptidases in human peripheral blood lymphocytes. Int. J. Biochem. 17, 1143–1148
4. Taylor, A. (1993) Aminopeptidases: towards a mechanism of action. TIBS 18, 167–172
5. Sanderink, G.-J., Artur,Y., Siest, G. (1988) Human aminopeptidase: A review of the literature. J. Clin. Chem. Biochem. 26, 795–807
6. Riemann, D., Göhring, B., Langner, J. (1994) Expression of aminopeptidase N/CD13 in tumor-infiltrating lymphocytes from human renal cell carcinoma. Immunol. Lett. 42, 19–23
7. Razak, K., Newland, A.C. (1992) The significance of aminopeptidase and haematopoetic cell differentation. Blood Rev. 6, 243–250
8. Look, A.T. (1989) Human myeloid plasma membrane glycoprotein CD13 (gp 150) is identical to aminopeptidase N. J.Clin.Invest. 83, 1299–1307
9. Mattern, T., Ansorge, S., Flad, H-D., Ulmer, A. (1993) Anti-CD26 monoclonal antibodies can reversibly arrest human T lymphocytes in the late G_1 phase of the cell cycle. Immunbiol. 188, 36–50
10. Reinhold, D., Bank, U., Bühling, F., Kähne, T., Kunz, D., Faust, J., Neubert, K., Ansorge, S. (1994) Inhibitors of Dipeptidyl peptidase IV (DPIV, CD26) specifically suppress proliferation and modulate cytokine production of strongly CD26 expressing U937 cells. Immunbiol. 192, 121–136
11. Murata, M., Kubota, Y., Tanaka, T., Iida-Tanaka, K., Takahara, J., Irino, S. (1994) Effect of ubenimex on the proliferation and differentation of U937 human histiocytic lymphoma cells. Leukemia 8, 2188–2193
12. Shibuya, K., Chiba, S., Hino, M., Kitamura, T., Miyagawa, K., Takaku, F., Miyazano, K. (1991) Enhancing effect of ubenimex (Bestatin) on proliferation and differentiation of hematopoietic progenitor cells, and the suppressive effect on proliferation of leukemic cell lines via peptidase regulation. Biomed. Pharmacother. 45, 71–80

13. Hass, R., Meinhardt, G., Hadam, M., Bartels, H. (1994) Characterization of human TUR leukemia cells: continued cell cycle progression in the presence of phorbol esters is associated with the resistance to apoptosis. Eur. J. Cell. Biol. 65, 408–416
14. Hütter, H.J. (1984) Characterization of human alanine aminopeptidases. Symposia Biologica Hungaria 25
15. MacIntyre, E.A. ,Roberts, P.J., Jones, M., van der Schoot, C.E., Favalaro,E.J., Tidman, N., Linch, D.C. (1989) Activation of human monocytes occurs on cross-linking monocytic antigens to an Fc receptor. J. Immunol. 142, 2377–2383
16. MacIntyre, E.A. ,Roberts, P.J.,Abdul-Gaffar, R., O'Flynn, K., Pilkington, G.R., Farace, F., Morgan, J., Linch, D.C. (1988) Mechanism of human monocytes activation via the 40-kDa Fc-receptor for IgG. J. Immunol. 141, 4333–4343
17. Hollenberg, M. D. (1996) Protease-mediated signalling: new paradigms for cell regulation and drug develoment. TIPS 17, 3–6

13

REGULATION OF THYMIC DEVELOPMENT BY NEPRILYSIN INHIBITION

Sandrine Guérin and Patrick Auberger*

CJF INSERM 9605
Faculté de Médecine
Avenue de Vallombrose 06107, Nice cédex 02, France

1. ABSTRACT

Development of T lymphocyte is regulated by both thymocyte-stromal cell interactions and production of soluble factors such as cytokines, peptides and hormones. The local concentration of active biological peptides is regulated by a specialized family of enzymes expressed at the cell surface, the ectopeptidases. We found that treatment of fetal thymic organ cultures (FTOC) with the specific CD10 (endopeptidase 24.11) inhibitor, (N-(3-[(hydroaxyamino)carbonyl]-2-benzylidene-1-oxopropyl]-N-glycine), RB25 results in a marked delay in thymocyte differentiation. RB25 causes a significant decrease in the number of DP ($CD4^+CD8^+$) cells in favor of the TN ($TcR\alpha\beta^-CD4^-CD8^-$) population. RB25 also blocks T lymphocyte differentiation in FTOC when preinjected into pregnant mice. Finally, RB25 was found to essentially affect the $CD44^+CD25^-$ and $CD44^-CD25^-$ thymocytes in "in vitro" and "in vivo" experiments after 2 days FTOC. Thus, a selective and stable endopeptidase 24.11 inhibitor impairs T cell development, an observation in agreement with the involvement of the CD10 antigen in early T cell development.

2. INTRODUCTION

CD10 (Neprilysin, endopeptidase 24.11, EC 3. 4. 24. 11) is a type II membrane glycoprotein that belongs to the NECK family of zinc metalloendopeptidases (1,2). In the hematopoietic system, CD10 is present on the majority of acute lymphoblastic leukemia and additional lymphoid malignancies (3) and is also expressed by normal lymphoid progeni-

* Tel. (33) 4. 93. 37. 76. 78 Fax: (33) 4. 93. 81. 54. 84. e-mail : auberger@unice.fr.

tors of human B and T cell lineage. CD10 cleaves small peptides such as enkephalins, on the amino terminal side of hydrophobic amino acids (4,5). It is implicated in the down-regulation of peptide-mediated response by reducing the local concentration of specific peptidic substrates. For noticeable example, CD10 has been involved in enkephalin-induced analgesia (4). CD10 is particularly abundant on thymic epithelial cells and bone marrow-derived stromal cells (3) and regulates the growth and the differentiation of B cell progenitors (6,7). We have recently brought evidence that CD10 is expressed on human immature thymocyte subpopulations (8,9) and human (S. Guérin in preparation) and murine thymic epithelial cells (10). This interesting observation suggests that CD10 may exert a similar function during early steps of thymic development. In this study, we used a selective and stable inhibitor of neprilysin to investigate the possible role of CD10 in thymic development. We show that CD10 plays an important physiological role in early thymocyte development, most probably by regulating the local concentration of an unidentified peptide involved in this process.

3. MATERIAL AND METHODS

3.1. Animals

NMRI mice were purchased from CERJ (Le Genest St Isle, France). Fetal mice were obtained from females impregnated during overnight matings. The day the males were removed was designated as day 1. In some experiments, pregnant mice were injected intraperitoneally with various doses of RB 25 at day 7, 10 and 14.

3.2. Fetal Thymic Organ Culture (FTOC)

Thymic lobes were excised from fetal mice at day 15 under binocular microscopy and placed on cellulose ester filters, which sat upon gelfoam sponges in media, in a 6-well plate. DMEM supplemented with 10% FCS, penicillin, streptomycin, L-glutamine, and sodium pyruvate (1 mM) was used as complete medium. The culture was carried out at 37°C in 10% CO2 in humidified air. The CD10 inhibitor was added, at the indicated concentrations, in the media at FTOC day 0, 2 and 5. Thymic lobes were collected at indicated times, and thymocytes prepared as described below. For each condition 5–8 thymic lobes were used.

3.3. Cell Suspensions

Fetal thymus cells from cultured organs were prepared by gently mincing the lobes in DMEM, in a potter and washed in PBS 0.1% BSA. Viable cells were systematically enumerated by trypan blue dye exclusion.

3.4. Cell Surface Staining and Flow Cytometry

FITC-anti-mouse CD8 MoAb, PE-anti-mouse CD4 MoAb , FITC-anti-mouse-CD44 MoAb, PE-anti-mouse CD25 MoAb and Cy-Chrome-anti-mouse αβTCR MoAb were used for cell surface staining. Briefly, cells were incubated for 30 min. at 4°C with saturating concentrations of MoAb in phosphate buffer saline (PBS) supplemented with 0.1%

BSA. Thymocytes were washed twice in PBS and resuspended in PBS containing 0.37% formol. Stained cells were analysed with a flow cytometer FACScan gated to exclude non-viable cells.

3.5. Endopeptidase 24.11 Assay

Cells were extensively washed in phosphate buffer saline (PBS) and resuspended in the same buffer. Endopeptidase 24.11 activity was determined in triplicate in 96 wells microtiter plates as previously described (8,11). Briefly, cells were incubated with 1 mM Suc-Ala-Ala-Phe-pNa in the presence of 10 μg/ml exogenous aminopeptidase N in the absence or the presence of phosphoramidon. Specific endopeptidase 24.11 activity is the phosphoramidon (10 μM) inhibitable activity.

3.6. Materials

MoAbs were purchased from Pharmingen (San Diego, CA). RB25, a stable analog of the antinociceptive agent kelatotorphan (N-(3-[(hydroaxyamino)carbonyl]-2-benzylidene-1-oxopropyl]-N-glycine) was kindly provided by Dr M.C Fournié-Zaluski (INSERM U166, Paris, France).

4. RESULTS

4.1. RB25 Delays Thymocyte Differentiation

To examine the role of CD10 in thymocyte differentiation we performed fetal thymus organ culture (FTOC) derived from fetal day 15 mouse embryos. Thymocytes phenotyped at day 15 were triple negative (TN, TcR $\alpha\beta^-CD4^-CD8^-$). These TN precursors proceed along different stages of differentiation and after 5 days in culture most of the cells were DP ($CD4^+CD8^+$), 10–15% ($CD4^+$ SP), 10% ($CD8^+$ SP) and only 10% TN ($TcR\alpha\beta^-CD4^-CD8^-$) (not shown). We investigated the effect of a selective CD10 inhibitor, RB25, on thymic maturation after 5 or 8 days FTOC. RB25 is derived from the antinociceptive agent kelathorphan and has been shown to inhibit endopeptidase 24.11 (IC50 = 1.5 nM), dipeptidyl-peptidase IV (CD26) (IC50 = 21 nM) and aminopeptidase N (CD13), (IC50 = 2000 nM) when assessed for enkephalin degradation (12). We found that half-maximal inhibition of endopeptidase 24.11 on the surface of human thymic epithelial cells measured on the chromogenic substrate Suc-Ala-Ala-Phe-pNA was obtained by pretreatment with 1 nM of inhibitor (not shown). A range of concentrations of RB25 were thus added "in vitro" to FTOC from day 15 embryos and thymic development was assessed 5 and 8 days later. As depicted in Figure 1, addition of 3 nM RB25 to FTOC provoked a small decrease in the number of DP thymocytes after 5 days of culture. At day 8 of FTOC the block of thymocyte differentiation was clearly visible, since 52% of thymocytes were TN as compared to 23% in untreated FTOC. Concentrations of RB25 as low as 0.3 nM also decreased thymic maturation (not shown), while doses up to 30 nM were less potent than 3 nM in blocking thymic differentiation. Thus at concentrations which only affected endopeptidase 24.11 activity, we observed maximal inhibitory effect of RB25 in FTOC. At higher concentrations of RB25 that also affected dipeptidylpeptidase IV activity the inhibitory effect was less pronounced (not shown). CD44 and CD25 represent two early

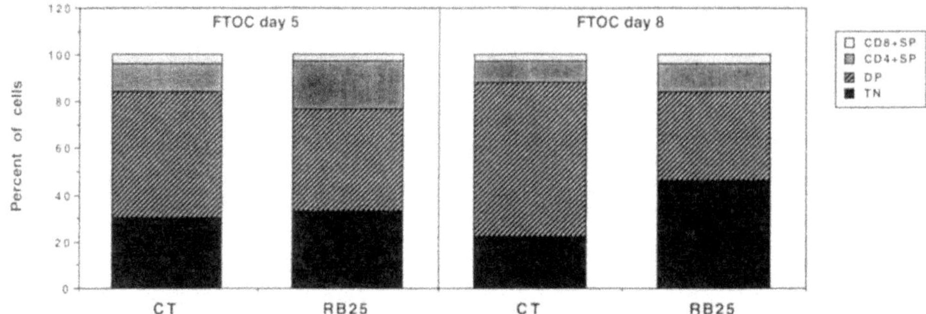

Figure 1. RB25 inhibits thymocyte differentiation in FTOC "in vitro". CD4 and CD8 phenotype of cells from day 5 and 8 of FTOC untreated (control) or treated with the kelatorphan analog RB25 (3.10^{-9} M). Thymocyte suspensions were stained with anti-CD4 and anti-CD8 monoclonal antibodies. Results are representative of at least 3 experiments.

lymphoid markers which define several TN subsets at different stages of differentiation (13). We established FTOC at day 15 of gestation and then looked for the effect of RB25, 2 days later. Figure 2 clearly shows that RB25 caused a 50% reduction in the number of total thymocytes as compared to untreated cells. CD10 inhibitor affected preferentially the CD25 negative subpopulations causing a 60% decrease in the number of both $CD44^-CD25^-$ and $CD44^+CD25^-$ thymocytes.

4.2. RB25 Intraperitoneal Injection of Pregnant Mice also Affects Thymic Development

The effect of RB25 was assessed after intraperitoneal injection of 2 μg (0.1 mg/kg) of the inhibitor or vehicle alone (0.025% EtOH) to pregnant mice. FTOC were then established at day 15 of gestation. RB25 markedly inhibits thymocyte maturation at day 2 of

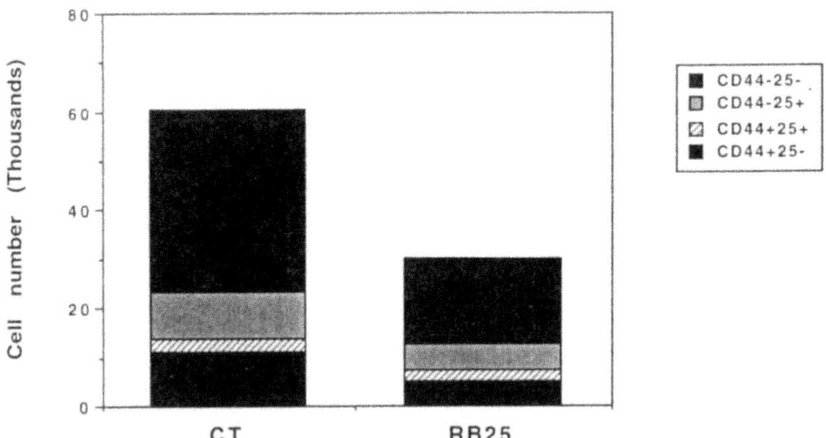

Figure 2. Effect of CD10 inhibition "in vitro" on cell yield of different CD44/CD25 thymocyte subsets from day 2 FTOC. FTOC were established from day 15 mouse embryos. FTOC were untreated or treated with RB25 (3.10^{-9} M) for 2 days. Thymocytes were dissociated and stained with anti-CD44 and anti-CD25 monoclonal antibodies. Data are expressed as the number of cells per lobe in culture.

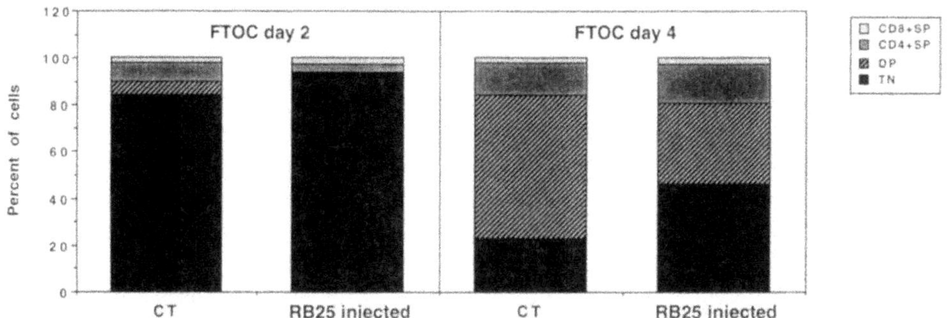

Figure 3. Intraperitoneal administration of RB25 to pregnant mice induces a significant block of thymocyte differentiation in FTOC. RB25 (2 µg, 0.1mg/kg) was given intraperitoneally to pregnant mice. FTOC were set up at day 15 of gestation and incubated for 2 or 4 days. CD4 and CD8 phenotype were determined as described above. Results are representative of 3 experiments.

FTOC (no DP thymocytes versus 6% in vehicle injected mice) (Figure 3). At day 5 of culture, 61% of the cells were DP in untreated mice while only 35% DP thymocytes were observed in RB25 injected mice (figure 3). The block in thymic differentiation was in favor of TN thymocytes (23% in control mice versus 46% in RB25 treated mice). As observed "in vitro", higher doses of RB25 (5 to 20 µg, 0.25 to 1 mg/kg) were found less potent than 2 µg (not shown). Finally, analysis of CD44 and CD25 expression indicated that RB25 treatment also decreased thymocyte number at day 2 and 5 of FTOC (33 and 40 % inhibition respectively) when preinjected "in vivo" to pregnant mice. As "in vitro" RB25 administration was shown to affect essentially the $CD44^{+}25^{-}$ and $CD44^{-}25^{-}$ subpopulations causing a 60% decrease in these particular subsets (Figure 4).

5. DISCUSSION

CD10 is known to regulate the growth and maturation of B cell progenitors "in vitro" and "in vivo" (6,7). In both models, addition or injection of the inhibitor signifi-

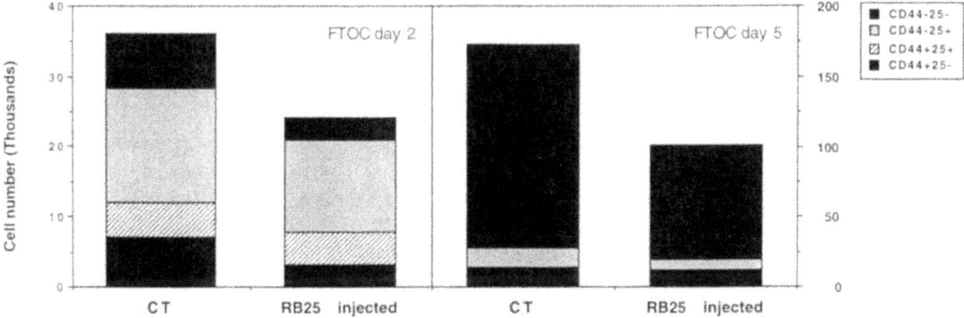

Figure 4. Effect of RB25 administration "in vivo" on early thymic development. RB25 (2 µg, 0.1 mg/kg) was given intraperitoneally to pregnant mice at day 7, 10 and 14 of gestation. FTOC were established at day 15 and thymocytes were dissociated, 2 days latter and stained with anti-CD44 and anti-CD25 monoclonal antibodies. Data are expressed as the number of cells per lobe in culture.

cantly increased B cell precursors growth and maturation, indicating that inhibition of CD10 activity may protect a yet unidentified stimulatory peptide from degradation which in turn may favor B cell differentiation. The high expression of CD10 on bone marrow stromal cells and thymic epithelial cells (S. Guérin personal observation) prompted us to examine the effect of CD10 inhibition in FTOC "in vitro" and "in vivo". In this report we show that a selective CD10 inhibitor, RB25, induced a drastic inhibition of thymic maturation "in vitro". The doses used to inhibit thymic differentiation "in vitro" are in the range of those which totally block CD10 activity on human TEC (not shown). Intraperitoneal injection of low concentrations of RB25 (2 to 10 μg , 0.1 to 1 mg/kg) was also found to significantly reduce thymic maturation "in vivo". The doses effective in inhibiting thymic maturation are those shown to exert an analgesic effect after intracerebroventricular injection to the mouse (0.4 to 4 mg/kg) (12). RB25 seems to affect early lymphoid development since it blocked the accumulation of $CD44^{+}25^{-}$ and $CD44^{-}25^{-}$ TN progenitors both "in vitro" and "in vivo". Taken together our results demonstrate that inhibition of CD10 enzymatic activity blocked lymphocyte differentiation in FTOC. This observation is particularly interesting since numerous biologically important peptides are present in the thymus or produced by thymocytes and thymic microenvironment (14). It is conceivable that such peptides may regulate the growth and maturation of thymocytes and TEC. The main function of CD10 is to down-regulate peptide-mediated response by reducing the local concentration of specific peptidic substrates (3–5), it appears possible that this enzyme could exert a stimulatory role in T cell development by inactivating a peptide that inhibits T cell differentiation. In this line, it has been recently shown that cAMP induced a block of T cell differentiation in FTOC (15). Thus it is reasonable to propose that CD10 positively regulates T cell development by reducing the local concentration of a peptide hormone whose receptor is coupled to adenylate cyclase stimulation and cAMP production. Finally, our results agree well with those of Salles et al. who recently suggested a role for CD10 in early B cell development (6,7). However, it appears that the nature of the peptidic substrates should be different in both cases (inhibitory and stimulatory respectively). In conclusion, our results strongly suggest that CD10 regulates early thymic development in mice by inactivating an inhibitory peptide.

6. REFERENCES

1. Shipp, M. A., Vijayaraghavan, J., Schmidt, E. V., Masteller, E. L., D'Adamio, L., Hersh, L. B. & Reinherz, E. L. (1989) . *Proc. Natl. Acad. Sci. USA* **86,** 297–301
2. Xu, D., Emoto, N., Giaid, A., Slaughter, C., Kaw, S., deWitt, D. & Yanagisawa, M. (1994) .*Cell* **78,** 473–485
3. Shipp, M. & Look, T. (1993) . *Blood* **82,** 1052–1070
4. Roques, B. P., Noble, F., Daugé, V., Fournié-Zaluski, M. C. & Beaumont, A. (1993). *Pharmacol. Rev.* **45,** 87–146
5. Erdös, E. & Skidgel, R. (1989) . *FASEB. J.***3,** 145–151
6. Salles, G., Rodewald, H.-R., Chin, B., Reinherz, E.-L. & Shipp, M.-A. (1993) I. *Proc. Natl. Acad. Sci. USA* **90,** 7618–7622
7. Salles, G., Chen, C. Y., Reinherz, E. L. & Shipp, M. A. (1992) *Blood* **80,** 2021–2029
8. Mari, B., Breittmayer, J.-P., Guérin, S., Belhacène, N., Peyron, J.-F., Deckert, M., Rossi, B. & Auberger, P. (1994).*Immunology* **82,** 833–838
9. Guérin, S., Mari, B., Belhacène, N., Rossi, B. & Auberger, P. (*Cell. Immunol.in press)*
10. Small, M., Kaiser, M., Tse, W., Heimfeld, S. & Blumberg, S. (1996) . *Eur. J. Immunol.* **26,** 961–964
11. Mari, B., Checler, F., Ponzio, G., Peyron, J.-F., Manié, S., Farahi Far, D., Rossi, B. & Auberger, P. (1992). *EMBO. J.* **11,** 3875–3885

12. Hernandez, J., Soleilhac, J., Roques, B. & Fournié-Zaluski, M. (1988) *J. Med. Chem.* **31,** 1825--1831
13. Godfrey, D., Kennedy, J., Suda, T. & Zlotnik, A. (1993) . *J. Immunol.* **150,** 4244–4252
14. Dardenne, M. & Savino, W. (1994) . *Immunol. Today* **15,** 518–551
15. Lalli, E., Sassone-Corsi, P. & Ceredig, R. (1995) *EMBO J.* **15,** 528–537

14

PROTEASES OF ISOLATED PANCREATIC ACINAR CELLS AFTER CAERULEIN HYPERSTIMULATION

Walter Halangk,[1] Dagmar Kunz,[2] Rainer Matthias,[2] Abidat Schneider,[2] Jörg Stürzebecher,[3] Hans-Ulrich Schulz,[1] and Hans Lippert[1]

[1]Centre of Surgery
Department of Experimental Surgery
[2]Institute of Clinical Chemistry
University Magdeburg, Leipziger Str. 44, D-39120 Magdeburg, Germany
[3]Centre of Vascular Biology and Medicine Erfurt
University Jena, Nordhäuser Str. 78, 99089 Erfurt, Germany

1. INTRODUCTION

Acute pancreatitis has been considered for a long time as autodigestive disease of the pancreatic gland. However, the pathophysiology of this disease is still poorly understood. Therefore, some aspects of the pathogenesis may be investigated in experimental animal models. Supramaximal doses of the cholecystokinin analogue caerulein induces a mild edematous pancreatitis in rats [1]. As main intracellular alterations, an impaired stimulus-secretion coupling, a formation of vacuolic structures, a diminished energy metabolism, and an activation of zymogens occur. A morphological feature of caerulein pancreatitis is the loss (or shedding) of the apical part of the plasma membrane. In the present study, this model was used to study the role of proteolytic enzymes to the pathological alterations of the pancreatic acinar cell. To this end, the activities of the potential cell surface peptidases dipeptidyl peptidase IV (DP IV) and aminopeptidase N (AP N) were measured in the course of caerulein pancreatitis and the following regeneration period. In a second set of experiments, the possible role of trypsin for initiating proteolytic actions within the acinar cell was investigated. For this purpose, we used rhodamine 110-based fluorogenic substrates to register proteolytic activities in intact cells.

2. MATERIALS AND METHODS

Caerulein pancreatitis was induced by up to 4 s.c. injections of 20 μg caerulein per kg b.w. in hourly intervals. The preparation of pancreatic acini from untreated and

Cellular Peptidases in Immune Functions and Diseases, edited by Ansorge and Langner
Plenum Press, New York, 1997

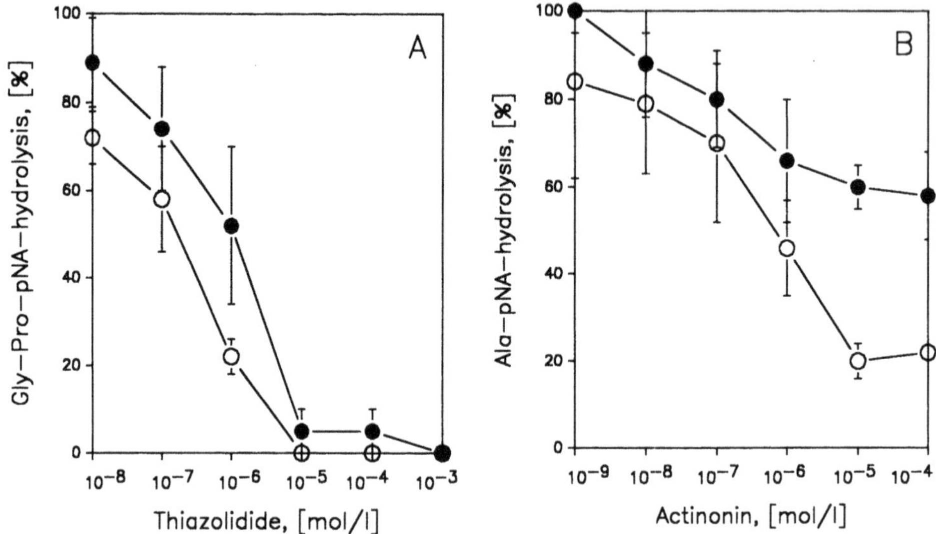

Figure 1. Inhibition of Gly-Pro-pNA (A) and Ala-pNA (B) hydrolysis by intact (•) and sonified (o) pancreatic acinar cells.

caerulein-treated rats was performed with purified collagenase [2]. Serine protease activity was measured in intact cells with the substrate (CBZ-arginine)$_2$-rhodamine 110 [3] (Molecular Probes Inc., Eugene). DP IV was determined with glycyl-prolyl-p-nitroanilide [4], AP N with alanyl-p-nitroanilide as described by Kunz et al. [5], and γ-glutamyltransferase (γ-GT) with γ-glutamyl-3-carboxy-p-nitroanilide. These activities were measured in intact and sonified cells, and in the soluble and the particulate fractions after centrifugation of sonified cells at 100.000 g for 1 h. To specify the (CBZ-arginine)$_2$-rhodamine 110 cleaving activity, the benzamidine derivative (BA) Nα-(2-naphthylsulfonyl)-3-amidinophenylalanine-carboxy methylpiperazide as inhibitor of serine proteases, trans-epoxysuccinyl-L-leucylamido-(4-guanidino) butane (E 64) as cysteine protease inhibitor, and soybean trypsin inhibitor (SBTI) were used. The Ba inhibitor was kindly provided by Dr P. Wikström (Pentapharm Ldt., Basel). Synthesis and determination of the Ki-values for inhibition of trypsin and related enzymes were described by Stürzebecher et al. [6] .Data are given as means ± S.D. or as representative experiments (Figs. 2–4) of at least 3 experiments.

3. RESULTS

3.1. Activities of Cell Surface Peptidases

The activities of DP IV and AP N were measured in suspensions of intact pancreatic acinar cells. Within 1 day of the caerulein treatment, the total Gly-Pro-pNA and Ala-pNA hydrolysis by the cells declined to about 60 to 70 % of untreated cells, and then these activities are restored to the initial levels within one week. Because the ectopeptidases may be interesting marker enzymes to follow the apical membrane shedding and the reappearance in the regeneration period, these activities were further characterized. As shown in Fig. 1, the Ala-pNA-splitting activity was only partially sensitive to actinonin, a specific

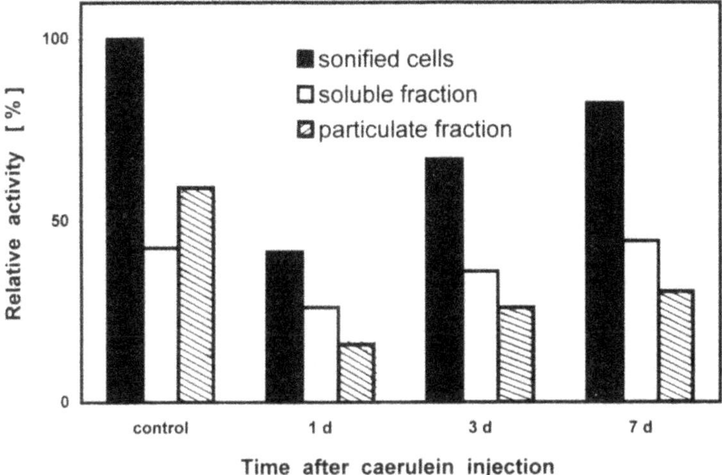

Figure 2. DP IV activity in fractions of acinar cells in the course of caerulein treatment of rats (4 injections of 20 μg/kg b.w. as indicated in Material and Methods).The activity of sonified control cells amounted to 22.7 pkat/10^6 cells.

AP N inhibitor [7]. The effect of the inhbitor was somewhat stronger to sonified cells. The incomplete inhibition indicates that a part of the total Ala-pNA hydrolysis is not caused by AP N. Ap N is localized on the cell surface as well as in the cytosolic compartment [3]. In fractionation studies it was found that the Ala-pNA-splitting activity was membrane-associated by only about 15–25 %. On the other hand, DP IV was strongly inhibited by the specific inhibitor Lys-[Z(NO2)]-thiazolidide [4] (Fig. 1). DP IV was predominantly found in the membrane fraction (Fig. 2), and this portion of activity was more affected by the caerulein treatment than the total activity. The restoration of the membrane-associated DP IV activity was not completed within one week. A similar time-course was found for the plasma membrane marker enzyme γ-GT (not shown). The preliminary data suggest a possible role of the membrane peptidases DP IV and γ-GT in the postpancreatitis regeneration period. Although acinar cells represent the major portion of the prepared cell population, the exact localization of the enzymes during caerulein pancreatitis and regeneration has to be found out [8].

3.2. Intracellular Activation of Serine Proteases

Investigating the pathophysiology of acute pancreatitis, the evidence of active trypsin is of special interest, because this enzyme is able to activate the other pancreatic proenzymes. In the experimental model of caerulein-induced pancreatitis, several groups determined trypsin activity in pancreas homogenates using more or less specific substrates [9–12]. In the present study, we used a bis-arginine-substituted rhodamine 110 [3] to measure active trypsin in isolated intact acini.

In order to evaluate to what extent this substrate can be cleaved by various pancreatic proteases, we measured the activities of purified enzymes with this substrate (Table 1). This selection was focused to pancreatic enzymes which can affect this cleavage site or which may exhibit noxious effects if they occur in their active forms. Although

Table 1. Relative cleavage of $(Bis\text{-}CBZ\text{-}Arg)_2$-Rhodamine-110 by proteinases

Enzyme(source)	pH	Relative activity
Trypsin(bovine pancreas)	7.4	100
Kallikrein(porcine pancreas)	7.4	0.02
Cathepsin B(bovine spleen)	7.4	0.006
	5.5	0.011
Chymotrypsin(bovine pancreas)	7.4	<0.02
Elastase(porcine pancreas)	7.4	<0.02

Enzymes activities were determined with 4 μM $(Bis\text{-}CBZ\text{-}Arg)_2$-R-110 at 37 °C. Fluorescence measurements were performed at 495 /521 nm using Perkin Elmer LS 50B.

these activities were compared on a molar and not on an active site basis, the data show that trypsin is mainly registered with this substrate.

In Fig. 3, the concentration-dependent action of caerulein on acini isolated from untreated rats are shown for the release of amylase and the $(CBZ\text{-}Arg)_2$-R110-cleaving activity. Amylase secretion showed the typical biphasic response with an inhibition at caerulein concentration > 0.1 nmol/l. Only at these high concentrations, a significant increase of protease activity vs. the basal level occurred. It was found in additional experiments that the $(CBZ\text{-}Arg)_2$-R110-cleaving activity was enhanced to 120.3±9.1% (n=7) at 0.1 nmol caerulein/l and to 192.4±16.3% (n=10) at 10 nmol caerulein/l. In order to specify the enzymatic activity observed under various conditions of caerulein stimulation, inhibitors of proteinases were applied. As inhibitor of trypsin, a benzamidine derivative (BA) was used

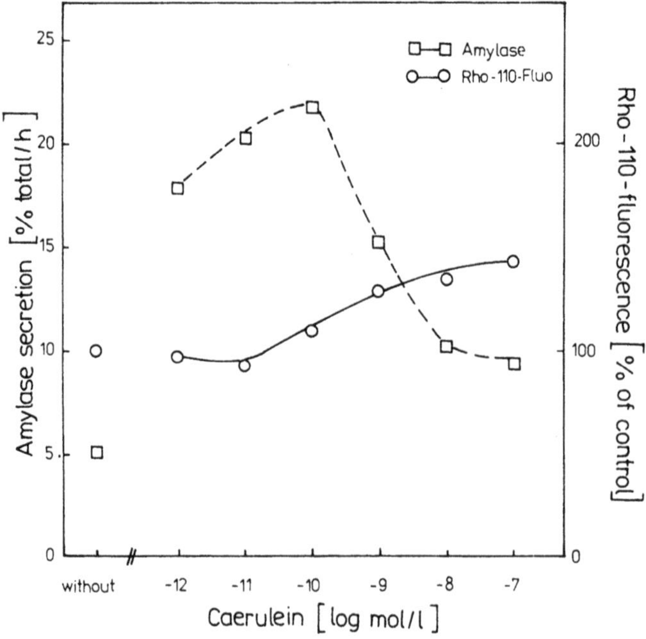

Figure 3. Amylase secretion and $(CBZ\text{-}Arg)_2$-R110 cleavage of acini in response to caerulein treatment.

which affects trypsin in the submicromolar range [4]. Testing its action on purified serine proteases in the rhodamine 110 assay, it was found that trypsin is inhibited at micromolar concentrations at which kallikrein or elastase were only marginally affected (data not shown). This inhibitor was also selected because it rapidly penetrates into the intact cell. Furthermore, E 64 as cysteine proteinase inhibitor was used to affect cathepsin B which is discussed to be able to activate trypsinogen. Although this lysosomal enzyme cleaves peptide-bonds with basic amino acid residues, its activity towards $(CBZ\text{-}Arg)_2$-R110 was low as shown in Table 1. Nevertheless, its activity as active lysosomal enzyme may contribute to the substrate cleavage. Therefore, we measured the inhibitor effects towards cathepsin B at pH 5.5, because the rhodamine substrate should reach the acid compartments of the acinar cells. The data in Fig. 4 show that trypsin was effectively inhibited by BA at 1 μmol/l, at which cathepsin B was unaffected. On the other hand, cathepsin B was completely inhibited by 1 μmol/l of E 64 to which trypsin is relatively insensitive up to 30 μmol/l. The 21 kDa SBTI was used in the experiments with intact acini in order to differentiate an intra- from an extracellular substrate cleavage.

We investigated the effect of these inhibitors in order to characterize the $(CBZ\text{-}Arg)_2$-R110-cleaving activity of untreated and caerulein-stimulated cells. As indicated in the dose-response experiments (cf. Fig. 3), pancreatic acini possess a substrate-cleaving activity in the unstimulated state, too. In Fig. 5 is shown that this activity was partially, but immediately inhibited by BA in low concentrations. This suggests that trypsin is likely responsible for this part of the substrate-cleaving activity. The lacking effect of low concentrations of E 64 indicates that cathepsin B does not participate in this activity. Addition of supramaximal concentrations of caerulein to the acini enhanced the protease activity about 2fold. This caerulein-induced activity showed a similar response towards benzamidine and E 64. The lacking effect of SBTI indicates that the substrate cleavage occurred intracellularly.

In the previous experiments the inhibitors were added after the induction of protease activity by caerulein. The observed increase of activity should be the result of a preceding

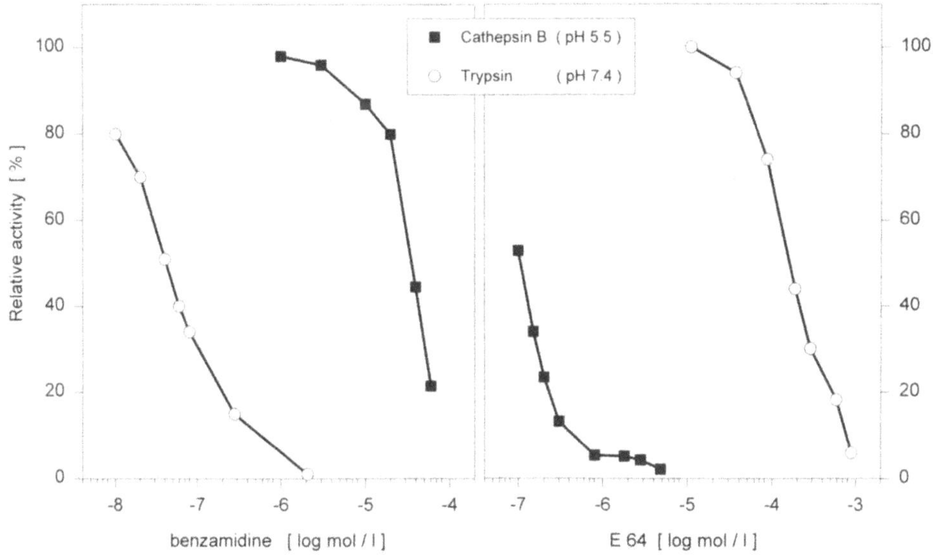

Figure 4. Inhibition of trypsin and cathepsin B by benzamidine and E 64. Both enzymes were measured with the substrate $(CBZ\text{-}Arg)_2$-R110.

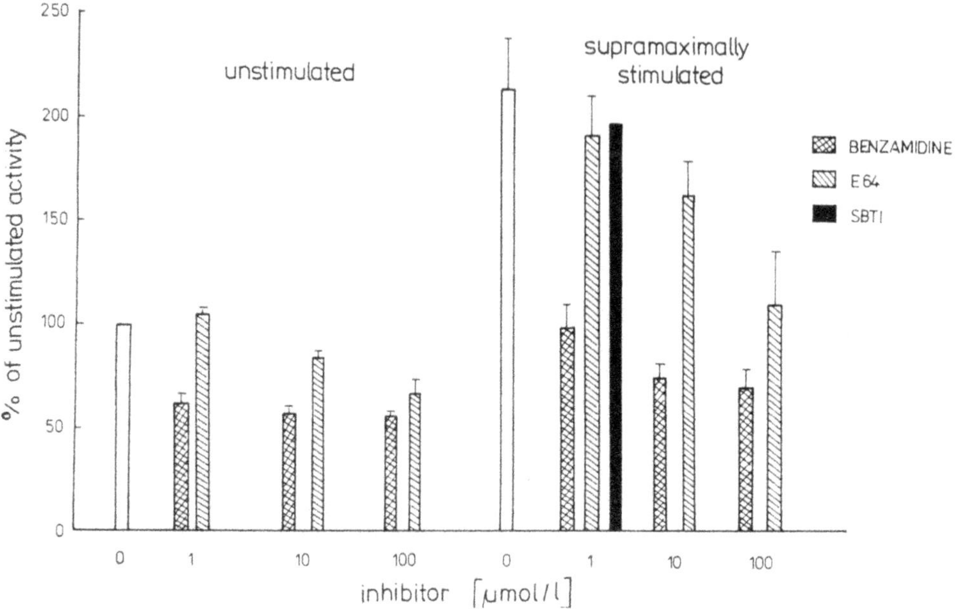

Figure 5. Effect of inhibitors on the $(CBZ\text{-}Arg)_2$-R110 cleaving activity of unstimulated and supramaximally stimulated (caerulein $1.4x10^{-8}$ mol/l) acini.

proenzyme activation by a proteolytic enzyme. Therefore, we tested the effect of the previously used inhibitors to characterize the formation process of the $(CBZ\text{-}Arg)_2$-R110-cleaving activity. To this end, the isolated acini were preincubated with the inhibitors, and then the acini were stimulated by supramaximal concentrations of caerulein. As shown in Fig. 6, the appearence of the caerulein-induced activity was strongly inhibited by BA, but not by E 64. This indicates that the activation of the substrate-cleaving enzyme is not caused by cathepsin B. Rather, the formation is dependent on an active serine protease. SBTI did not show an inhibiting effect to this process indicating that the formation of the protease activity as well as the substrate cleavage are intracellular events.

The question arose, whether the caerulein-induced increase of serine protease activity in vitro is qualitatively and quantitatively comparable to the caerulein effect in vivo. Rats were treated by caerulein in vivo, and acini were prepared from these animals. The cells were incubated with $(CBZ\text{-}Arg)_2$-R110 (Fig. 7). The order of magnitude of the measured substrate-cleaving activity was comparable to that of in vitro-stimulated acini, although there were remarkable individual deviations. Nevertheless, the response to the inhibitors used was identical to that of in vitro-stimulated acini.

4. DISCUSSION

Summarizing the presented results, it may be stated that rhodamine 110-based fluorogenic protease substrates are useful tools to characterize intracellular protease activites in pancreatic cells. In combination with protease inhibitors, it was found that the treatment of isolated acini with supramaximal concentrations of caerulein leads to an increase of serine protease activity. This confirmed results of Leach et al. [13] who could

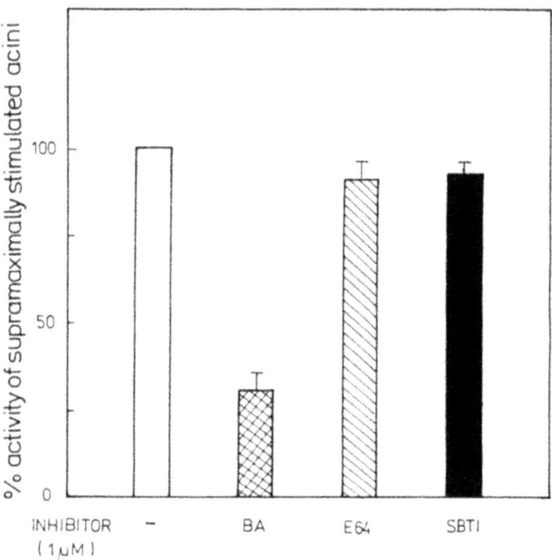

Figure 6. Effect of protease inhibitors on the formation of the $(CBZ\text{-}Arg)_2$-R110 cleaving activity of supramaximally stimulated acini. Acini were preincubated in the presence of inhibitors for 15 min.

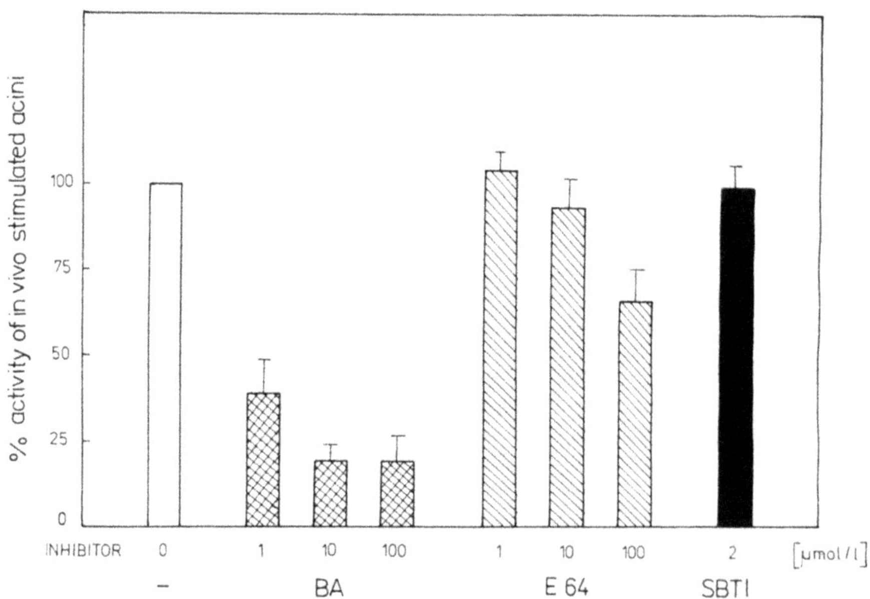

Figure 7. Inhibition of the $(CBZ\text{-}Arg)_2$-R110 cleaving activity of acini which were isolated from caerulein-treated rats (2 injections of 20 μg/kg b.w. as indicated in Material and Methods).

demonstrate the action of a serine protease in isolated pancreatic acini stimulated by high dosages of CCK measuring the conversion of procarboxypeptidase A1 into carboxypeptidase A1.

Suggesting that the caerulein-induced increase of serine protease activity is preceded by a proenzyme conversion, the sensitivity to BA indicates the action of a serine protease as activating enzyme. A cysteine protease, as discussed for cathepsin B, seems to be not involved in this activation process in vitro. Caerulein hyperstimulation of rats induces a serine protease activity in pancreatic acinar cells which was found to be in the order of magnitude as in the in vitro-stimulated acini. This findings suggest that the serine protease proenzyme conversion is an intracellular event also under in vivo-conditions. The available results favour a concept of an autocatalytic activation of trypsinogen.

This work was supported by the Deutsche Forschungsgemeinschaft SFB 387/B2.

5. REFERENCES

1. Gorelick, F. S., G. Adler, and H. F. Kern. Cerulein-induced pancreatitis. In: The pancreas: Biology, pathobiology, and disease. Go, V. L. W., E. P. DiMagno, J. D. Gardner, E. Lebenthal, H. A. Reber, and G. A. Scheele (eds.) Raven Press, New York, 501–526, 1993
2. Amsterdam, A., and J. D. Jamieson. Structural and functional characterization of isolated pancreatic exocrine cells. Proc. Natl. Acad. Sci. U.S.A. 69: 3028–3032, 1972
3. Rothe, G., S. Klingel, I. Assfalg-Machleidt, W. Machleidt, C. Zirkelbach, R. B. Banati, W. F. Mangel, and G. Valet. Flow cytometric analysis of protease activities in vital cells. Biol. Chem. Hoppe-Seyler 373: 547–554, 1992
4. Schön, E., I. Born, H.-U. Demuth, J. Faust, K. Neubert, T. Steinmetzer, A. Barth, and S. Ansorge. Dipeptidyl peptidase IV in the immune system. Biol. Chem. Hoppe-Seyler 372: 305–311, 1991
5. Kunz, D., F. Bühling, H.-J- Hütter, T. Aoyagi, and S. Ansorge. Aminopeptidase N (CD 13, EC 3.4.11.2) occurs on the surface of resting and concanavalin A-stimulated lymphocytes. Biol. Chem. Hoppe-Seyler 374: 291–296, 1993
6. Stürzebecher, J., D. Prasa, P. Wikström, H. Vieweg, and J. Hauptmann. Synthesis and structure-activity relationships of potent thrombin inhibitors: piperazides of 3-amidinophenylalanine. J. Med. Chem. (1996) in press
7. Tieku, S. and N. M. Hooper. Inhibition of aminopeptidases N, A and W. Biochem. Pharm. 44: 1725–1730, 1992
8. Heymann, E., R. Mentlein, I. Nausch, C. Erlanson-Albertsson, T. Yoshimoto, and A. C. Feller. Processing of pro-colipase and trypsinogen by pancreatic dipeptidyl peptidase IV. Biomed. Biochem. Acta 45: 575–584, 1986
9. Yamaguchi, H., T. Kimura, K. Kimura, and N. Nawata. Activation of proteases in caerulein-induced pancreatitis. Pancreas 4: 565–571, 1989
10. Bialek, R., S. Willemer, R. Arnold, and G. Adler. Evidence of intracellular activation of serine proteases in acute caerulein-induced pancreatitis in rats. Scand. J. Gastroenterol. 26: 190–196, 1991
11. Lüthen, R., C. Niederau, and J. H. Grendell. Intrapancreatic zymogen activation and levels of ATP and glutathione during caerulein pancreatitis in rats. Am. J. Physiol. 268: G592–604, 1995
12. Grady, T, A. Saluja, A. Kaiser, and M. Steer. Edema and intrapancreatic trypsinogen activation precede glutathione depletion during caerulein pancreatitis. Am. J. Physiol. 271: G20-G26, 1996
13. Leach, S. D., I. M. Modlin, G. A. Scheele, and F. S. Gorelick. Intracellular activation of digestive zymogens in rat pancreatic acini. Stimulation by high doses of cholecystokinin. J. Clin. Invest. 87: 362–366, 1991

15

STRUCTURE OF CD26 (DIPEPTIDYL PEPTIDASE IV) AND FUNCTION IN HUMAN T CELL ACTIVATION

Martin Hegen, Junichi Kameoka, Rui-Ping Dong, Chikao Morimoto, and Stuart F. Schlossman

Division of Tumor Immunology
Dana-Farber Cancer Institute and Department of Medicine
Harvard Medical School
44 Binney Street, Boston, Massachusetts 02115

1. INTRODUCTION

1996 was the 10th anniversary of the establishment of the cluster of differentiation antigen, CD26 by the Second Workshop and Conference on Human Leukocyte Differentiation Antigens [1]. First described as a T cell activation antigen, CD26 was found to be strongly expressed on subsets of both activated CD4 and CD8 T cells [2]. Subsequent work showed the expression of CD26 on activated Nk cells, B cells and B cell lines and demonstrated CD26 expression on a variety of different cell types, most abundantly on epithelia of the intestine, prostate and the proximal tubuli of the kidneys [3,4,5]. Furthermore, it was demonstrated that CD26 possesses enzymatic activity, dipeptidyl peptidase IV (DPP IV; EC 3.4.14.5), in its extracellular domain [4] and is identical to the previously described adenosine deaminase (ADA, EC 3.5.4.4) binding protein [6,7]. Recent studies indicate that CD26 plays a seminal role in T cell costimulation and may be involved in severe combined immunodeficiency and HIV infection [8,9].

2. THE STRUCTURE OF CD26

The cDNA encoding the human CD26 antigen was cloned in 1992 [10,11]. The isolated cDNA predicts a protein of 766 amino acids, having type II membrane topology with only a 6 amino acid cytoplasmic region and a 22-residue hydrophobic transmembrane region. The extracellular region has 10 potential sites for N-glycosylation, and differences in post-translational processing may account for the different molecular weight forms found with different CD26 mAbs. The extracellular region can be divided into 3 regions: the N-terminal glycosylated region starting with a flexible 'stalk' region; a cysteine-rich region; and a

Cellular Peptidases in Immune Functions and Diseases, edited by Ansorge and Langner
Plenum Press, New York, 1997

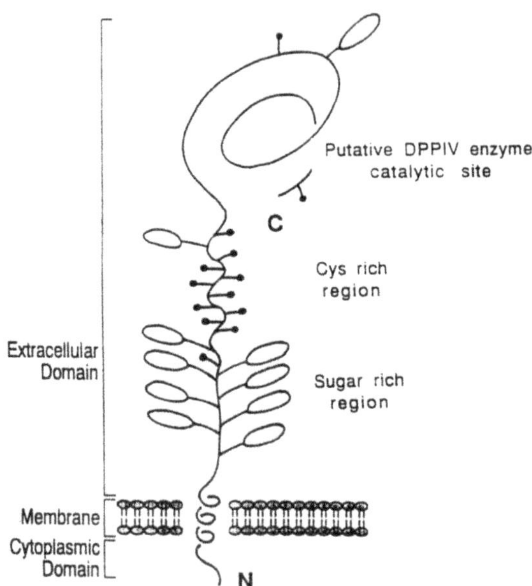

Figure 1. Schematic primary structure of the human CD26/DPP IV protein. The protein is predicted to have type II transmembrane topology containing 6 N-terminal amino acids in the cytoplasm.

260 amino acid C-terminal region containing the catalytic triad [10]. The predicted amino acid sequence showed around 85% homology with the rat DPP IV enzyme [12,13] and mouse thymocyte activating molecule (THAM) [14], the mouse homologue of the human CD26 antigen. Figure 1 shows the schematic structure of the human CD26 protein encoded by the CD26 cDNA.

CD26 is a member of the prolyl oligopeptidase family which is defined by the requirement of the catalytic triad in the unique order Ser, Asp, His [15]. The predicted CD26 amino acid sequence around the serine catalytic residue is Gly-Trp-Ser-Tyr-Gly at position 628–632. The serine 630 was confirmed by oligonucleotide directed mutagenesis to be the active-site serine [16]. Although the natural substrate for the DPP IV has not been established, the enzyme can cleave amino-terminal dipeptides from polypeptides with either L-proline or less efficiently L-alanine at the penultimate position [17]. It should also be emphasized that many biologically active cytokines or chemokines may be targets for the enzyme and all available evidence suggests that the enzyme lacks endopeptidase activity [18].

3. DIRECT ASSOCIATION OF ADENOSINE DEAMINASE WITH HUMAN CD26

A molecule of 43-kDa protein can be coprecipitated by anti-CD26 antibody from ^{125}I-labeled T cells, phytohemagglutinin (PHA) activated blast cells, and from CD26-transfected Jurkat cell lines [6,19]. The 43-kDa protein band obtained from CD26 transfected Jurkat cells was consistently more dense in SDS-PAGE than the band obtained from peripheral T cells and PHA blasts. Therefore, the CD26-transfected Jurkat cells was used to identify the 43-kDa protein by immunoaffinity chromatography [6]. The purified protein was separated by SDS-polyacrylamide gel electrophoresis (PAGE), transferred to a nitrocellulose membrane, and stained by Ponceau S, resulting in a single band of 43-kDa in addition to the CD26 110-kDa protein. The 43-kDa protein was then digested with trypsin, separated by reversed-phase high-pressure liquid chromatography (rpHPLC), and subjected to amino acid sequencing. According to the homology search, the amino acid sequences of the two peptides derived from the 43-kDa were completely identical to those of the human adenosine deaminase (ADA) [20].

ADA is a 41-kDa protein, expressed in all tissues (highest expression in lymphocytes), that catalyzes the conversion of adenosine and deoxyadenosine to inosine and deoxyinosine, respectively. ADA is present on the cell surface, as well as in the cytoplasm, of human fibroblasts, rabbit renal tubular cells, and human mononuclear blood cells [21,22]. Despite its wide distribution, ADA deficiency causes severe combined immunodeficiency disease (SCID) in humans [23]. ADA was coexpressed with CD26 on the Jurkat CD26 transfectants, and *in vitro* binding assay showed that the binding was through the extracellular domain of CD26. These studies show for the first time that ADA and CD26 interact on the T cell surface and suggest that this interaction may provide a clue to the pathophysiology of SCID caused by ADA deficiency. Recent studies indicate that the ADA bound to CD26 is from an extracellular source and may have an immunoregulatory function in human T cell activation [24,25].

4. FUNCTION OF CD26 IN HUMAN T CELL ACTIVATION

The CD3/TcR complex plays a central role in T cell activation and function. In general, specific peptide antigen or antibodies against the antigen-specific T cell receptor (TcR)/CD3 complex alone cannot induce T cell proliferation and lymphokine secretion. Rather, T cells require a second costimulatory signal which can be provided by a number of accessory molecules, including CD26, expressed on the T cell surface [26]. The unique memory/helper population of human CD4 cells ($CD45RO^{+}CD29^{+}$) which expresses CD26 is the only one that can respond to recall antigens, induce B-cell immunoglobulin synthesis and activate MHC restricted cytotoxic T cells [27]. $CD4^{+}$ T cells lacking CD26 cannot be triggered to elicit helper functions but can respond to mitogens and alloantigens.

Initially, anti-CD26 antibody (anti-Tp103) was shown to induce T cell activation and enhanced cytotoxic activity when cross-linked via the FcR on the surface of accessory or target cells [28]. Subsequently, a mAb, anti-1F7 was developed, which defines a functionally unique epitope of CD26 distinct from that recognized by anti-Ta1 [27]. Interestingly, mAb anti-1F7 but not mAb anti-Ta1 inhibited both tetanus-toxoid induced T cell proliferation and pokeweed mitogen-driven B cell IgG synthesis [27]. Although mAb anti-Ta1 can only mediate T cell activation in the presence of accessory cells bearing high-density FcR and exogenous interleukin-2 (IL-2) [29], solid-phase immobilized mAb anti-1F7 can mediate a co-mitogenic effect on human T cell proliferation via the CD3 and CD2 pathways or after treatment with phorbolester in the absence of either exogenous IL-2 or accessory cells [30]. Moreover, anti-1F7 treatment leads to a decrease in the surface expression of the CD26 antigen via the internalization, and such modulation results in an enhanced proliferative response to anti-CD3 or anti-CD2 [31].

To further characterize the function of human CD26, a CD26 negative human T cell line, Jurkat, was transfected with CD26 cDNA [10,32]. Transfected Jurkat cells now express a 110-kDa molecule similar to that found on peripheral blood T cells and the isolated Ag has dipeptidyl peptidase IV (DPP IV) activity. Further study showed that Jurkat cells expressing CD26 have enhanced IL-2 production when triggered with solid-phase immobilized anti-CD26 (1F7) and anti-CD3 antibodies, similar to that reported for peripheral $CD4^+$ T cells. In addition, mAb-mediated cross-linking of the CD26 and CD3 antigens results in markedly enhanced $[Ca^{2+}]i$ mobilization in CD26 transfected Jurkat cells. These results directly support the notion that the CD26 plays a key role in T cell costimulation.

5. SIGNAL TRANSDUCTION OF CD26

The earliest biochemical events seen in stimulated T lymphocytes activated through the engagement of the TCR is the tyrosine phosphorylation of a panel of cellular proteins [33]. It has been demonstrated that $CD4^+$ T cells lacking CD26 cannot be triggered to elicit a memory T cell response [34] and previous studies indicate the involvement of tyrosine phosphorylation in CD26 mediated T cell proliferation [35]. Therefore, we investigated the CD26-mediated signal transduction pathway in order to determine in more detail which pathways are involved. We demonstrate that antibody-induced cross-linking of CD26 in CD26-transfected Jurkat cells induced tyrosine phosphorylation of several intracellular proteins with a similar pattern to that seen after TCR/CD3 stimulation [36]. Herbimycin A, an inhibitor of the *src* family protein tyrosine kinases dramatically inhibited this CD26-mediated effect on tyrosine phosphorylation. Major tyrosine phosphorylated proteins were identified by immunoblotting, and are identical with $p56^{lck}$, $p59^{fyn}$, zeta associated protein-tyrosine kinase of 70-kDa (ZAP-70), mitogen-activated protein (MAP) kinase, c-Cbl, and phospholipase Cγ (Figure 2). CD26-induced tyrosine phosphorylation of MAP kinase correlated with increased MAP kinase activity. In addition, CD26 was costimulatory to CD3 signal transduction since co-cross-linking of CD26 and CD3 antigens induced prolonged and increased tyrosine phosphorylation in comparison with CD3 activation alone providing a putative mechanism for the enhancement of the response to recall antigens mediated through CD26. We therefore conclude that CD26 is a true costimulatory entity which can up regulate the signal transducing properties of the TCR.

6. ASSOCIATION MOLECULES OF CD26 AND ITS SIGNAL INDUCING ACTIVITY

According to the amino acid sequence predicted from the cloned cDNA sequence, the CD26 antigen has a cytoplasmic region consisting of only six amino acids, which may be too short to fully explain its signal-transducing activity. ADA, a cytoplasmic enzyme has been shown to associate with CD26 [6,7]. Therefore, ADA may be a possible signal transducing candidate of CD26. However, as the cDNA sequence of ADA does not show a transmembrane or cytoplasmic region, and alternative spliced variants have not been reported [20], it is unlikely that ADA is directly involved in CD26 signaling. Another molecule which associates with CD26 is CD45. CD26 was comodulated on the T cell surface with CD45RO, a known membrane-linked protein tyrosine phosphatase and that anti-CD26 was capable of precipitating CD45 from T cell lysates [37]. Moreover, we demonstrated that modulation of CD26 from T cell surface induced by anti-CD26 led to enhanced phosphorylation of CD3 ζ tyrosine residues and increased CD4 associated $p56^{lck}$ tyrosine kinase activity [37]. Thus, one possible mechanism for this costimulatory activity involves the association of CD26 with the membrane-linked protein-tyrosine-phosphatase CD45. This association may account for the increased Lck kinase activity, increased tyro-

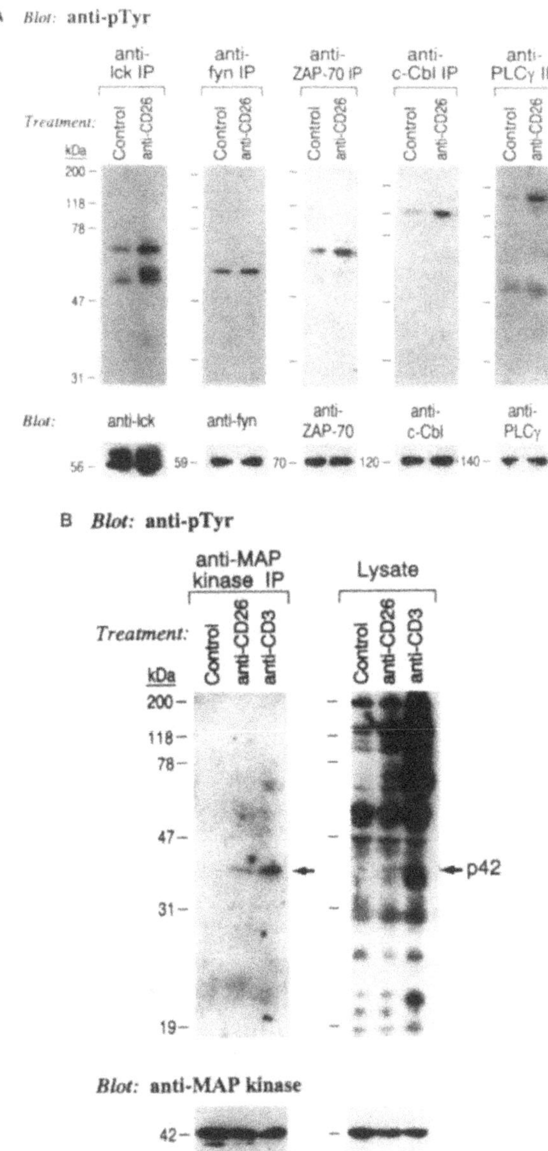

Figure 2. CD26 ligation induces tyrosine phosphorylation of $p56^{lck}$, $p59^{fyn}$, ZAP-70, c-Cbl, PLCγ, and MAP kinase in CD26 transfected Jurkat cells (J.CD26). **A.** J.CD26 cells were untreated or incubated with anti-CD26 mAb and stimulated by cross-linking with goat anti-mouse Ab for 15 min. Lysates of stimulated cells were immunoprecipitated with anti-lck conjugated beads, anti-fyn polyclonal Ab, anti-ZAP-70 mAb, anti-c-Cbl polyclonal Ab and anti-PLCγ polyclonal Ab, and the immunoprecipitates were analyzed by anti-phosphotyrosine immunoblotting. The blots were stripped and reprobed with the specific Abs followed by horse radish peroxidase (HRP)-conjugated secondary Abs and bound Abs were detected by enhanced chemiluminescence technique. **B.** J.CD26 cells were untreated or incubated with anti-CD3 mAb or anti-CD26 mAb and stimulated by cross-linking with goat anti-mouse Ab. Immunoprecipitates with anti-ERK 2 conjugated beads (left panel) and whole lysates (right panel) were immunoblotted with anti-phosphotyrosine mAb, stripped and reprobed with anti-MAP kinase Ab followed by HRP-conjugated secondary Abs and bound Abs were detected by enhanced chemiluminescence technique.

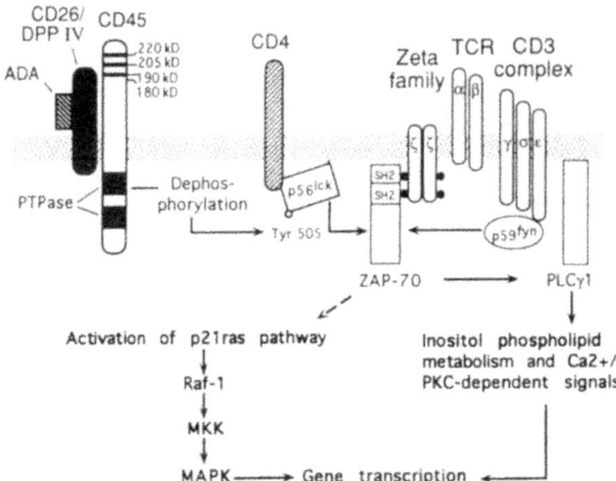

Figure 3. Model of CD26/DPP IV signal transduction and associated molecules, CD45, a membrane-linked protein tyrosine phosphatase and adenosine deaminase (ADA). The association of CD26 with CD45 may explain the enhanced tyrosine phosphorylation of several intracellular proteins by antibody-induced cross-linking of CD26. The protein-tyrosine kinase activity of the *src* kinases $p56^{lck}$ and $p59^{fyn}$ can be upregulated by CD45-mediated dephosphorylation of tyrosine 505 (Tyr 505). The $p56^{lck}$ kinase is able to phosphorylate the immunoreceptor tyrosine-based activation motifs (ITAMs) of the CD3 ζ chain, thus enabling the recruitment and activation of SH_2-carrying proteins such as ZAP-70 kinase. Subsequently, the activation by tyrosine phosphorylation of phospholipase Cγ (PLCγ) leads to the generation of inositol phosphate (IP) production and increase in intracellular free calcium concentration on one hand, and to diacylglycerol synthesis and protein kinase C activation on the other. Another major downstream event is the activation of the RAS-GTPase activating (RAS-GAP) pathway including the activation of the mitogen-activated protein (MAP) kinase.

sine phosphorylation and intracellular calcium mobilization after binding of anti-CD26 antibody to the antigen (Figure 3).

7. CONCLUSIONS

CD26 is a 110-kDa T cell activation antigen. It has been shown to have DPP IV enzyme activity which cleaves amino-terminal dipeptide with either L-proline or less efficiently L-alanine at the penultimate position. Recent studies have shown that CD26 plays an integral role in T cell activation. A partial explanation of the mechanism of CD26-mediated T cell signaling appears to be its association with CD45 tyrosine phosphatase as well as adenosine deaminase. In addition, it has been proposed that CD26 can interact with extracellular matrix proteins such as fibronectin or collagen. Moreover, it may be linked to the pathogenesis of AIDS [9]. Thus, CD26 is a multifunctional molecule that may have an important role in various components of the immune system.

ACKNOWLEDGMENT

This work is supported by grants NIH AI12609, AR33713 and CA55601.

REFERENCES

1. Knowles, R. W. 1986. In *Leukocyte typing II* (ed. E. L. Reinherz, B. F. Haynes, L. M. Nadler, and I. D. Bernstein), Springer-Verlag, New York, pp. 259–88.
2. Fox, D. A., Hussey, R. E., Fitzgerald, K. A., Acuto, O., Poole, C, Palley, L., Daley, F. J., Schlossman, S. F. and Reinherz, E. L. 1984. Ta1, a novel cell surface molecule involved in helper function of CD4 cells. *Journal of Immunology* **133,** 1250–6.
3. Stein, H., Schwarting, R. and Niedobitek, G. 1989. In *Leukocyte typing IV*. (ed. W. Knapp, B. Dörken, W. R. Gilks, E. P. Rieber, R. E. Schmidt, H. Stein, and A. E. G. Kr. von dem Borne), Oxford University Press, Oxford, pp.412–5.
4. Hegen, M., Niedobitek G., Klein, C. E., Stein, H. and Fleischer, B. 1990. The T cell triggering molecule Tp103 is associated with dipeptidyl aminopeptidase IV activity. *Journal of Immunology* **144,** 2908–14.
5. Bühling, F., Junker, U., Reinhold, D., Neubert, K., Jäger, L. and Ansorge, S. 1995. Functional role of CD26 on human B lymphocytes. *Immunology Letters* **45,** 47–51.
6. Kameoka, J., Tanaka, T., Nojima, Y., Schlossman, S. F. and Morimoto, C. 1993. Direct association of adenosine deaminase with a T cell activation antigen, CD26. *Science* **261,** 466–9.
7. Morrison, M. E., Vijayasaradhi, S., Engelstein, D., Albino, A. P., Houghton, A. N. 1993. A marker for neoplastic progression of human melanocytes is a cell surface ectopeptidase. *Journal of Experimental Medicine* **177,** 1135–43.
8. Fleischer, B. 1994. CD26: a surface protease involved in T cell activation. *Immunology Today* **15,** 180–4.
9. Morimoto, C. and Schlossman, S. F. 1994. CD26, a key costimulatory molecule on CD4 memory T cells. *The Immunologist* **2,** 4–7.
10. Tanaka, T., Camerini, D., Seed, B., Torimoto, Y., Dang, N. H., Kameoka, J., Dahlberg, H. N., Schlossman, S. F. and Morimoto, C. 1992. Cloning and functional expression of the T cell activation antigen CD26. *Journal of Immunology* **149,** 481–6.
11. Darmoul, D., Lacasa, M., Baricault, L., Marguet, D., Sapin, C., Trotot, P. Barbat, A. and Trugnan, G. 1992. Dipeptidyl peptidase IV (CD26) gene expression in enterocyte-like colon cancer cell lines HT-29 and Caco-2. Cloning of the complete human coding sequence and changes of dipeptidyl peptidase IV mRNA levels during cell differentiation. *Journal of Biological Chemistry* **267,** 2200–8.
12. Hong, W. and Doyle, D. 1987. cDNA cloning for a bile canaliculus domain-specific membrane glycoprotein of rat hepatocytes. *Proceedings of the National Academy of Science USA*. **84,** 7962–65.
13. Ogata, S., Misumi, Y. and Ikehara, Y. 1989. Primary structure of rat liver dipeptidyl peptidase IV deduced from its cDNA and identification of the NH_2-terminal signal sequence as the membrane-anchoring domain. *Journal of Biological Chemistry* **264,** 3596–601.
14. Marguet, D., Bernard, A.-M., Vivier, I., Darmoul, D., Naquet, P. and Pierres, M. 1992. cDNA cloning for mouse thymocyte-activating molecule. *Journal of Biological Chemistry* **267,** 2200–8
15. Abbott, C. A., Baker, E., Sutherland, G. R. and McCaughan, G. W. 1994. Genomic organisation, exact localization, and tissue expression of the human CD26 (dipeptidyl peptidase IV) gene. *Immunogenetics* **40,** 331–8.
16. Tanaka, T., Kameoka, J., Yaron, A., Schlossman, S. F. and Morimoto, C. 1993. The costimulatory activity of the CD26 antigen requires dipeptidyl peptidase IV enzyme activity. *Proceedings of the National Academy of Science USA*. **90,** 4586–90.
17. Kenny, A. J., Booth, A. G., George, S. G., Ingram, J., Kershaw, D., Wood, E. J. and Young, A. R. 1976. Dipeptidyl peptidase IV, a kidney brush-border serine peptidase. *Biochemical Journal* **155,** 169–82.
18. Yaron, A. and Naider, F. 1993. Proline-dependent structural and biological properties of peptides and proteins. *Critical Review Biochemical Molecular Biology* **28,** 31–81.
19. Torimoto, Y., Dang, N. H., Tanaka, T., Prado, C., Schlossman, S. F. and Morimoto, C. 1992. Biochemical characterization of CD26 (dipeptidyl peptidase IV): Functional comparison of distinct epitopes recognized by various monoclonal antibodies. *Molecular Immunology* **29,** 183–92.
20. Daddona, P. E., Shewach, D. S., Kelley, W. N., Argos, P., Markham, A.F. and Orkin, S. H. 1984. Human adenosine deaminase: cDNA and complete primary amino acid sequence. *Journal Biological Chemistry* **259,** 12101–6.
21. Schrader, W. P., Miczek, A.D., West, C. A. and Samsonoff, W. A. 1988. Evidence for receptor-mediated uptake of adenosine deaminase in rabbit kidney. *Journal Histochemistry Cytochemistry* **36,** 1481–7.
22. Aran, J. M., Colomer, D., Matutes, E., Vives-Corrons, J. L. and Franco R. J. 1991. Presence of adenosine deaminase on the surface of mononuclear blood cells: Immunochemical localization using light and electron microscopy. *Journal Histochemistry Cytochemistry* **39,** 1001–8.

23. Hirschhorn R. 1990. In Immunodefiency Reviews. (eds. Rosen, F.S. and Seligmann, M.). Howard, London. Vol. 2. pp. 175–98.
24. Dong, R.-P., Kameoka, J., Hegen, M., Tanaka, T., Xu, Y, Schlossman, S. F. and Morimoto, C. 1996. Characterization of adenosine deaminase binding to human CD26 on T cells and its biologic role in immune response. *Journal of Immunology* **156,** 1349–55.
25. Martin, M., Huguet, J., Centelles, J. J. and Franco, R. 1995. Expression of ecto-adenosine deaminase and CD26 in human T cells triggered by the TCR-CD3 complex. Possible role of adenosine deaminase as costimulatory molecule. *Journal of Immunology* **155,** 4630–43.
26. Geppert, T. D., Davis, L. S., Gur, H., Wacholtz, M. C. and Lipsky, P. E. 1990. Accessory cell signals involved in T cell activation. *Immunological Reviews* **117,** 5–66.
27. Morimoto, C., Torimoto, Y., Levinson, G., Rudd, C. E., Schrieber, M., Dang, N. H.. Letvin, N. L. and Schlossman, S. F. 1989. 1F7, a novel cell surface molecule involved in helper function of CD4 cells. *Journal of Immunology* **143,** 3430–9.
28. Fleischer, B. 1987. A novel pathway of human T cell activation via a 103KD human T cell activation antigen. *Journal of Immunology* **138,** 1346–9.
29. Dang, N. H., Torimoto, Y., Deusch, K., Schlossman, S. F. and Morimoto, C. 1990. Comitogenic effect of solid-phase immobilized anti-1F7 on human CD4 T cell activation via CD3 and CD2 pathways. *Journal of Immunology* **144,** 4092–100.
30. Dang, N. H., Torimoto, Y., Sugita, K., Daley, J. F., Schow, P., Prado, C., Schlossman, S. F. and Morimoto, C. 1990. Cell surface modulation of CD26 by anti-1F7 monoclonal antibody: Analysis of surface expression and human T cell activation. *Journal of Immunology* **145,** 3963–71.
31. Hegen, M., Camerini, D. and Fleischer B. 1993. Function of dipeptidyl peptidase IV (CD26, Tp103) in transfected human T cells. *Cellular Immunology* **146,** 249–60.
32. Dang, N. H., Hafler, D. A., Schlossman, S. F. and Breitmeyer, J. B. 1990. FcR-mediated crosslinking of Ta1 (CDw26) induces T cell activation. *Cellular Immunology* **125,** 42–57.
33. Chan, A. C., Desai, D. M. and Weiss, A. 1994. The role of protein tyrosine kinases and protein tyrosine phosphatases in T cell antigen receptor signal transduction. *Annual Review Immunology* **12,** 555–92.
34. Hafler, D. A., Fox, D. A., Benjamin, D. and Weiner, H. L. 1986. Antigen rective memory T cells are defined by Ta1. *Journal of Immunology* **147,** 414–8.
35. Munoz, E., Blazquez, M. W., Madueno, J. A., Rubio, G. and Pena, J. 1992. CD26 induces T cell proliferation by tyrosine phosphorylation. *Immunology* **77,** 43–50.
36. Hegen, M., Kameoka, J., Dong, R.-P., Schlossman, S. F. and Morimoto, C. 1997. Cross-linking of CD26 by antibody induces tyrosine phosphorylation and activation of mitogen-activated protein kinase. *Immunology* **90,** 257–64.
37. Torimoto, Y., Dang, N. H., Vivier, E., Tanaka, T., Schlossman, S. F. and Morimoto, C. 1991. Coassociation of CD26 (dipetidyl peptidase IV) with CD45 on the surface of human T lymphocytes. *Journal of Immunology* **147,** 2514–7.

MOLECULAR ASSOCIATIONS REQUIRED FOR SIGNALLING VIA DIPEPTIDYL PEPTIDASE IV (CD26)

Bernhard Fleischer, Christiane Steeg, Jochen Hühn, and Arne von Bonin

Bernhard Nocht Institute for Tropical Medicine
D-20359 Hamburg, Germany

1. INTRODUCTION

CD26, or dipeptidyl peptidase IV, is a widely distributed cell surface glycoprotein of approximately 110 kD molecular weight. It is constitutively expressed on a variety of different cell types, particularly on epithelial cells of the intestine, prostate and kidney-proximal tubules[1-3]. On T cells the expression of CD26 is regulated more stringently. The molecule is absent from the majority of human resting peripheral blood T lymphocytes and is expressed weakly on a fraction of T cells in the blood. Expression increases within two days of T cell activation and increases further after culture *in vitro* [1,2]. It has been shown to be a suitable marker for T cells activated *in vivo* [4] and memory T cells have been shown to reside in the $CD26^+$ T cell fraction[5]. CD26 has several unique properties that distinguish this molecule from related surface proteases:

i. the expression of CD26 is tightly regulated in different tissues,
ii. it is associated with the differentiation and activation status of different cells,
iii. CD26 is highly conserved among different species[6].

CD26 is associated with other molecules on the cell surface and within the cells. It binds adenosine desaminase (ADA)[7] and co-precipitates with CD45 on T lymphocytes[8]. Modulation of CD26 on T cells leads to an enhanced phosphorylation of the CD3ζ-chain and increases the activity of the lck-kinase[9]. CD26 also binds specifically to the *tat* protein of HIV. It has been reported that the enzymatic activity is inhibited by this protein[10].

Specific substrate analogues have been used to selectively block the enzymatic activity of CD26 and have been found to block T cell activation by antigens and by lectin and can even suppress an immune response *in vivo* [11-13]. Moreover, certain antibodies against CD26 inhibit T cell proliferation induced by antigen and the mitogen-driven T cell-dependent synthesis of immunoglobulins by B cells[14]. Furthermore, antigen-induced T cell proliferation can be enhanced by soluble CD26[15]. Taken together, these studies suggest that CD26 may have biologically important immuno-modulatory activity.

Cellular Peptidases in Immune Functions and Diseases, edited by Ansorge and Langner
Plenum Press, New York, 1997

2. ACTIVATION OF T CELLS VIA CD26

More than ten years ago we have first demontrated that CD26 may be involved in T cell activation, when we found that a monoclonal antibody directed against a T cell activation antigen of approximately 103 kD (Tp103) induced re-directed lysis in human cytotoxic T lymphocyte clones and interleukin-2 (IL-2)-secretion and proliferation in human preactivated $CD4^+$ and $CD8^+$ T cells[16, 17]. Later we found that Tp103, CD26 and dipeptidyl peptidase IV are identical[18]. A similar activating property was described for mouse dipeptidyl peptidase IV[19]. T cell triggering via CD26 appears to be unique because other cell surface proteinases like CD10 and CD13 that are also type II membrane proteins fail to induce T cell activation. Moreover, stimulation of CD26 leads to the activation of all functional programs of the T cell, including cytotoxicity and granule exocytosis. The signalling capacity of CD26 can be transferred by transfected CD26 molecules indicating that the stimulation of T cells via CD26 does not require additional activation-dependent molecules[20]. It is remarkable that this stimulatory signal of CD26 on T cells has to be induced by crosslinking monoclonal anti-CD26-antibodies via Fc-receptors on accessory cells and cannot be induced by plastic-immobilized anti-CD26-antibodies. So far a variety of different monoclonal antibodies against CD26 directed against different epitopes of the CD26 molecules can all induce this stimulation. The stimulatory signal of CD26 is regulated by the CD4 and CD8 molecules indicating the use of physiological pathways in CD26-mediated signalling[21]. We and others have also been able to show that CD26 can deliver a so-called "co-stimulatory signal" to suboptimal stimulation of the T cell receptor (TCR)[20, 22]. This finding indicates that a physiological role of CD26 could be the binding of a ligand on the antigen-presenting cell for enhancing a signal given via antigen/MHC complexes.

3. ASSOCIATION WITH THE T CELL RECEPTOR IS CRITICAL FOR SIGNALLING

The intra-cytoplasmic tail of CD26 is only six amino acids in length[1-3]. It is therefore likely that CD26 is unable to signal by itself and has to utilize other signal-transducing molecules to confer the incoming signal. In contrast to the CD2 molecule that is able to signal in T cells expressing the CD3 complex as well as in NK cells lacking the TCR/CD3 complex, but expressing CD16 associated with γ-chains, CD26 signals only in T cell receptor expressing $\alpha\beta TCR^+$ and $\gamma\delta TCR^+$ T cells[23]. Moreover, modulation of the T cell receptor complex abrogates the capacity of CD26 to signal in modulated T cells, although CD26 does not co-modulate with the TCR. Mutant Jurkat cells that do not express a functional T cell receptor are non-responsive to CD26-mediated signalling[20]. However, in these cells also the CD2-induced activation is impaired indicating the dependence of this stimulation on components of the TCR[23]. Thus, it is likely that the response of NK cells stimulated by CD2 is due to the presence of a ζ-like chain of the Fc-receptor for IgG (CD16)[24].

4. THE ROLE OF THE CD3 ζ-CHAIN IN THE STIMULATION VIA CD26

Signal transduction from the T cell receptor after recognition of antigen/MHC complexes or after triggering by TCR-specific antibodies is dependent on the presence of so-called "immune-receptor tyrosine-based activation motif" (ITAM) in the TCR-associated CD3-chains[25]. The first step in T cell activation is the tyrosine phosphorylation of these ITAM[26]. The CD3 ζ-chain is unusual in having three motifs, whereas the CD3 δ-, CD3-ε and the CD3-γ-subunits have only one ITAM. The TCR complex consists of at least two autonomous transduction molecules because the deletion of the ITAM in the ζ-chain still allows signalling after antigen-recognition[27].

To analyze the role of the ζ-chain in CD26-mediated signal-transduction we made use of two different cell types. Both cells lines are derived from the thymoma cell line BW5147[27, 28]. The MM16, MM17 and MM18 TCR-negative variants lack α-, β-, γ- δ- and ζ-chain proteins. They express a chimeric CD25/CD3-ζ-chain-protein on the cell surface, where the extracellular portions of the human CD25-molecule were fused to the cytoplasmic tail of the ζ-protein[28]. The MM16 cells express full length cytoplasmic tails of the ζ-chain containing all three functional ITAM whereas the MM17 express a mutated ζ-chain with the tyrosine residues at positions 72, 111 and 142 having been exchanged to phenylalanine residues destroying the ζ-specific functional ITAM. The MM18 cell lines contain a tyrosine- phenylanaline substitution at the position 153 inactivating the third ITAM.

The second set of BW5147-derived cells are positive for TCR-α-, TCR-β-, CD3γ- and CD3ε-chains mRNA molecules, but not for CD3δ- and CD3δ-chains[27]. The cells were transfected with the entire wild-type CD3δ-gene, and the resultant cell line BWδ was transfected with plasmids encoding cDNAs for different forms of the CD3ζ-chain: wild-type ζ-chain, containing three ITAM (leading to the cell line BWδζ); a truncated ζ-chain in which amino acids 66 to 114 were deleted, containing one functional ITAM (BWδζD66-114) and a ζ-chain in which amino acids 66 to 157, i.e. all three ITAM, were deleted (BWδζD66-157). Both sets of cell lines were supertransfected with either expression vectors encoding the CD2- or the CD26-coreceptor molecule[29].

BWδ cells expression no ζ-chain have only very little TCR on the cell surface and respond poorly when stimulated with anti-CD3-specific antibodies. The three cell lines containing different forms of the ζ-chain showed identical expression of the TCR. All supertransfected cells expressed human CD26 or human CD2. Several isolates of each triple-transfectant were used for the experiments described. To prove the identity of the transfectant the presence of the CD3-ζ-mRNA of the predicted size was verified by RT-PCR. CD26-transfected cell lines but not the untransfected cells expressed dipeptidyl peptidase IV enzymatic activity.

All ζ-chain-transfectants were equally responsive to monoclonal antibodies against the CD3-ε molecule. The reactivity against the superantigen staphylococcal enterotoxin A (SEA) differed between the transfected cells. BWδζD66-114 cells with one functional ITAM were as efficient responders to SEA as the wild-type cells, whereas the BWδζD66-157 cells lacking all three ITAM required 10 to 100 times higher concentrations of SEA to reach optimal secretion of IL-2.

We then investigated which of the mutant TCR was able to provide stimulatory capacity to the transfected CD26 molecule. Our results showed that the complete deletion of all three ITAM from the ζ-chain completely abolished the activating capacity of the CD26 molecule. Cells expressing TCR with ζ-chains that contain only one ITAM still could respond to anti-CD26 although a 100fold higher concentration of the antibody was required.

Interestingly, stimulation via CD2 in all three ζ-transfectants resulted also in a significant IL-2-production. This is in contrast with earlier observations demonstrating that the ζ-chain is a prerequisite for effective CD2-mediated signalling (30). Obviously, other proteins of the CD3 complex compensate for the lack of endogenous ζ-proteins after stimulation with the anti-CD2-specific antibodies in the CD2-transfectants.

We then asked the question whether the CD25-ζ-chimeric proteins in MM16 CD26 cells were able to provide stimulatory capacity to the transfected CD26 molecule. These "minimal receptors" were able to activate the cells after stimulation with antibodies against the extracellular CD25 part of the chimeric proteins. In contrast, none of the CD26 transfectants was able to respond to anti-CD26-crosslinking. As another control served the MM16 cells transfected with the CD2 molecule. These cells could easily be stimulated after anti-CD2-crosslinking to release IL-2 (Table 1).

Taken together, these results suggest that the CD3-ζ-chain is required but not sufficient for stimulation via CD26 whereas it is sufficient but not required for stimulation via CD2.

5. ROLE OF THE ENZYMATIC ACTIVITY OF CD26 IN SIGNALLING

Since signalling via CD26 is likely to occur in a complex with other molecules this interaction might be mediated through the catalytic site of CD26. To assess the role of dipeptidyl peptidase IV-enzymatic activity several groups including our own used specific inhibitors of dipeptidyl peptidase IV-activity to block specifically the enzymatic activity without effecting the structure of CD26[11, 12]. It cannot be excluded, however, that these inhibitors have side effects beyond their action on the catalytic center of dipeptidyl peptidase IV that might effect T cell activation independent of their specific activity on the enzyme. In fact, we have found that CD26-negative Jurkat cells are as susceptible to inhibition by specific inhibitors of dipeptidyl peptidase IV as CD26-expressing transfected Jurkat cells[31].

In a second experimental approach we used CD26 transfectants which lack dipeptidyl peptidase IV-enzymatic activity by a mutation of residue 630 from serine to alanine[32]. Jurkat cells transfected with such CD26 mutant molecules have been reported to respond much less to CD26-mediated co-stimulation than Jurkat cells transfected with the wild-type enzyme[33]. We therefore tested a large number of different transfectants for the signalling capacity of the mutant molecules compared to the wild-type molecule (Table 2). The response of transfected clones to direct stimulation with anti-CD26-antibodies and to

Table 1. Response of T cells expressing different forms of the TCR to various stimuli

		Stimulation via/by				
Mutant	No. ITAM	CD3	CD25	SEA	CD2	CD26
Bwδζ (BW1)	0	(+)	–	–	–	–
BWδζ (BW2)	3	+++	na	+++	+++	+++
BWδζD66-114 (BW4)	1	+++	na	+++	+	+
BWδζD66-157 (BW3)	0	+++	na	+++	–	–
MM16	3	na	+++	na	++	–
MM17	0	na	–	na	–	–
MM18	2	na	nd	na	nd	–

Different cell lines expressing wildtype and truncated forms of the ζ-protein were tested for their ability to produce IL-2 after having been treated with different stimuli (na = not applicable, nd = not done). As shown for CD26 signalling, triggering requires a ζ-chain with at least one functional ITAM. Again, the sole expression of the ζ-protein, demonstrated with the different "MM"-cell lines, does not result in effective IL-2 secretion. Interestingly, the stimulation of BWδζD66-157 (BW3-CD2) resulted also in IL-2 secretion indicating a possible role for non ζ-chain signal transducing modules in CD2-induced signalling.

co-stimulation via CD26 was variable and not clearly dependent on the amount of CD26 and TCR expressed on the T cells. Several mutant transfectants were more easily triggered via CD26 than cells transfected with the wild-type molecule. It appears that several parameters affect the response to anti-CD26. The expression of the TCR at an appropriate level is probably very important. Since signalling via CD26 depends directly on the presence of a functional TCR, it is likely that a low-level expression of CD3 will affect the signalling capacity of both mutant and wild-type CD26 molecules. High TCR expression may even be more important than the level of expression of CD26 itself. In addition there appeared to be clonal differences in the overall capacity to respond to stimuli or to produce IL-2 among the transfected clones. These points could explain the discrepancy between our and results reported by others. Taken together we can conclude that the enzymatic activity of CD26 is not required for signalling via this molecule.

6. COMPARISON OF THE REACTIVITY OF CD26-POSITIVE AND CD26-NEGATIVE T CELL LINES

To assess whether the expression of CD26 might confer a better reactivity to T cells we compared the response of two different cell lines, either transfected with CD26 or mock-transfected with vector alone, to a stimulation via TCR. CD26 negative and positive Jurkat transfectants were compared with respect to their reactivity to the superantigen staphylococcal enterotoxin E . As shown in Figure 1, the dose response curve of these cells was identical, irrespective of the presence of CD26. The same results were obtained using CD26-transfected or mock-transfected mouse cell lines. These results show that the expression of CD26 does not confer a specific advantage to the expressing T cell.

7. CONCLUDING REMARKS

Here we have described a somewhat reductionistic approach to analyze the function of CD26 in T cell stimulation. The complex function of CD26 in T cell activation was reduced to the serach for the molecular requirements for signalling via a transfected CD26-molecule. We

Table 2. Response of transfectants expressing wildtype and mutant CD26 molecules to anti-CD26 and anti-CD3

	Expression of[a]		IL-2 production after stimulation[b]		
Cell line	CD26	CD3	Medium	CD26	CD3
M7	67 (17)	132 (85)	100	200	18400
M15	85 (90)	205 (100)	100	26200	9000
M19	82 (61)	81 (90)	200	2600	4500
M21	64 (25)	193 (100)	400	23600	26300
M22	232 (91)	42 (25)	100	9700	15900
M24	75 (90)	105 (100)	50	13800	8600
WT1	125 (91)	164 (100)	300	14100	24600
WT2	100 (80)	190 (100)	100	8200	20500

Cells were assayed for surface expression of CD26 and CD3 by flowcytometry and for IL-2 production in the presence of 1 μg/ml CB.1 or 0.1 μg/ml 14-2C11 in the presence of A20/J cells.
[a]mean fluorescence intensity (% positive cells)
[b]cpm 3H-Thymidine incorporated by CTLL cells

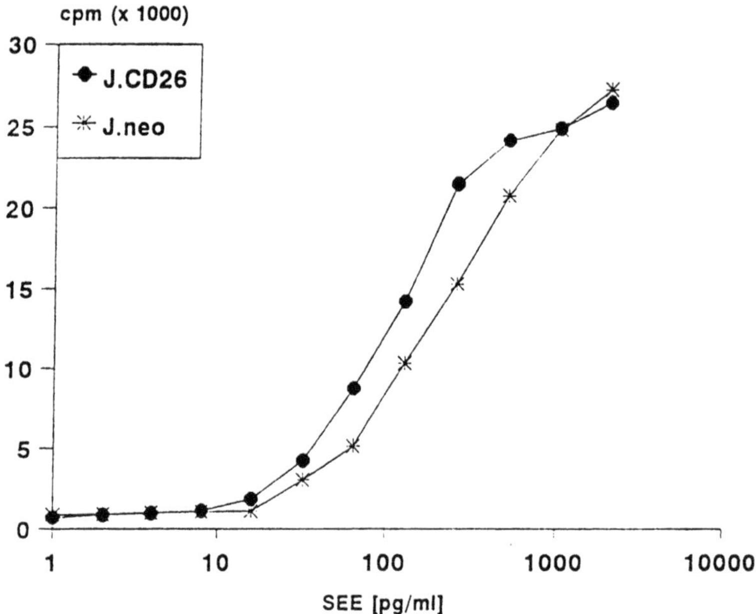

Figure 1. Jurkat cells that express a Vβ8 TCR reactive with staphylococcal enterotoxin E (SEE) were transfected with a CD26 cDNA-encoding expression vector (o) or with the vector without CD26 (*). The transfectants were stimulated with SEE at the concentrations indicated in the presence of Raji cells as superantigen-presenting cells. The IL-2 produced upon stimulation was measured in a bioassay using IL-2 dependent CTLL-2 cells[34]. Data are given as cpm 3H-Thymidin incorporated by CTLL-2 cells.

have used cloned cell lines transfected with a cDNA encoding CD26 to probe the molecular associations. There are several advantages in using such cells as experimental system:

1. they represent a cloned homogenous cell population.
2. they produce but do not consume IL-2. Thus, IL-2-production is directly proportional to the number of cells stimulated.
3. non-transfected and mock-transfected counterparts without expression of CD26 are available.
4. transfection with a cDNA-encoding CD2 can be used as a control.
5. mutant cell lines with defined deletions or substitutions of signal-transducing molecules are available.

Our data show that for transmitting an effective signal to the T cell via CD26 the T cell receptor is absolutely required. Moreover, it is clear that the cytoplasmic tail of the ζ-chain is a prerequisite for T cell activation via CD26, although the ζ-chain is not sufficient for conducting these signals. This shows that CD26 does not interact directly with the ζ-chain of the TCR but probably requires an additional so far undefined "protein X" that couples the incoming signals from CD26 to the ζ-chain of the TCR. Appearently the complete TCR complex is required for this association of CD26 plus "protein X" to the ζ-chain (Figure 2).

Our results also show that the enzymatic activity of CD26 is not required, neither for the direct signal transduction via CD26 after crosslinking with antibodies nor for its co-stimulatory function, i.e. as an additional signal in concert with the TCR. These results do not exclude that the enzymatic activity of CD26 might play an important role in the regulation of the immune response.

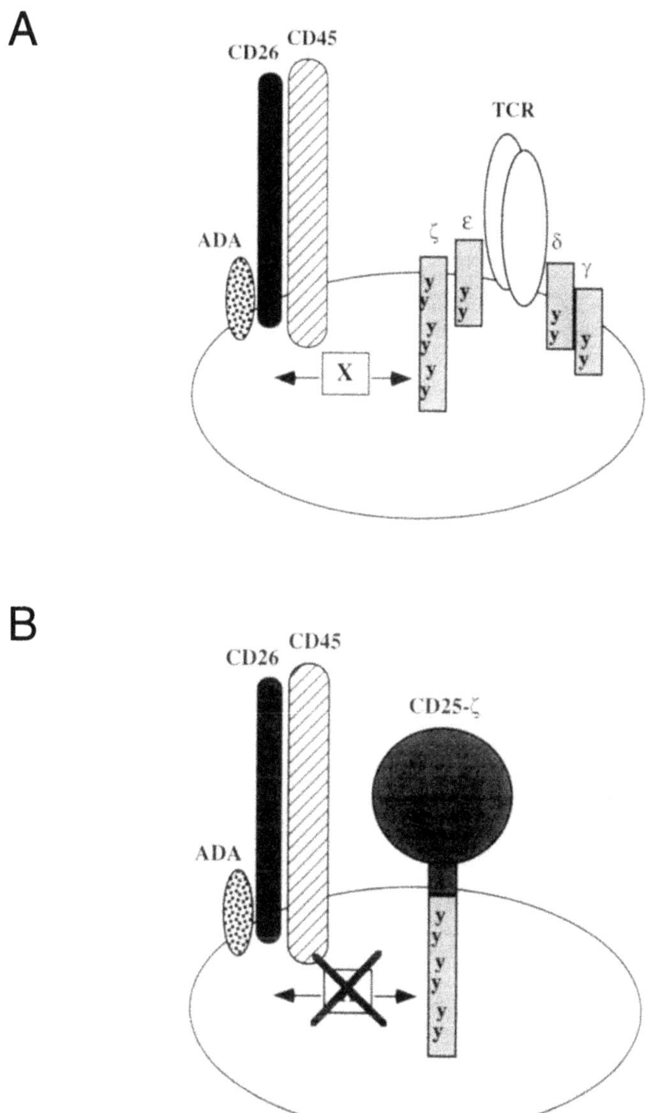

Figure 2. Hypothetical model of CD26-mediated signalling in TCR-expressing and TCR non-expressing cells. Figure 2A shows a possible situation for CD26-induced signalling in TCR-expressing cells. Since CD26 expresses only a very short cytoplasmic tail incapable of transmitting signals, CD26 must interact with other proteins to trigger IL-2 production. Two coprecipitating partners have been identified, adenosine desaminase (ADA) and the tyrosinephosphatase CD45. However, both proteins do not interact directly with the ζ-chain, making it likely that additional proteins are involved in the signalling events induced after anti CD26-crosslinking. Figure 2B depicts the situation in TCR$^-$ cells. So far no CD26-dependent stimulation could be observed in cells expressing no TCR. This may be due to the fact that expression of the ζ-protein at the cell surface without the rest of the TCR complex leads to an incomplete signalling module insufficient for CD26-dependent signalling. Therefore we propose that a signalling module for CD26 must consist of a ζ-chain with at least one functional ITAM in combination with the TCR and a bridging protein denoted as "X". For coupling the incoming signals from CD26 to the ζ-chain signalling module the "X"-protein requires obviously the expression of the complete TCR complex on the cell surface.

8. REFERENCES

1. Fleischer, B. 1995. Dipeptidylpeptidase IV (CD26) in metabolism and the immune response. *Molecular Biology Intelligence Unit. R.G. Landes Co., Austin/Springer Verlag, Berlin, 1-202.*
2. Fleischer, B. 1994. CD26: a surface protease involved in T-cell activation. *Immunol. Today 15:180-184.*
3. Yaron, A. and N., F. 1993. Proline dependent structural and biological properties of peptides and proteins. *Crit. Rev. Biochem. Mol. Biol. 28:31-81.*
4. Enssle, K. H., and B. Fleischer. 1990. Absence of Epstein-Barr virus-specific, HLA class II-restricted CD4+ cytotoxic T lymphocytes in infectious mononucleosis. *Clin. Exp. Immunol. 79:409-415.*
5. Hafler, D. A., D. A. Fox, and D. Benjamin. 1986. Antigen reactive memory T cells are defined by Ta1. *J. Immunol. 137:414-418.*
6. Marduet, D., A. M. Bernard, I. Vivier, D. Darmoul, P. Naquet, and M. Pierres. 1992. cDNA cloning for mouse thymocyte-activating molecule. A multifunctional ecto-peptidylpeptidase IV (CD26) included in a subgroup of serine proteases. *J. Biol. Chem 267:2200-2208.*
7. Kameoka, J., T. Tanaka, Y. Nojima, S. F. Schlossmann, and C. Morimoto. 1993. Direct association of adenosin desaminase with a T cell activation antigen, CD26. *Science 261:566-469.*
8. Torimoto, Y., N. H. Dang, E. Vivier, T. Tanaka, S. F. Schlossmann, and C. Morimoto. 1991. Coassociation of CD26 (dipeptidyl peptidase IV) with CD45 on the surface of human T lymphocytes. *J. Immunol. 147:2514-2517.*
9. Dang, H. H., Y. Torimoto, K. Sugita, J. F. Daley, P. Schow, C. Prado, S. F. Schlossmann, and C. Morimoto. 1990. Cell surface modulation of CD26 by anti 1F7 monoclonal antibody. Analysis of surface expression and human T cell activation. *J. Immunol. 145:3963-3971.*
10. Subramanyam, M., W. G. Gutheil, W. W. Bachovchin, and B. T. Huber. 1993. Mechanism of HIV-1 tat induced inhibition of antigen-specific T cell responsiveness. *J. Immunol. 150:2544-2553.*
11. Schön, E., S. Jahn, S. T. Kiessig, H. U. Demuth, K. Neubert, A. Barth, R. von Baehr, and S. Ansorge. 1987. The role of dipeptidylpeptidase IV in human T lymphocyte activation. Inhibitors and antibodies against dipeptidyl peptidase IV suppress lymphocyte proliferation and immunglobulin synthesis in vitro. *Eur. J. Immunol. 17:1821-1826.*
12. Flentke, G. R., E. Munoz, B. T. Huber, A. G. Plaut, C. A. Kettner, and Bachovchin W.W. 1991. Inhibition of dipeptidyl aminopeptidase IV (DP-IV) by Xaa-boroPro dipeptides and use of these inhibitors to examine the role of DP-IV in T-cell function. *Proc.Natl.Acad.Sci. USA 88:1556-1559.*
13. Kubota, T., G. R. Flentke, W. W. Bachovchin, and B. D. Stollar. 1992. Involvement of dipeptidyl peptidase IV in an in vivo immune response. *Clin. Exp. Immunol. 89:192-197.*
14. Morimoto, C., Y. Torimoto, G. Levinson, C. E. Rudd, M. Schrieber, N. H. Dang, N. L. Letvin, and S. F. Schlossmann. 1989. 1F7, a novel surface molecule, involved in helper function of T cells. *J. Immunol 143:3430-3439.*
15. Tanaka, T., Duke-Cohan, J.S., Kameoka, J., Yaron, A., Lee, I., Schlossmann, S.F., and C. Morimoto. 1994. Enhancement of antigen induced T cell proliferation by soluble CD26 (DPP-IV). *Proc. Natl. Acad. Sci.* USA 91:3082-3086.
16. Fleischer, B. 1987. A novel pathway of human T cell activation via a 103 kD T cell activation antigen. *J. Immunol. 138:1346-1349.*
17. Fleischer, B., Schendel, D., D. von Steldern. 1986. Triggering of the lethal hit in human cytotoxic T lymphocytes: a functional role for a 103 KD T cell activation antigen. *Eur. J. Immunol. 16:741-746.*
18. Hegen, M., G. Niedobitek, C. E. Klein, H. Stein, and B. Fleischer. 1990. The T cell triggering molecule Tp103 is associated with dipeptidyl aminopeptidase IV activity. *J. Immunol. 144:2908-2914.*
19. Vivier, I., D. Marguet, P. Naquet, and e. al. 1991. Evidence that thymocyte-activating molecule is mouse CD26 (dipeptidyl peptidase IV). *J. Immunol. 147:447-454.*
20. Hegen, M., D. Camerini, and B. Fleischer. 1993. Function of dipeptidyl peptidase IV (CD26, Tp103) in transfected human T cells. *Cell. Immunol. 146:249-260.*
21. Schrezenmeier, H., and B. Fleischer. 1988. A regulatory role for the CD4 and CD8 molecules in T cell activation. *J. Immunol. 141:398-403.*
22. Dang, N. H., Y. Torimoto, K. Deusch, S. F. Schlossmann, and C. Morimoto. 1990. Comitogenic effect of solid phase immobilized anti-1F7 on human CD4 T cell activation via CD3 and CD2 pathways. *J. Immunol. 144:4092-4100.*
23. Fleischer, B., E. Sturm, J. E. de Vries, and H. Spits. 1988. Triggering of cytotoxic T lymphocytes and NK cells via Tp103 pathway is dependent on the expression of the T cell receptor/CD3 complex. *J. Immunol. 141:1103-1107.*

24. Howard, F. D., P. Moingeon, U. Moebius, D. J. McConkey, B. Yandava, T. E. Gennert, and E. L. Reinherz. 1992. The CD3 ζ cytoplasmic domain mediates CD2-induced T cell activation. *J.Exp.Med. 176:139-145.*
25. Reth, M. 1989. Antigen receptor clue. *Nature 338:383-384.*
26. Samelson, L. E., M. D. Patel, A. M. Weissman, J. B. Harford, and R. D. Klausner. 1986. Antigen activation of murine T cell induces tyrosine phosphorylation of a polypeptide that is associated with the T cell antigen receptor. *Cell 46:1083-1090.*
27. Wegener, A. M. K., F. Letourneur, A. Hoeveler, T. Brocker, F. Luton, and B. Malissen. 1992. The T cell receptor/CD3 complex is composed of at least two autonomous transduction modules. *Cell 68:83-95.*
28. Donnadieu, E., A. Trautmann, M. Malissen, J. Trucy, B. Malissen, and E. Vivier. 1994. Reconstitution of CD3 ζ coupling to calcium mobilization via genetic complementation. *J. of Biological Chemistry 269:32828-32834.*
29. Mittrücker, H. W., C. Steeg, B. Malissen, and B. Fleischer. 1995. The cytoplasmic tail of the T cell receptor ζ chain is required for signaling via CD26. *Eur. J. Immunol. 25:295-297.*
30. Moingeon, P., J. L. Lucich, D. J. McConkey, F. Letourneur, B. Malissen, J. Kochan, H. C. Chang, H. R. Rodewald, and E. L. Reinherz. 1992. CD3 ζ dependence of the CD2 pathway of activation in T lymphocytes and natural killer cells. *Proc.Natl.Acad.Sci. USA 39:1492-1496.*
31. Hegen, M., H. W. Mittrücker, R. Hug, and B. Fleischer. 1993. Enzymatic activity of CD26 (dipeptidyl peptidase IV) is not required for its signalling function in T cells. *Immunobiol. 189:483-493.*
32. Steeg, C., U. Hartwig, and B. Fleischer. 1995. Unchanged signaling capacity of mutant CD26/Dipeptidyl peptidase IV molecules devoid of enzymatic activity. *Cell. Immunol. 164:311-315.*
33. Tanaka, T., J. Kameoka, A. Yaron, S. F. Schlossmann, and C. Morimoto. 1993. The costimulatory activity of the CD26 antigen requires dipeptidyl peptidase IV enzymatic activity. *Proc. Natl. Acad. Sci. USA 90:4586-4590.*
34. Grabstein, K., J. Eisenmann, D. Modizuki, P. Schanebeck, P. Cordon, T. Hopp, C. March, and S. Gilles. 1986. Purification to homogeneity of B cell stimulating factor. *J. Exp. Med. 163:1405-1414.*

17

CD26/DIPEPTIDYL PEPTIDASE IV IN LYMPHOCYTE GROWTH REGULATION

Siegfried Ansorge, Frank Bühling, Thilo Kähne, Uwe Lendeckel, Dirk Reinhold, Michael Täger, and Sabine Wrenger

Institute of Experimental Internal Medicine
Department of Internal Medicine
Otto-von-Guericke-University
Magdeburg, Germany

1. INTRODUCTION

T lymphocyte activation is controlled by the T cell antigen receptor-CD3 complex and a number of accessory molecules on the surface of lymphocytes which are triggered by special ligands outside of the cell. During the last years, there has been a considerable progress in the identification of these accessory structures and the mechanisms of their actions.

One of them has been identified to be a proteolytic enzyme, the dipeptidyl peptidase IV (DP IV, EC 3.4.15.5). This 766 amino acid type II integral membrane protein was described for the first time by Hopsu-Havu and Glenner[1] as glycylprolyl-ß-naphthylamide hydrolyzing enzyme and 10 years later by A. J. Kenny et al.[2] as a constituent of the plasma membrane. At the Fourth International Workshop on Human Leukocyte Differentiation Antigens in Boston, 1984, this peptidase was classified as an activation marker on lymphocytes under the entry CD26[3–6]. Since that time DP IV / CD26 has been admitted to the family of immunologically relevant molecules. The growing knowledge on DP IV / CD26 was reviewed and summarized in the book "Dipeptidyl Peptidase IV (CD26) in Metabolism and the Immune Response" edited by Bernhard Fleischer[7].

The available data of the recent years taken together, verify that the DP IV / CD26 system is involved in diverse fundamental functions of the cell. Most of them, however, are still not understood with respect to the detailed molecular and cellular mechanisms (Table 1).

In this paper we intend to concentrate on the question as to what way DP IV / CD26 is involved in the regulation of lymphocyte functions. For this purpose we shall review our results of recent years on the role of DP IV / CD26 in the control of lymphocyte growth, in the production and release of various cytokines and the molecular signal transduction events of T cells.

Cellular Peptidases in Immune Functions and Diseases, edited by Ansorge and Langner
Plenum Press, New York, 1997

Table 1. What do we really know about DP IV / CD26?

CD26 / DP IV Property / Involvement	Detailed molecular mechanism
• Ectoenzyme component / soluble peptidase	
• Different molecular forms	
• Catalyzing NH_2 - X - Pro release	Serine peptidase
• Peptide inactivation (e.g. SP)	Hydrolysis
• Lymphocyte activation structure	?
• Adhesion molecule	?
• **Growth regulation (inhibitors, Abs, nat. ligands)**	**?**
• **Cytokine production**	**?**
• **T cell signalling**	**?**
• ADA binding	?
• Peptide transport	?
• Virus receptor	?
• Glucose regulations in vivo	?

2. DP IV / CD26 IN LYMPHOCYTE GROWTH REGULATION

2.1. DP IV Inhibitors, Effects on Lymphocyte Functions

One of the most common approaches to examining the cellular function of an enzyme is to use different effectors of the catalyst. The first hint that DP IV has a function in lymphocyte growth came from Ekkehard Schön et al.[8] in our laboratory who could show that N-Ala-Pro-O-(nitrobenzoyl-)hydroxylamine which irreversibly blocks the enzymatic activity of DP IV, is capable of suppressing the DNA synthesis of human lymphocytes stimulated by the mitogens phythemagglutinin or concanavalin A in vitro (Figure 1). There was found a significant decrease in ^{3}H-thymidine incorporation due to the action of the DP IV inhibitor although incubation was performed in presence of 17 % human blood serum in the cell culture medium which contains considerable amounts of soluble DP IV.

In the mean time a multitude of different inhibitors of DP IV[9] (Table 2) have been studied under different stimulation conditions in different cell systems. Most of the inhibitors used in our laboratory were designed by the groups of Alfred Barth, Hans-Ulrich Demuth and Klaus Neubert from the Martin-Luther-University in Halle. The most potential inhibitors used in our lab were Lys[Z(NO_2)]-piperide or Lys[Z(NO_2)]-thiazolidide which cause a 50 % inhibition of DP IV in the range of 2 - 3μM. A 50 % suppression of DNA synthesis after 3 days of incubation could be reached at an inhibitor concentration around 10μM[9].The various stimulation conditions included the usage of the mitogens PHA, ConA and PWM, antigens as tetanus antigen, anti-CD3 antibodies, different cytokines, e.g. IL-1, IL-2, IL-6 and special monoclonal anti-CD26 antibodies alone or in combination with anti-CD3 antibodies (cf. below). All of them were found to induce an increase in DNA synthesis in mononuclear cells which can be suppressed by DP IV inhibitors (Figure 2).

These data suggest that DP IV / CD26 occupies a central place in the regulation of DNA synthesis of lymphocyte irrespective of the special triggering signal. Moreover, recently we could show that this phenomenon is not restricted to T lymphocytes[10,11]. CD26 was found to be expressed also on 40 - 50 % of stimulated B lymphocytes of healthy do-

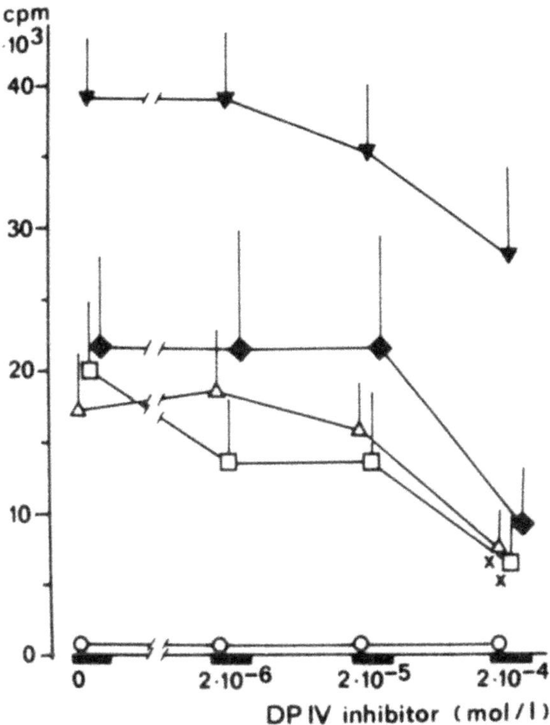

Figure 1. Influence of DP IV inhibitor on mitogen induced T lymphocyte proliferation. N-Ala-Pro-O-(nitrobenzoyl-)hydroxylamine was used as DP IV inhibitor in cultures containing no mitogen (O), PHA in a final dilution of 1 : 3000 (Δ) and 1 : 300 (▼), or Con-A at 8μg/ml (□) and 40μg/ml (◆). Mononuclear cells of human peripheral blood were cultured under standard conditions in presence of 17 % human serum. After 76 h incubation, the cultures were pulsed with ^{3}H-thymidine.

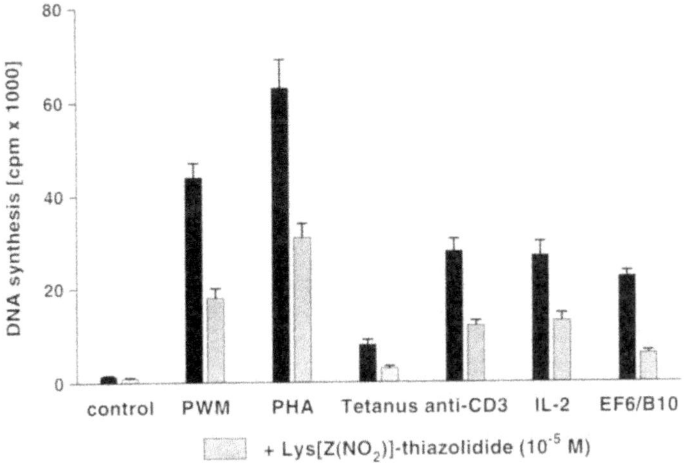

Figure 2. Suppression of DNA synthesis by DP IV inhibitors in differently stimulated peripheral blood mononuclear cells. Human PBMC were cultured under serum free conditions in presence of 2μg/ml PWM, 1μg/ml PHA, 0.1μg/ml Tetanus antigen, 5 μg/ml anti-CD3 antibodies, 10 U IL-2 or 5μg/ml anti-CD26 antibody EF6/B10 without and with 10^{-5} M DP IV inhibitor Lys-[Z(NO$_2$)]-thiazolidide. Cells were labeled with ^{3}H-thymidine 18 hours before harvesting.

Table 2. DP IV inhibitors (reprinted from[9], with kind permission from Walter de Gruyter, Berlin & New York)

Compound	DP IV activity IC_{50}	DNA synthesis	
		IC_{50}	Residual activity at 5×10^{-4} mol/l
	[μmol/l]	[μmol/l]	[% of contr.]
1 Lys[Z(NO_2)]-pyrrolidide	2.0 ± 0.2	[a]	
2 Lys[Z(NO_2)]-thiazolidide	2.7 ± 0.3	8.0 ± 1.5	
3 Ile-thiazolidide	2.8 ± 0.2	47 ± 4	
4 Val-pyrrolidide	6.0 ± 0.3	[b]	54 ± 14
5 Ile-Pro-Ile (diprotin A)	8.0 ± 0.8	[b]	69 ± 20
6 DIFP	10 ± 2	[c]	
7 Ile-pyrrolidide	17 ± 1	400 ± 190	
8 Phe-Pro-NHO-CO-$-C_6H_4-NO_2$	20 ± 2	[a]	
9 Ala-Pro-NHO-CO-$-C_6H_4-NO_2$	30 ± 3	50 ± 23[d] 370 ± 50	
10 Leu-pyrrolidide	38 ± 3	[b]	61 ± 11
11 Phe-pyrrolidide	54 ± 4	260 ± 48	
12 Ala-thiazolidide	87 ± 6	320 ± 150	
13 Leu-Pro-Gly-Val-Gly [lymphotoxin-(1–5)-peptide]	320 ± 60	[b]	61 ± 18
14 Tyr-Pro-Phe-Pro-NH_2 (morphiceptin)	500 ± 60	[e]	
15 Tyr-Pro-Phe-Pro-Gly [β-casomorphin-(1–5)-peptide]	530 ± 70	[e]	
16 Ile-Pro	620 ± 110	[e]	
17 Lys[Z(NO_2)]-Pro	700 ± 60	200 ± 90[d] 250 ± 40	
18 Lys[Cap]-Pro	820 ± 50	[e]	
19 Val-Pro-Leu (diprotin B)	920 ± 160	[e]	
20 Ala-Ala	[f]	[b]	69 ± 8

[a] Toxic to lymphocyte cell cultures.
[b] Less than 50% inhibition at highest concentration tested (5×10^{-4}M).
[c] Not applicable in cell culture.
[d] Concanavalin A used as mitogen (10 μg/ml).
[e] No effect (evaluation limit 80% of the DNA synthesis rate of the control culture) up to a concentration of 500μM.
[f] No effect up to 1mM.

nors and of CVID patients[10] as well as of a population of human NK cells after in vitro treatment with IL-2[11]. Interestingly enough, treatment of activated B and NK cells with specific competitive DP IV inhibitors leads to a significant inhibition of DNA synthesis in a concentration-dependent manner[10, 11] (Figure 3). Flow cytometric DNA analyses as well as cellular activation marker studies with DP IV inhibitors and with anti-CD26 antibodies revealed that DP IV / CD26 has a function in overriding the cell cycle restriction point at the late G1 phase before DNA synthesis, G1/S[11]. From these data we conclude that DP IV/CD26 is also involved in regulation of pro-liferation of NK cells and B cells and that this phenomenon might be a more common one in the hematopoietic cells.

Suppressive effects of DP IV inhibitors strongly correlate with the expression of CD26 / DP IV on the surface of cells. This could be proven on U937 human histiocytic lymphoma cell clones with low or high expression of CD26 / DP IV[12]. In U937 cells strongly expressing CD26, DP IV inhibitors were shown to suppress DNA synthesis significantly whereas in CD26-negative U937 cells these inhibitors had no effect on DNA synthesis (Figure 4). These results also indicate that the inhibitors used are highly DP IV specific and that the effects measured are not due to toxic activities by them. Moreover, it could be excluded that DP IV inhibitors may induce apoptosis in the various cell systems. For this reason we used different methods for measuring apoptosis: the DNA ladder assay, annexin V labeling, terminal deoxynucleotidyl transferase (TdT)-mediated dUTP nick end labeling (TUNEL), and the cell death detection assay. With none of these techniques, apoptosis could be detected in presence of DP IV inhibitors in the cell culture assays.

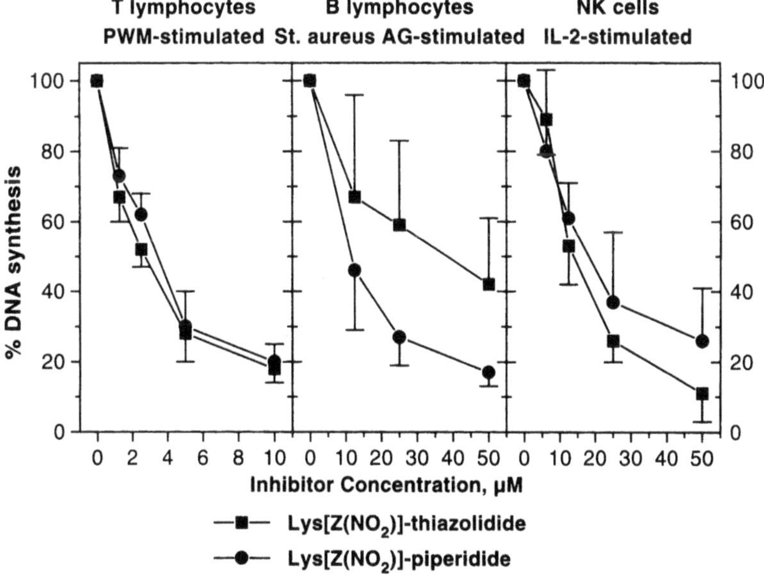

Figure 3. Suppression of DNA synthesis by DP IV inhibitors in different lymphocyte lineages. Isolated T cells, B cells or NK cells were incubated with pokeweed mitogen (2µg/ml), Staphylococcus aureus protein (0.004 %, V/W), and IL-2 (1000 U/ml), resp., in presence and absence of the DP IV inhibitors Lys-[Z(NO_2)]-thiazolidide or -piperidide for 3 days under standard conditions. Cells were labeled with ^{3}H-thymidine 16 hours before harvesting.

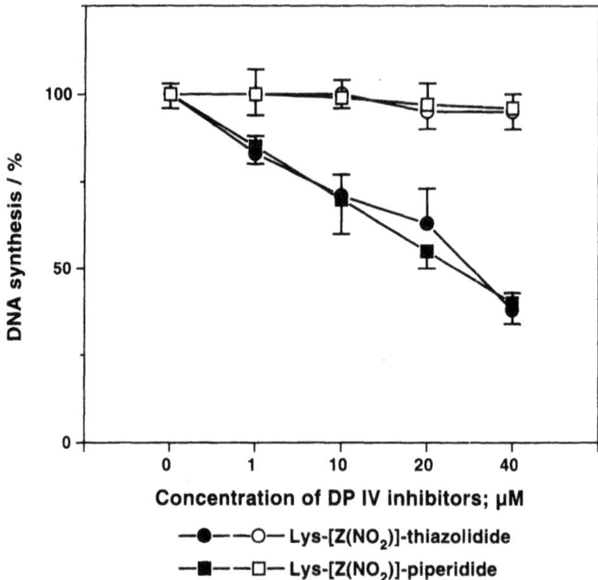

Figure 4. Suppression of DNA synthesis by DP IV inhibitors in CD26 positive U937 cells but not in CD26 negative U937 cells. U937 $CD26^+$ cells and U937 $CD26^-$ cells were incubated with different concentrations of Lys-[Z(NO_2)]-thiazolidide or -piperidide. After 42 hours the cells were pulsed with ^{3}H-methyl-thymidine for further 6 hours.

2.2 The Action of CD26 / DP IV Antibodies on Lymphocyte Functions

An alternative approach to studying the role of DP IV / CD26 independently of its active site is to employ specific antibodies separately or in combination with other effectors. We have investigated different monoclonal antibodies from our laboratory against purified DP IV or DP IV on U937 cells[13]. Biochemical characterization of these antibodies indicates that they have a unique epitope specificity. They are capable of recognizing pig kidney DP IV, they were found to react with the protein backbone of CD26 and were incapable of reacting with the carbohydrate moieties.Moreover, they recognize a more restricted pattern of human lymphocytic CD26 forms separated by isoelectric focusing[14]. Particularly the anti-CD26 antibodies EF6/B10 and EF5/A3 were found to be capable of strongly stimulating the DNA synthesis as well as the IL-2, IL-10 and IFNγ production of human T cells[13] (Figure 5) and, with respect to DNA synthesis, of mononuclear cells[15]. The full activation of purified T cells by soluble anti-CD26 antibodies requires the addition of submitogenic concentrations of phorbol myristate acetate (PMA, 2 ng/ml). However, similar results were obtained when low amounts of monocytes were used instead of PMA in the stimulation assay. As expected, addition of anti-CD3 antibodies or interleukin-2 in concentrations of 5µg/ml and 10 U/ml, respectively increases the DNA synthesis of T cells in a concentration dependent manner. Both the stimulatory effects of anti-CD26 antibodies that induce increased DNA synthesis on their own as well as the increased DNA synthesis due to the costimulatory action of CD26 antibodies and anti-CD3 antibodies or IL-2 could be strongly suppressed by Lys[Z(NO_2)]-thiazolidide or -piperidide[16]. These data demonstrate that T cells can be stimulated directly by CD26 / DP IV antibodies alone or in combination with classical stimulatory agents and that the enzymatic activity or at least the catalytic domain of DP IV is important for mediating the anti-CD26 effects as well as the anti-CD3 and IL-2 action on DNA synthesis.

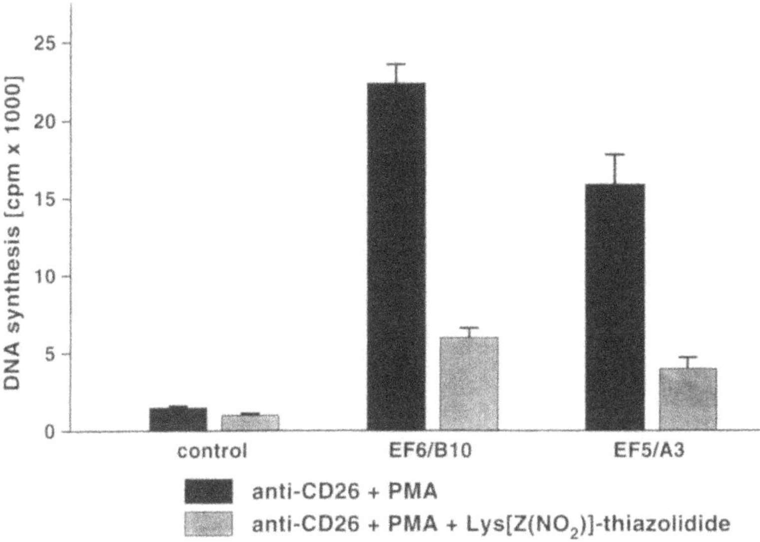

Figure 5. Suppression of DNA synthesis by DP IV / CD26 inhibitors in anti-CD26 antibody-stimulated T cells. Purified T lymphocytes from human peripheral blood were cultured under serum free conditions for 3 days in presence of a submitogenic concentration (2 ng/ml) of phorbol myristate and of the anti-CD26 antibodies EF6/B10 or EF5/A3 (5ng/ml) with or without Lys-[Z(NO_2)]-thiazolidide (2 x 10^{-5} M). The DP IV inhibitor was added at time O. Cells were labeled with ^{3}H-thymidine 6 hours before harvesting.

2.3. Potential Natural Ligands of DP IV / CD26

Considering the specificity of DP IV, the question arises, what could be the natural ligands and/or substrates of it in the lymphocyte function? Many cytokines, as for instance human IL-1ß, IL-2, IL-3, IL-5, IL-6, IL-13, TNF-ß, G-CSF, GM-CSF, ILGF-IA and ILGF-IB, have N-terminal sequences with proline in the penultimate position which makes these polypeptides to potential DP IV substrates[16]. In the case of other cytokines, e.g. IL-1α, interferons and recombinant cytokines with N-terminal methionine, proline is localized in the third or later position of the N-terminus.

In these instances, a joint action of aminopeptidase N / CD13 or other aminopeptidases and DP IV / CD26 could be theoretically possible for the removal of the N-terminal part and the generation of proline containing dipeptides. Torsten Hoffmann et al.[17] have investigated in our laboratory the capability of soluble DP IV and/or aminopeptidase N (AP-N) to modify the N-terminal part of some cytokines with susceptible sequences. Using the method of capillary electrophorsis, we could show that oligopeptides (chain length 4 to 24 amino acids) with sequences identical to the N-terminal part of human IL-2, IL-1ß, TNF-ß and murine IL-6 were hydrolyzed by purified DP IV / CD26 and AP-N / CD13. The role of DP IV-catalyzed hydrolysis of these peptides was negatively correlated with their chain length. Using glycosylated IL-2 oligopeptides, it could be shown that glycosylation has only a slight negative effect on the velocity of hydrolysis of these oligopeptides. However, intact native cytokines investigated so far, e.g. IL-1α, IL-1ß, IL-2 and G-CSF, have been found to be resistent to degradation by these peptidases under in vitro conditions used. Our data suggest that processing of cytokines by DP IV and other exopeptidases possibly needs one or more preceding endoproteolytic steps by ectoproteases, and

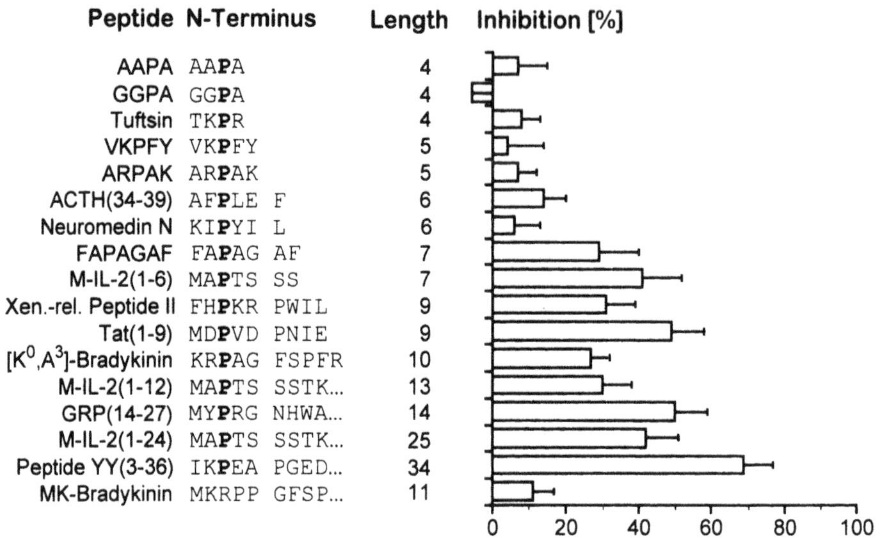

Figure 6. X-X-Pro peptides are capable of inhibiting DP IV enzymatic activity. DP IV activity was measured by IL-2 (1–12) degradation (release of X-Pro) using the capillary electrophoresis. The N-terminal amino acid sequences are written in single-letter code.

that DP IV and AP-N could play a role in the modification of the generated N-terminal peptides with the conserved proline (alanine) in the second position. Alternatively it could be possible that the susceptibility of cytokines to exo-peptidase-catalyzed processing could be changed by conformational changes due to the binding of these factors to their specific receptors and that membrane-bound DP IV or AP-N have another conformation than the soluble forms and another capability of handling receptor-bound cytokines. The essential reaction for cell activation via DP IV / CD26 in all systems discussed could be the generation of X-Pro(X-Ala) peptides on the surface of cells.

Another category of functional ligands of the catalytically active structure are natural inhibitors. Recently evidence has been provided[18,19] that X-X-Pro peptides with proline in the third position are capable of inhibiting the DP IV catalyzed degradation of IL-2 (1–12) (Figure 6). Importantly, the Tat protein of the human immunodeficiency virus type 1 (HIV-1) belongs to this group of molecules[18,19,20].

Interestingly enough, DNA synthesis of PWM stimulated T cells[19] and U937 cells expressing CD26 were strongly suppressed by HIV-1 Tat (1–86) whereas the N-terminally rhodamine-substituted R-HIV-1 Tat (1–72) had no effect at all on DNA synthesis of T cells (Figure 7). A number of cytokines and cytokine precursors contain many proline residues near the N-terminus. Erythropoietin, IL-6 and TGF-ß1 prepropeptide are X-X-Pro peptides and G-CSF, IL-1α-precursor, IL-1ß-precursor, LIF, lymphotoxin-precursor and IL-13 are putative substrates for DP IV that release X-X-Pro peptides after cleavage of the N-terminal X-Pro- or X-Ala-dipeptide. Our results raise the possibility that DP IV mediated effects on the proliferation of activated T lymphocytes and other cells expressing DP IV / CD26 could be controlled by such endogenous peptides. Furthermore, we obtained evidence that the immuno-suppressive effects of HIV-1 Tat protein were mediated, at least in part, by DP IV / CD26. Thus, CD26-Tat interactions may play a crucial role in the development and progression of HIV mediated AIDS disease.

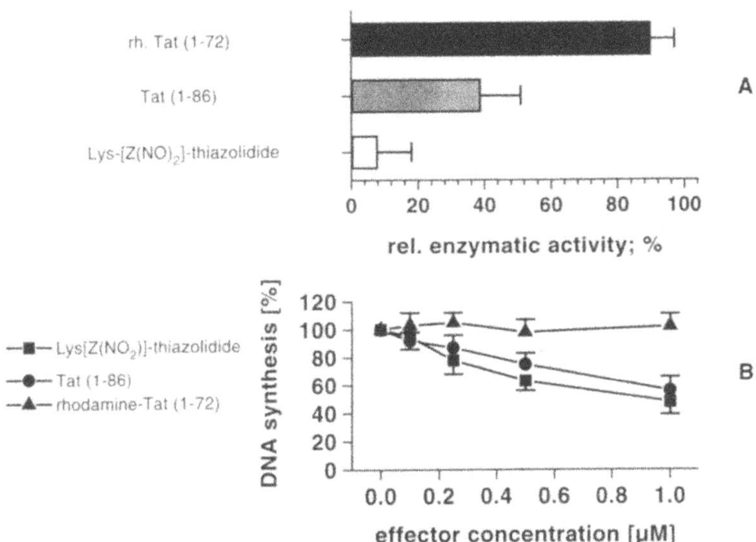

Figure 7. The N-terminal X-X-Pro motif of HIV-1 Tat mediates suppression of DNA via inhibition of DP IV enzymatic activity. (A) IL-2 (1–12) cleavage by DP IV in presence of 40µM rhodamine-Tat (1–72), Tat (1–86), and Lys-[Z(NO_2)]-thiazolidide was measured by free zone capillary electrophoresis. The error bars indicate the standard deviation of 2 different experiments each being carried out in triplicate. (B) Purified T cells were stimulated with PWM (2µg/ml) in presence of Tat (1–86), rhodamine-Tat (1–72), or Lys-[Z(NO_2)]-thiazolidide in the indicated concentrations. After 90 h, the cultures were pulsed with ([^{3}H]TdR) for an additional 6 h. [^{3}H]TdR incorporations in cpm are indicated as mean ± S.D. from 4 different experiments. The values are expressed as % [^{3}H]TdR incorporated related to control cultures without effector.

3. ROLE OF DP IV / CD26 IN CYTOKINE PRODUCTION

In addition to DP IV inhibitors acting on the DNA synthesis, they also exhibit strong suppressive effects on the production of most cytokines (Table 3). This could be shown for different cell systems under different stimulation conditions[12,13,16,21–23]. Interestingly enough, in U937 cells expressing strongly DP IV / CD26, IL-1ß and TNFα production is also markedly diminished by Lys[Z(NO_2)]-thiazolidide and -piperidide, whereas the generation of IL-1-RA and TGF-ß1 is significantly increased under the same conditions[12,24]. TGF-ß1 is a multifunctional cytokine, strongly suppressing the DNA synthesis and the production of IL-2, IL-6, IL-10 and IL-12 by PWM stimulated peripheral blood mononuclear cells[24,25]. Recently we also found that the membrane expression of CD26 on T cells is suppressed by TGF-ß1 at physiological concentrations[24]. On the other hand, DP IV inhibitors such as Lys[Z(NO_2)]thiazolidide, -piperidide, and -pyrrolidide strongly stimulate the production and release of latent TGF-ß1 in a concentration dependent manner (Figure 8). This has been demonstrated for human peripheral blood mononuclear cells, for purified T lymphocytes and NK cells as well. Moreover, very recently evidence has been provided that DP IV inhibitors induce TGF-ß1 generation also on the transcriptional level (Figure 8, bottom).

These findings suggest that TGF-ß1 mediates at least in part some of the immune suppressive effects of DP IV inhibitors and might be directly involved in the DP IV dependent regulation of cell growth and production of various other cytokines. In other

Table 3. Effects of DP IV inhibitors on cytokine production by different cells in vitro

	Production suppressed	Production stimulated
PBMC	IL-2, IL-6, IL-10, IL-12	TGF-ß1
T lymphocytes	IL-2, IL-6, IL-10, IL-12	TGF-ß1
NK cells	TNF-alpha, GM-CSF, IFN-gamma	TGF-ß1
U 937	TNF-alpha, IL-1ß	TGF-ß1, IL-1RA

words, DP IV seems to play a crucial role in regulating the biological activity of TGF-ß1 in that it inhibits the generation of the active cytokine.

4. DP IV / CD26 IN SIGNAL TRANSDUCTION OF LYMPHOCYTES

DP IV / CD26 has been studied by different groups with respect to its involvement in signal transduction processes of lymphocytes (for reviews cf.[7,26,27]). Considering the amino acid sequence predicted from the cloned cDNA, DP IV has a cytoplasmic tail consisting of only six amino acids which seems to be too short to be directly involved in signal transduction. Interestingly enough, the adenosine deaminase[28,29] and the tyrosine phosphatase CD45R[30] were shown to be functionally associated with DP IV / CD26 (cf. also the contributions of B. Fleischer and M. Hegen in this book). The detailed molecular mechanism of the CD26 / DP IV-mediated signal transduction is still unknown. There is some evidence that it is associated with the CD3 ζ chain which was found to be phosphorylated after CD26 stimulation[31]. We have studied the effects of different types of our own and of commercial monoclonal antibodies against DP IV / CD26 as well as DP IV inhibitors with respect to their effects on Ca^{++} influx and protein phosphorylation of lymphocytes[16,32]. Here we will briefly summarize recent results on protein phosphorylation. Using a commercial tyrosine kinase assay (Boehringer) with a gastrine peptide as protein-tyrosine kinase substrate, we found different effects with different antibodies[16]. A strong increase in tyrosine phosphorylation within 10 minutes after treatment of human T cells was measured with the anti-CD26 antibodies EF6/F11 and EF7/A10 from our laboratory but not with Ta1 and CB.1. Obviously, special epitopes of DP IV / CD26 are involved in protein-tyrosine phosphorylation by anti-CD26 antibodies. To obtain more detailed information about these processes, we studied a possible functional relationship between DP IV / CD26 and the protein-tyrosine kinase $p56^{lck}$. Among other members of the src gene family ($p59^{fyn}$, $p62^{yes}$), this kinase is one of the major protein tyrosine kinases expressed in T lymphocytes. The $p56^{lck}$ itself is regulated by reversible phosphorylation and is associated with the inner face of the plasma membrane due to a myristylated moiety of the protein.

Therefore, it is in an ideal position to pass on and amplify signals generated by surface molecules. Triggering of the TCR-CD3 complex, CD2 or CD28 surface molecules, respectively, results in a hyperphosphorylation of $p56^{lck}$ and a concomitant shift in electrophoretic

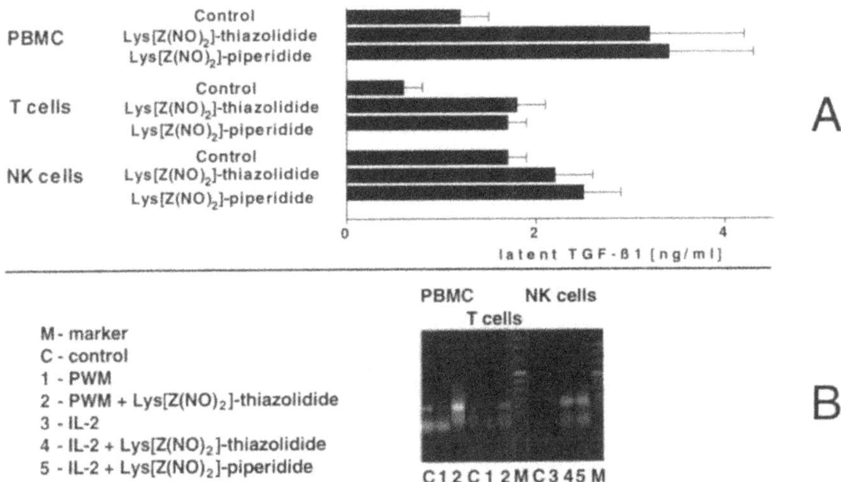

Figure 8. Induction of TGF-ß1 by DP IV inhibitors in different cell systems. (A) PBMC, T cells and NK cells (10^6 cells/ml) were incubated with PWM (2µg/ml) or IL-2 (1000 U/ml, NK cells) and the DP IV inhibitors Lys-[Z(NO$_2$)]-thiazolidide and Lys-[Z(NO$_2$)]-piperidide (10^5 M). After 4 hours, supernatants were harvested, stored at -70 °C, and the TGF-ß1 concentrations were measured with an enzyme immunoassay. Samples were tested after transient acidification (latent TGF-ß1). Results are expressed as mean ± SD of three independent experiments. (B) Enzymatic amplification of TGF-ß1 cDNA derived from untreated PBMC or T cells (C), cells stimulated with PWM (1), cells stimulated with PWM in the presence of Lys-[Z(NO^2)]-thiazolidide (2), IL-2 stimulated NK cells (3), and NK cells stimulated with IL-2 and Lys-[Z(NO$_2$)]-thiazolidide (4) or Lys-[Z(NO$_2$)]-piperidide (5).

mobility of p56^{lck} to p60^{lck}. We have investigated the action of DP IV-specific inhibitors (Lys[Z(NO$_2$)]-thiazolidide and -piperidide) on phorbol 12-myristate 13-acetate (PMA)-induced hyperphosphorylation of p56^{lck} in human T cells[32]. It could be found that this hyperphosphorylaton was strongly suppressed by both inhibitors in a dose-dependent manner (Figure 9). Removal of these competitive inhibitors fully restored the hyperphosphorylation. Therefore, these effects could be considered to be reversible. The data provide further evi-

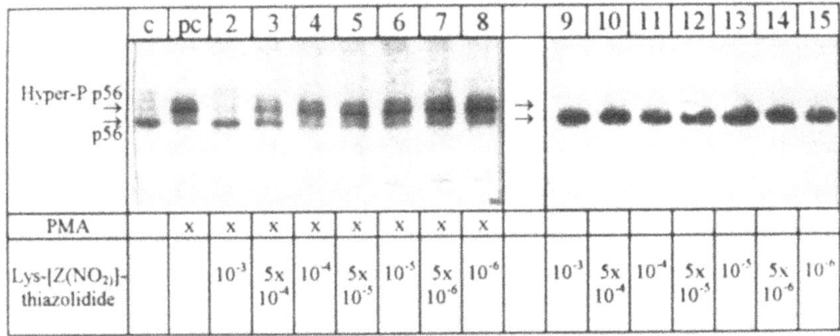

Figure 9. PMA-induced hyperphosphorylation of the tyrosine kinase p56^{lck} is suppressed by DP IV inhibitors. Purified human T lymphocytes were incubated for 30 minutes with different concentrations of the DP IV inhibitor Lys-[Z(NO$_2$)]-thiazolidide. After this time, the hyperphosphorylation of p56^{lck} was induced by PMA (8 x 10^{-10} M). After further 30 minutes of incubation at 37 °C cell lysates were analyzed by SDS-polyacrylamide electrophoresis. p56^{lck} was detected by means of rabbit anti-p56^{lck} immunoglobuline (reprinted from[32], with kind permission from Elsevier Science Ireland Ltd).

dence that DP IV / CD26 is directly involved in early processes of T cell activation and signalling. Furthermore, these findings strongly support the hypothesis that the growth regulating and signalling function of DP IV / CD26 requires its enzymatic activity or at least its active molecular site. Very recently, we have been studying the direct effect of DP IV inhibitors on PMA induced tyrosine phosphorylation of different proteins in human T cells. We found that the PMA-induced tyrosine phosphorylation of most of the potentially phosphorylated proteins is significantly reduced in presence of DP IV inhibitors (Figure 10). Moreover, some new tyrosine phosphorylated proteins appeared in presence of peptidase inhibitors. These data again demonstrate that DP IV plays a central role in regulating tyrosine phosphorylation in the process of T cell activation. Moreover, these data suggest an involvement of enzymatic DP IV activity or at least the active site of this peptidase in signal transducing processes. The results presented here as well as those of other groups have provided a number of hints that it is improbable that DP IV / CD26 influences separate signal transduction events in a direct way. The available data obtained with anti-CD26 and anti-CD3 antibodies as well with DP IV-Inhibitors suggest that DP IV / CD26 does have a more indirect function, obviously in tuning up the main streams of signal transduction of lymphocyte after activation.

SUMMARY/CONCLUSION

- DP IV / CD26 is involved in regulation of DNA synthesis and proliferation as well as production of cytokines of hematopoietic cells under various conditions.
- Inhibition of DNA synthesis in T lymphocytes, B lymphocytes, NK cells and myelomonocytic cells as well as of the production of IL-2, IL-6 TNFα, IL-1, IL-10, IL-12, IL-13, IFN-γ, GM-CSF are not due to apoptosis of these cells.

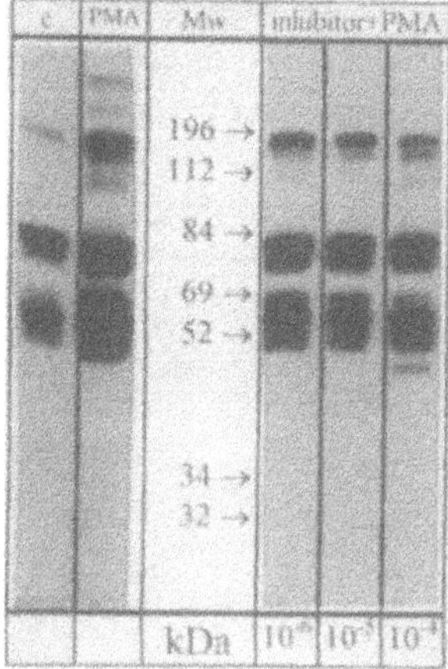

Figure 10. PMA-induced protein tyrosine phosphorylation is strongly changed by DP IV inhibitors. Aliquots of 2.5×10^6 resting human T cells were incubated with the DP IV inhibitor Lys-[Z(NO_2)]-thiazolidide) at concentrations as listed above for 30 minutes. After this time, PMA (50ng/ml) was added to each sample followed by a further incubation step of 15 minutes. Cell lysates were subsequently separated in a 10 % SDS-polyacrylamide gel. After transfering the proteins onto nitrocellulose membranes tyrosine-phosphate containing proteins were detected by means of anti-P-Tyr (4G10-biotin) and streptavidin-POD/SuperSignal-chemiluminescent substrate (Pierce, U.S.A.). c = untreated control, PMA = control without DP IV inhibitor.

- DP IV / CD26 inhibitors induce TGF-ß1 mRNA synthesis and latent protein release demonstrating a crucial role of TGF-ß1 in mediating CD26 function.
- X-X-Pro peptides as HIV-Tat protein strongly inhibit DP IV enzymatic activity and suppress DNA synthesis. This group of peptides may represent a class of natural DP IV / CD26 ligands and effectors, respectively.
- Hyperphosphorylation of $p56^{lck}$ as well as protein tyrosine phosphorylation of a number of proteins in T lymphocytes can be modulated by DP IV inhibitors. These data suggest that enzymatic activity or, at least in part, the active site of DP IV are both essential for its regulatory function in lymphocytes.
- Further work is required to determine the natural ligands, i.e. substrates and effectors, which are play the central role in DP IV / CD26 action in T cell growth and to understand the molecular mechanism of the early steps of this fundamental process.

ACKNOWLEDGMENTS

The authors wish to express his appreciation to the colleagues who were involved in this work, Dr. Torsten Hoffmann, Dr. Ekkehard Schön, Prof. Klaus Neubert and Dr. Jürgen Faust (Halle), Prof. Hans-Dieter Flad, Prof. Artur-J. Ulmer and Dr. Taila Matern (Borstel) and Dr. Rainer Frank and Dr. Margot Kraft (Heidelberg). This work was supported by the Deutsche Forschungsgemeinschaft, SFB 387 and the Fonds der Chemischen Industrie im Verband der Chemischen Industrie e.V.

6. REFERENCES

1. Hopsu-Havu VK, Glenner GG (1966) A new peptide naphtylamidase hydrolyzing glycylprolyl-naphtylamide. Histochemistry 7:197–201
2. Kenny AJ, Booth AG, George SG (1976) Dipeptidyl peptidase IV, a kidney brush border serine peptidase. Biochem J 157:169–182
3. Stein H, Schwarting R, Niedobitek G (1989) Cluster report: CD26. In: Leukocyte Typing IV, Oxford Press, Oxford, pp 412–418
4. Mattern T, Flad H-D, Feller AC, Heymann E, Ulmer AJ (1989) Anti-DPIV and anti-Ta1 but not IOT15 should be clustered in CDw26. In: Leukocyte Typing IV, edited by Knapp W, Dorken B, Gilks WR. Oxford Press, Oxford, pp 416
5. Ulmer AJ, Mattern T, Feller AC, Heymann E, Flad HD (1990) CD26 antigen is a surface dipeptidyl peptidase IV (DPPIV) as characterized by monoclonal antibodies clone TII-19-4-7 and 4EL1C7. Scand J Immunol 31:429–435
6. Hegen M, Niedobitek G, Klein CE, Stein H, Fleischer B (1990) The T cell triggering molecule Tp103 is associated with dipeptidyl aminopeptidase IV activity. J Immunol 144:2908–2914
7. Fleischer B (1995) Dipeptidyl peptidase IV (CD26) in metabolism and the immune response, R. G. Landes Company Austin, Texas, USA, pp 202
8. Schön E, Mansfeld HW, Demuth HU, Barth A, Ansorge S (1985) The dipeptidyl peptidase IV, a membrane enzyme involved in the proliferation of T lymphocytes. Biomed Biochim Acta 44 (2):K 9-K 15
9. Schön E, Born I, Demuth HU, Faust J, Neubert K, Steinmetzer T, Barth A, Ansorge S (1991) Dipeptidyl Peptidase IV in the immune system. Effects of specific inhibitors on activity of dipeptidyl peptidase IV and proliferation of human lymphocytes. Biol Chem Hoppe Seyler 372:305–311
10. Bühling F, Kunz D, Reinhold D, Ulmer AJ, Ernst M, Flad HD, Ansorge S (1994) Expression and functional role of dipeptidyl peptidase IV (CD26) on human natural killer cells. Nat Immunity 13:270–279
11. Bühling F, Junker U, Reinhold D, Neubert K, Jäger L, Ansorge S (1995) Functional role of CD26 on human B lymphocytes. Immunology Letters 45:47–51

12. Reinhold D, Bank U, Bühling F, Kähne T, Kunz D, Faust J, Neubert K, Ansorge S (1994) Inhibitors of dipeptidyl peptidase IV (DPIV, CD26) specifically suppress proliferation and modulate cytokine production of strongly CD26 expressing U937 cells. Immunobiology 192:121–136
13. Kähne T, Reinhold D, Lendeckel U, Bühling F, Täger M, Faust J (1997) The effect of anti-CD26 antibodies on DNA synthesis and cytokine production (IL-2, IL-10 and IFN-γ) depends on enzymatic activity of DPIV / CD26. In: Leukocyte Typing VI, edited by Kishimoto T. Garland Publishing, Cambridge, in press
14. Kähne T, Kröning H, Thiel U, Ulmer AJ, Flad H-D, Ansorge S (1996) Alterations in structure and cellular localization of molecular forms of DPIV/CD26 during T cell activation. Cell Immunol 170:63–70
15. Poppema S, Visser L (1996) Some CD26 reagents directly activate T cells. Tissue Antigens 48:451(Abstract)
16. Ansorge S, Bühling F, Hoffmann T, Kähne T, Neubert K, Reinhold D (1995) DPIV/CD26 on human lymphocytes: Functional roles in cell growth and cytokine regulation. In: Dipeptidyl peptidase IV (CD26) in metabolism and the immune response, edited by Fleischer B. Springer-Verlag, Heidelberg, Germany, pp 163–184
17. Hoffmann T, Faust J, Neubert K, Ansorge S (1993) Dipeptidyl peptidase IV (CD26) and Aminopeptidase N (CD13) catalyzed hydrolysis of cytokines and peptides with N-terminal cytokine sequences. FEBS Letters 336,Nr.1:61–64
18. Hoffmann T, Reinhold D, Kähne T, Faust J, Neubert K, Frank R, Ansorge S (1995) Inhibition of dipeptidyl peptidase IV (DP IV) by anti-DP IV antibodies and non-substrate X-X-Pro-oligopeptides ascertained by capillary electrophoresis. J Chromatography 716:355–362
19. Wrenger S, Reinhold D, Hoffmann T, Kraft M, Frank R, Faust J, Neubert K, Ansorge S (1996) The N-terminal X-X-Pro sequence of the HIV-1 Tat protein is important for the inhibition of dipeptidyl peptidase IV (DP IV/CD26) and the suppression of mitogen-induced proliferation of human T cells. FEBS Letters 383:145–149
20. Reinhold D, Wrenger S, Bank U, Bühling F, Hoffmann T, Neubert K, Kraft M, Frank R, Ansorge S (1996) CD26 Mediates the Action of HIV-1 Tat Protein on DNA Synthesis and Cytokine production in U937 Cells. Immunobiology 195:119–128
21. Ansorge S, Schön E (1987) Dipeptidyl peptidase IV in human T lymphocytes: an approach to the function of this peptidase in the immune system. In: Peptides and Proteases: Recent Advances, edited by Schowen RL, Barth A. Pergamon Press, Oxford, pp 3–10
22. Schön E, Demuth HU, Eichmann E, Horst HJ, Körner IJ, Kopp J, Mattern T, Neubert K, Noll F, Ulmer AJ, Barth A, Ansorge S (1989) Dipeptidyl peptidase IV in Human T-Lymphocytes Impaired Induction of Interleukin 2 and Gamma Interferon due to Specific Inhibition of Dipeptidyl peptidase IV. Scand J Immunol 29:127–132
23. Reinhold D, Bank U, Bühling F, Neubert K, Mattern T, Ulmer AJ, Flad H-D, Ansorge S (1993) Dipeptidyl peptidase IV (CD26) on human lymphocytes. Synthetic inhibitors of and antibodies against dipeptidyl peptidase IV suppress the proliferation of pokeweed mitogen-stimulated peripheral blood mononuclear cells, and IL-2 and IL-6 production. Immunobiology 188:403–414
24. Reinhold D, Bank U, Bühling F, Lendeckel U, Ulmer AJ, Flad H-D, Ansorge S (1994) Transforming growth factor beta1 (TGF-beta1) inhibits DNA synthesis of PWM-stimulated PBMC via suppression of IL-2 and IL-6 production. Cytokine 6:382–388
25. Reinhold D, Bank U, Bühling F, Lendeckel U, Ansorge S (1995) Transforming growth factor beta1 inhibits interleukin-10 mRNA expression and production in pokeweed mitogen-stimulated peripheral blood mononuclear cells and T cells. Journal of Interferon and Cytokine Research 15:685–690
26. Fleischer B (1994) CD26: a surface protease involved in T cell activation. Immunology Today 15:180–184
27. Morimoto C, Schlossman SF (1994) CD26. A key costimulatory molecule on CD4 memory T cells. Immunologist 2:4–7
28. Morrison ME, Vijayasaradhi S, Engelstein D, Albino AP, Houghton AN (1993) A marker for neoplastic progression of human melanocytes is a cell surface ectopeptidase. J Exp Med 177:1135–1143
29. Kameoka J, Tanaka T, Nojima Y, Schlossman SF, Morimoto C (1993) Direct association of adenosine deaminase with a T cell activation antigen, CD26. Science 261:466–469
30. Torimoto Y, Dang NH, Vivier E, Tanaka T, Schlossman SF, Morimoto C (1991) Coassociation of CD26 (dipeptidyl peptidase IV) with CD45 on the surface of human T lymphocytes. J Immunol 147:2514–2517
31. Mittrücker H-W, Steeg C, Malissen B, Fleischer B (1995) The cytoplasmic tail of the T cell receptor zeta chain is required for signalling via CD26. Eur J Immunol 25:295–297
32. Kähne T, Neubert K, Ansorge S (1995) Enzymatic activity of DPIV/CD26 is involved in PMA-induced hyperphosphorylation of p56lck. Immunol Lett 46:189–193

18

CD26 IS INVOLVED IN REGULATION OF CYTOKINE PRODUCTION IN NATURAL KILLER CELLS

F. Bühling,[1] D. Reinhold,[1] U. Lendeckel,[1] J. Faust,[2] K. Neubert,[2] and S. Ansorge[1]

[1]Institute of Experimental Internal Medicine
Otto-von-Guericke-University Magdeburg, Germany
[2]Dept. of Biochemistry and Biotechnology
Martin-Luther-University Halle-Wittenberg, Germany

1. INTRODUCTION

Dipeptidyl peptidase IV (DP IV, CD26) is a postproline cleaving enzyme which catalyses the degradation of polypeptides from the N-terminal part. DP IV is expressed on the surface of cells from several tissues. High levels of DP IV activity were found in the brush border membrane of the intestinal epithelium, in the proximal tubulus of the kidney, and in the serum [1,2]. In the immune system, DP IV was observed to be expressed on activated T cells, on activated NK cells, and on activated B lymphocytes [3,4]. DP IV was characterised as a key enzyme in the regulation of activation and proliferation of T lymphocytes. The mitogen- and antigen-induced DNA synthesis was found to be inhibited by specific DP IV inhibitors [5]. Earlier we described the expression of DP IV/CD26 on activated NK cells using enzymatic and immunologic techniques. Moreover, evidence was presented that DP IV is involved in the regulation of DNA synthesis and cell cycle progression, but not in that of the natural cytotoxic activity of NK cells against K 562 cells. The mechanisms of regulation of the DNA synthesis are unclear as yet. In T lymphocytes it could be shown that DP IV inhibitors specifically regulate the cytokine production of these cells and by that means affect the autocrine growth regulation [6]. It was shown that both resting and activated NK cells are capable of cytokine production [7,8]. These cytokines seem to play a role in regulation of NK cell activation as well as in their effector functions [9]. Other authors postulated a link between CD26 expression and apoptosis [10].

In the present study we could show that inhibition of enzymatic DP IV activity leads to a suppression of the production of the stimulatory cytokines interferon-γ (IFN-γ), tumour necrosis factor-α (TNF-α) and granulocyte/macrophage-colony stimulatory factor

Cellular Peptidases in Immune Functions and Diseases, edited by Ansorge and Langner
Plenum Press, New York, 1997

(GM-CSF). Remarkably, in contrast to these results, secretion of the immune-suppressive cytokine transforming growth factor-β1 was induced.

2. METHODS

2.1. Isolation and Culture of NK Cells

NK cells were isolated from non-adherent cells by depletion of T- and B-lymphocytes with help of anti-CD3- and anti-CD19-coupled magnetic beads (MACS system; Miltenyi Biotec), stimulated with rIL-2 (1000 U/ml) for 3 days [3] or cultured under limiting dilution conditions in presence of irradiated autologous PBL, 1 μg/ml PHA and 100 U/ml IL-2 [11].

2.2. Cytokine ELISA

TNF-α, IFN-γ and GM-CSF were measured with help of commercially available enzyme immune assays (R&D systems, Laboserve). The assay for measurement of TGF-β1 was established in our laboratory. The lowest measurable concentration was 50 pg/ml. Samples were tested after transient acidification (reduction of pH to 1.5 by addition of 5 N HCl for 30 min at 37 °C and neutralisation with 1.4 N NaOH in 0.7 M HEPES).

2.3. Amplification of Cytokine mRNA

Total RNA was isolated from stimulated cells (RNeasy isolation kit, Qiagen), reverse transcribed, and Polymerase Chain Reaction (PCR) amplified using primers specific for IFN-γ, TNF-α, TGF-β1, TGF-β2, IL-12, and IL-2 (Stratagene).

2.4. Detection of Apoptotic Cells

a. Nick end labelling of DNA strand beaks with BrDUTP and staining with anti-BrDU-FITC antibodies or
b. Labelling of surface phosphatidyl serine of intact cells with Annexin-V-FITC. Measurement with flow cytometer FACSCalibur (Becton Dickinson).

3. RESULTS AND DISCUSSION

3.1. Suppression of IFN-γ, TNF-α and GM-CSF Production

We have shown that DP IV inhibitors specifically suppress the IL-2-induced DNA synthesis of natural killer cells [3]. To determine the role of the enzymatic activity on the production of cytokines by IL-2 stimulated natural killer cells we incubated freshly isolated NK cells with 1000 U/ml rIL-2 and various concentrations of the competitive DP IV inhibitors Lys[Z(NO_2)]-thiazolidide and Lys[Z(NO_2)]-pyrrolidide. The concentrations of cytokines of IFN-γ, TNF-α, and GM-CSF released after 72 h were determined by enzyme immunoassays. As shown in Figure 1, the production of all three cytokines was suppressed in a dose dependent manner.

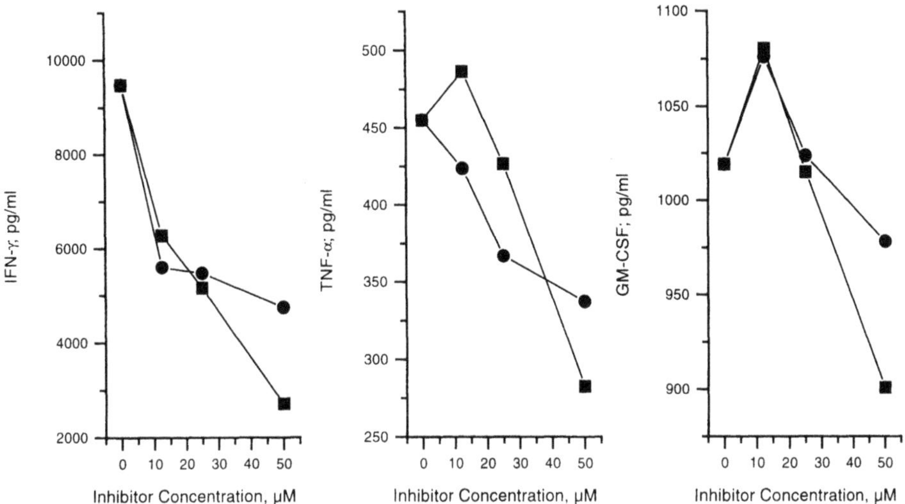

Figure 1. Dose dependent inhibition of the production of IFN-γ (A), TNF-α (B), and GM-CSF (C) by the competitive DP IV inhibitors Lys[Z(NO_2)]-thiazolidide (■) and Lys[Z(NO_2)]-pyrollidide (▼). Isolated NK-cells were incubated (3 d) with rIL-2 (1000 U/ml) and the indicated concentrations of the DP IV inhibitors. The cytokine concentrations in the supernatants were measured with enzyme immunoassays.

Furthermore, we incubated the $CD3^{-}CD56^{+}$ NK cell clone G4G10 with both DP IV inhibitors (10^{-5} M) for 2 days and measured the resulting cytokine concentrations in the supernatants. Again, we could show that the concentrations of IFN-γ, TNF-α and GM-CSF were decreased after suppression of enzymatic activity (Fig. 2).

PCR analyses of IL-2-stimulated NK cells showed a decrease in cytokine-specific mRNA of TNF-α, IFN-γ, and IL-12. We could not detect IL-2 or TGF-β2 mRNA (Fig. 3). Taken together, these data convincingly show that inhibition of enzymatic DP IV activity leads to a dose dependent decrease of cytokine production on the trancriptional as well as the translational level. LANGE et. al. have shown that the NK cells are capable of secreting both TNF-α and IFN-γ and that both cytokines are important regulators of proliferation and cytotoxic activity of these cells [7]. Other studies show that GM-CSF is also involved in the regulation of NK cell activity [12]. Our data show that DP IV has a crucial role in these processes and that this enzyme takes part in the regulation of the secretion of these cytokines.

3.2. Induction of TGF-β1 Production by DP IV Inhibitors

TGF-β1 has been demonstrated to modulate the immune response by affecting proliferation, the activation state, and the differentiation of immune cells. It has been also shown to inhibit the IL-2-dependent proliferation of NK cells [9,13]. To answer the question whether the growth inhibition by DP IV inhibitors is mediated by TGF-β1, we incubated isolated NK cells with IL-2 and different concentrations of DP IV inhibitors. After 3 days the supernatants were collected and the concentration of latent TGF-β1 was measured. As shown in Fig. 4, we found decreased levels of TGF-β1 after IL-2 stimulation without inhibitors and a dose dependent increase of latent TGF-β1 after incubation of the NK cells with DP IV inhibitors. Also the RT PCR analyses of the TGF-β1 specific mRNA

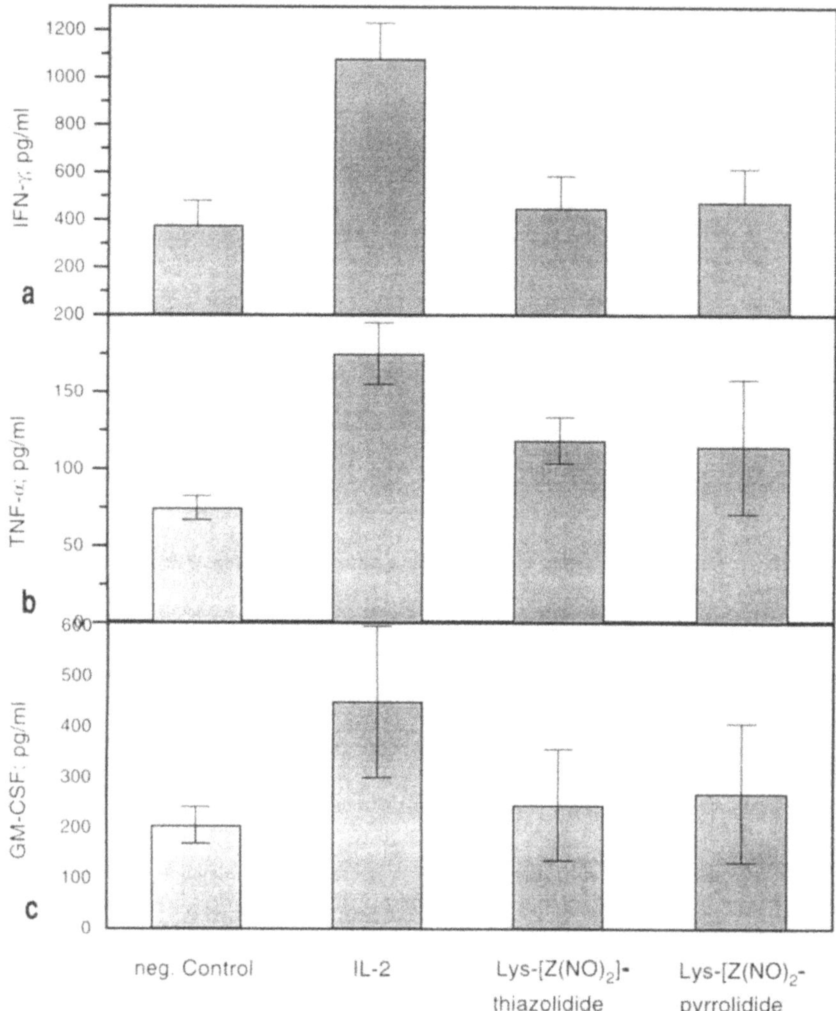

Figure 2. Inhibition of the production of IFN-γ (a), TNF-α (b), and GM-CSF (c) by the NK cell clone G4G10. The cells were incubated (2 d) in absence of IL-2, in presence of IL-2 (1000 U/ml), in presence of IL-2 and 5 x 10^{-5} M Lys[Z(NO_2)]-thiazolidide, and with IL-2 and Lys[Z(NO_2)]-pyrrolidide. The cytokine concentrations in the supernatants were measured with enzyme immunoassays.

showed a higher amount of TGF-β1 mRNA after incubation with DP IV inhibitors (5 x 10^{-5} M, Fig. 3). We could not find detectable amounts of TGF-β2 mRNA. These findings support that the suppressive effect of DP IV inhibitors on DNA synthesis may result, at least in part, from the induction of the inhibitory cytokine TGF-β1. They are in good correlation to the data of Reinhold et. al. showing a induction of TGF-β in U937 cells [14].

3.3. DP IV Inhibitors Do Not Induce Apoptosis

An other possibility of growth regulation of immune cells is the specific induction of apoptosis. It was shown that transfected cells expressing mutated CD26 without enzymatic

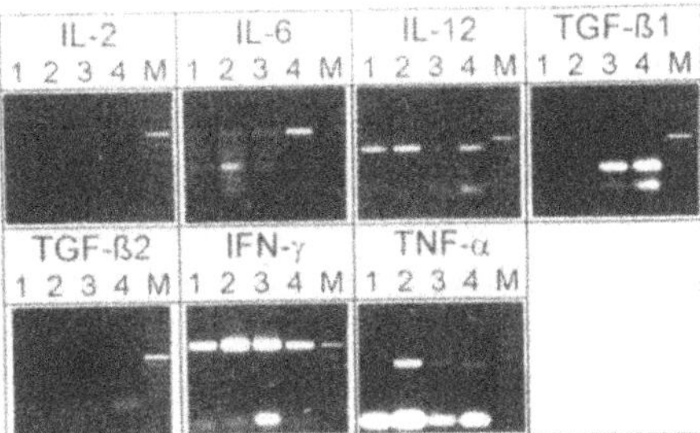

Figure 3. Cytokine mRNA expession after incubation of isolated NK-cells without IL-2 (1), with IL-2 (2), with IL-2 and 5 x 10^{-5} M Lys[Z(NO_2)]-thiazolidide (3), and with IL-2 and Lys[Z(NO_2)]-pyrrolidide (4). Equal amounts of cDNA were used for RT PCR analyses with cytokine specific primers. Equal amounts of the reaction mixtures and the molecular weight DNA standards (M) were loaded on an agarose gel, electrophoresed and then stained with ethidium bromide.

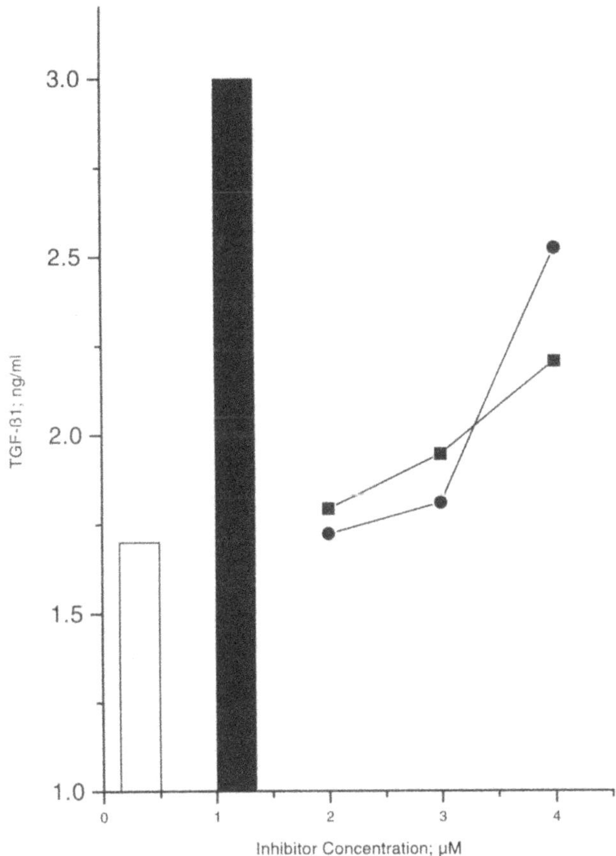

Figure 4. Dose dependent induction of the TGF-β1 production in isolated NK cells. The cells were incubated (3 d) in absence of IL-2 (open bars), in presence of IL-2 (1000 U/ml, solid bars), in presence of IL-2 and 5 x 10^{-5} M Lys[Z(NO_2)]-thiazolidide (■), and with IL-2 and Lys[Z(NO_2)]-pyrrolidide (●).

activity are more susceptible to the induction of apoptosis than cells expressing a molecule with enzymatic activity [10]. For further studies on the influence of the enzymatic DP IV activity on the induction of apoptosis, we incubated isolated NK cells with different concentrations of DP IV inhibitors. In control experiments the NK cells were incubated with Dexametason, which is known to be capable of inducing apoptosis. As shown in Fig. 5, DP IV inhibitors did not induce apoptosis in stimulated NK cells.

In summary, our findings show that CD26/DP IV plays an important role in the regulation of cytokine production but not in induction of apoptosis. At least some of the immunsuppressive effects of DP IV inhibitors could be mediated by inhibition of immunostimulatory cytokines (IFN-γ, GM-CSF, TNF-α) and particularly by an induction of the immunosuppressive cytokine TGF-β1.

5. ACKNOWLEDGMENTS

This work was supported by the Deutsche Forschungsgemeinschaft SFB387/A2/A4. Our thanks go to Marianne Blichmann for expert technical assistance.

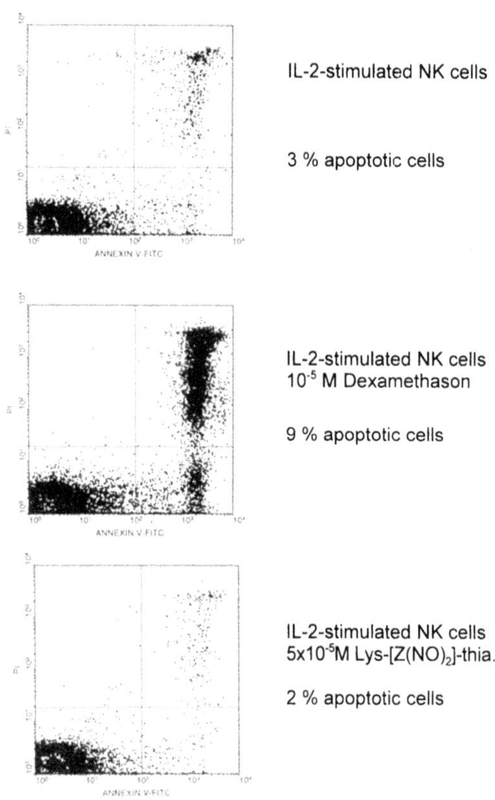

Figure 5. DP IV inhibitors do not induce apoptosis. Isolated NK cells were incubated with IL-2, with IL-2 and 10^{-5} M Lys[Z(NO_2)]-thiazolidide, or with IL-2 and 10^{-5} M Dexametason for 3 days. Apoptotic cells were stained with help of Annexin V-FITC.

4. REFERENCES

1. Krepela, E., Kraml, J., Vicar, J., Kadlecová, L. and Kasafírek, E. (1983) Demonstration of two molecular forms of dipeptidyl peptidase IV in normal human serum. Physiol. Bohemoslov. 32, 486–496.
2. Gossrau, R. (1979) Peptidase II. Zur Lokalisation der Dipeptidylpeptidase IV (DPP IV). Histochemische und biologische Untersuchung. Histochemistry 60, 231–248.
3. Bühling, F., Kunz, D., Reinhold, D., Ulmer, A.J., Flad, H.-D. and Ansorge, S. (1994) Expression and functional role of dipeptidyl peptidase IV (CD26) on human natural killer cells. Nat. Immun. Cell Growth Regul. 13, 270–279.
4. Bühling, F., Junker, U., Reinhold, D., Neubert, K., Jäger, L. and Ansorge, S. (1995) Functional role of CD26 on human B lymphocytes. Immunol. Lett. 45, 47–51.
5. Schön, E., Jahn, S., Kiessig, S.T., Demuth, H.-U., Neubert, K., Barth, A., vonBaehr, R. and Ansorge, S. (1987) The role of dipeptidyl peptidase IV in human T lymphocyte activation. Inhibitors and antibodies against dipeptidyl peptidase IV suppress lymphocyte proliferation and immunoglobulin synthesis in vitro. Eur. J. Immunol. 17, 1821–1826.
6. Reinhold, D., Bank, U., Bühling, F., Neubert, K., Mattern, T., Ulmer, A.J., Flad, H.-D. and Ansorge, S. (1993) Dipeptidyl peptidase IV (CD26) on human lymphocytes. Synthetic inhibitors of and antibodies against dipeptidyl peptidase IV suppress the proliferation of pokeweed mitogen-stimulated peripheral blood mononuclear cells, and IL-2 and IL-6 production. Immunobiology 188, 403–414.
7. Lange, A., Moniewska, A., Fetting, R., Ernst, M. and Flad, H.-D. (1995) rIL-2-activated killer cell generation is influenced by autologous TNF-α production. Nat. Immun. 14, 2–10.
8. Richards, S.J. and Scott, C.S. (1992) Human NK cells in health and disease: Clinical, functional, phenotypic and DNA genotypic characteristics. Leuk. Lymphoma 7, 377–399.
9. Bellone, G., Aste-Amezaga, M., Trinchieri, G. and Rodeck, U. (1995) Regulation of NK cell functions by TGF-β1. J. Immunol. 155, 1066–1073.
10. Morimoto, C., Lord, C.I., Zhang, C., Duke-Cohan, J.S., Letvin, N.L. and Schlossman, S.F. (1994) Role of CD26/dipeptidyl peptidase IV in human immunodeficiency virus type 1 infection and apoptosis. Proc. Natl. Acad. Sci. USA 91, 9960–9964.
11. Ciccone, E., Pende, D., Viale, O., Di Donato, C., Tripodi, G., Orengo, A.M., Guardiola, J., Moretta, A. and Moretta, L. (1992) Evidence of a natural killer (NK) cell repertoire for (allo) antigen recognition: Definition of five distinct NK-determined allospecificities in humans. J. Exp. Med. 175, 709–718.
12. Takahashi, K., Sone, S., Saito, S., Kamamura, Y., Uyama, T., Ogura, T., Monden, Y., van den Bosch, G., Preijers, F., Vreugdenhil, A., Hendriks, J., Maas, F. and De Witte, T. (1995) Granulocyte-macrophage colony-stimulating factor (GM-CSF) counteracts the inhibiting effect of monocytes on natural killer (NK) cells. . Jpn. J. Cancer Res. 101, 515–520.
13. Ortaldo, J.R., Mason, A.T., O'Shea, J.J., Smyth, M.J., Falk, L.A., Kennedy, I.C., Longo, D.L. and Ruscetti, F.W. (1991) Mechanistic studies of transforming growth factor-beta inhibition of IL-2-dependent activation of CD3- large granular lymphocyte functions. Regulation of IL-2R beta (p75) signal transduction. J. Immunol. 146, 3791–3798.
14. Reinhold, D., Bank, U., Bühling, F., Ulmer, A.J., Flad, H.-D. and Ansorge, S. (1992) Inhibitors of the dipeptidyl peptidase IV (DP IV, CD26) induce secretion of transforming growth factor-β (TGF-β) on immune cells in vitro. Immunobiology 186, 109(Abstract).

19

THE EFFECT OF ANTI-CD26 ANTIBODIES ON DNA SYNTHESIS AND CYTOKINE PRODUCTION (IL-2, IL-10 AND IFN-γ) DEPENDS ON ENZYMATIC ACTIVITY OF DP IV/CD26

D. Reinhold,[1] T. Kähne,[1] M. Täger,[1] U. Lendeckel,[1] F. Bühling,[1] U. Bank,[1] S. Wrenger,[1] J. Faust,[2] K. Neubert,[2] and S. Ansorge[1]

[1]Institute of Experimental Internal Medicine
Center of Internal Medicine
Otto-von-Guericke University Magdeburg
Leipziger Str. 44, D-39120 Magdeburg, Germany
[2]Department of Biochemistry and Biotechnology
Martin-Luther-University Halle-Wittenberg
Kurt-Mothes-Str. 3, 06120 Halle(Saale), Germany

1. INTRODUCTION

Dipeptidyl peptidase IV (DP IV, EC 3.4.14.3) is a transmembrane type II glycoprotein present on most mammalian cells [1,2].

In the immune system, this enzyme is found on T lymphocytes, but also on B lymphocytes and NK cells [1,3,4]. DP IV is a serine peptidase that catalyzes the release of N-terminal dipeptides from peptides and proteins with preferentially proline, hydroxyproline and alanine at the penultimate position [5]. In the plasma membrane, DP IV occurs as a dimer with a total molecular mass of 220–240 kD. At the 4th Workshop on Leukocytes Differentiation Antigens monoclonal antibodies recognizing DP IV were subsumed under the term CD26 [6-8].

DP IV/CD26 is also known to be an adenosine desaminase (ADA) binding protein [9] and is involved in CD3/T-cell receptor (TcR)-mediated signal transduction [10-12].

DP IV has been shown to be an activation marker for various types of immune cells. Surface expression of CD26 is upregulated after mitogenic, anti-CD3 or IL-2 stimulation of T cells, St. aureus protein stimulation of B cells and IL-2 stimulation of NK cells [3,4,13-15]. Data from several groups have been shown that DP IV plays an intergral role in the regulation of differentiation and growth of lymphocytes [13,14,16-23]. Using specific inhibitors of DP IV, it has

Cellular Peptidases in Immune Functions and Diseases, edited by Ansorge and Langner
Plenum Press, New York, 1997

been demonstrated that DP IV is involved not only in the regulation of DNA synthesis, but also in production of various cytokines by human CD26^{+}-immune cells [20,24-26].

Here we describe the functional effects on T cell proliferation and cytokine production of five new monoclonal CD26-antibodies. Biochemical characterisation of these antibodies indicates that they have a unique epitope specificity. Support for the view that CD26 / DP IV is directly involved in the process of T cell activation via its enzymatic activity comes from the observation that DP IV specific inhibitors strongly interfere with the activating properties of these antibodies.

2. RESULTS AND DISCUSSION

Numerous immunological investigations have been aimed at the clarification of the functional effects of monoclonal DP IV antibodies, which have been used as a substitute for the still unknown physiological effectors (ligands, substrates, inhibitors) of this enzyme. The data obtained are in part contradictory, and both inhibitory and stimulatory effects have been described [16,27,28]. In our laboratory, four new antibodies EF6/B10, EF5/A3, EF6/F11 and PEG2/C3 have been raised against DP IV of U937 cells. The hybridomas

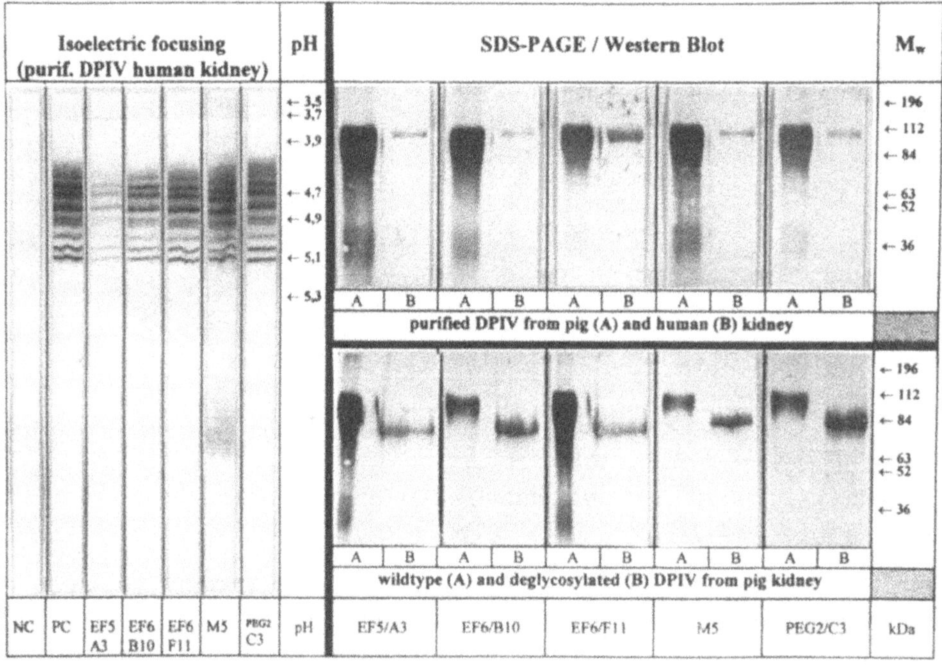

Figure 1. Comparison of immunoreactivity of native, denatured and deglycosylated pig and human DP IV. Isoelectric Focusing: DP IV was separated by thin-layer isoelectric focusing, followed by transfer of the proteins onto nitrocellulose. Blots were stained using anti-DP IV mAbs (except lane NC: negative control) and detected with goat anti-mouse alkaline phosphatase conjugate as the second antibody and NBT and BCIP as chromogenic phosphatase subtrates (PC: positive control; commercially available anti-CD26 mAb 4EL1C7, Ta1; Coulter Electronics). SDS-Page/Western Blot: Purified DP IV as well as enzymatically deglycosylated DP IV were separated by SDS gel electrophoresis according to the method of Laemmli and blotted onto nitrocellulose. Blots were stained using with anti-DP IV mAbs and detected with goat anti-mouse alkaline phosphatase conjugate as the second antibody and NBT and BCIP as chromogenic phosphatase subtrates.

have been screened on highly purified DP IV from pig kidney. Therefore, CD26 antibodies which were obtained recognize epitopes present on both human and pig DP IV (Fig. 1). The antibody M5 was raised against purified DP IV from pig kidney. As shown in Figure 1, this antibody recognize human DP IV as well.

Investigating most commercially available CD26 antibodies we have found that only the antibodies of our lab are capable of recognising pig kidney DP IV (data not shown). This property makes them unique among the CD26 antibodies available.

Binding studies using deglycosylated DP IV from human kidney characterised an epitope specificity of all five antibodies to the protein backbone of CD26.

Isoelectric focusing studies have clearly demonstrated that these two antibodies are immunoreactive to the whole pattern of molecular forms of purified pig kidney DP IV. In contrast to the strong immunoreactivity to all molecular forms of purified DP IV, both antibodies recognise a more restricted pattern of human lymphocytic CD26 forms separated by isoelectric focusing [29]. Whether these findings are of functional relevance is unclear.

Immunoprecipitation analysis performed as described in [30] has indicated that all five antibodies are capable of immunoprecipitating a 110 kDa protein, corresponding to the molecular weight of the CD26 monomer from the surface of human PHA-stimulated PBMC (Figure 2) [28].

The unique epitope specificity of these antibodies makes them a useful tool for studying functional effects resulting from binding cell surface CD26. We have found that all five antibodies strongly stimulated the DNA synthesis of human purified T cells (Fig. 3). The full activation of purified T cells by anti-CD26 required the addition of submitogenic amounts of phorbol myristate acetate (PMA, 2 ng/ml). However, similar results

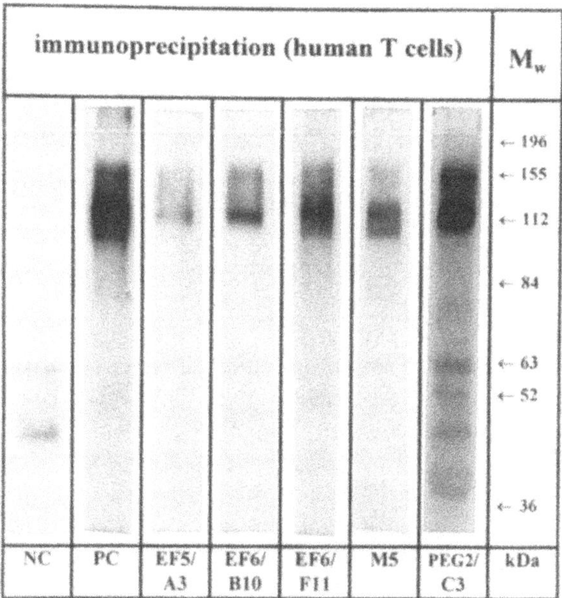

Figure 2. Immunoprecipitation of CD26 from lysates of PHA activated mononuclear cells. Human peripheral blood mononuclear cells were stimulated with PHA for 4 days. The surface proteins of these cells were biotinylated using Biotin-LC-Hydrazide. The cells were lysed and DP IV immunoprecipitated as described previously [29]. After separating the precipitates by SDS-polyacrylamide electrophoresis, they were transferred to nitrocellulose membranes and biotinylated proteins were detected by incubation with streptavidin-horseraddish peroxidase conjugates and a peroxidase chemiluminescent substrate. Membranes were exposed to X-ray film for 2 to 60 minutes.

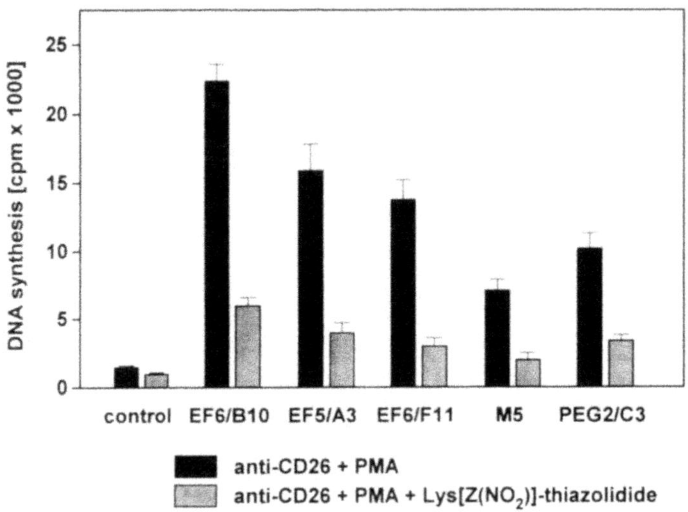

Figure 3. Influence of anti-CD26 antibodies and DP IV inhibitors on DNA synthesis of purified T cells. Purified T cells (separated from the venous blood of healthy donors; 10^5 cells/100µl), were incubated with the anti-CD26 antibodies EF6/B10, EF5/A3, EF6/F11, M5 and PEG2/C3 (5 µg/ml) in combination with submitogenic concentrations of phorbol myristate acetate (PMA, 2 ng/ml) and the synthetic inhibitor of DP IV, Lys[Z(NO_2)]-thiazolidide (10^{-5} M). After 90 hours the cultures were pulsed for additional 6 hours with ^{3}H-methyl-thymidine ([^{3}H]TdR, 0.2 µCi per well). Cells were harvested onto glass fibre filters and the incorporated radioactivity was measured by scintillation spectrometry.

have been obtained when small numbers of monocytes were used instead of PMA in the stimulation assay.

In further experiments we have shown that this anti-CD26-mediated activation leads to a strong increase in the production of IL-2 (Fig. 4). The same effect have been found concerning IL-10 and IFN-94Symbol"cproduction (data not shown). Interestingly, the competitive DP IV inhibitor Lys[Z(NO_2)]-thiazolidide significantly suppressed DNA synthesis and cytokine production on anti-CD26/PMA-stimulated purified T cells (Fig. 3 and Fig. 4). These findings strongly strengthen the view that CD26 is directly involved in early processes of T cell activation via its enzymatic activity or at least via its catalytic domain.

Furthermore, RT-PCR analyses have been carried out to clarify whether both effects described above are due to changes of cytokine-mRNA levels. Enzymatic amplification has been performed on cDNA generated from RNA of unstimulated T cells and from T cells stimulated with anti-CD26/PMA in presence and absence of 10^{-5} M Lys[Z(NO_2)]-thiazolidide. In unstimulated cells as well as in the presence of submitogenic amounts of PMA, the mRNA of the investigated cytokines was present at low levels only. However, strong induction of mRNA has been observed after stimulation with the anti-CD26 antibodies. In the presence of the DP IV inhibitor, levels of IL-2, IL-10 and IFN-94Symbol"c-mRNA were found to be decreased (data not shown).

In summary, the data presented here demonstrate that T cells can be stimulated directly by CD26/DP IV antibodies. The unique epitope specificity of the antibodies EF6/B10, EF5/A3, EF6/F11, M5 and PEG2/C3 may be the reason for their capability of stimulating T cells without crosslinking and without submitogenic amounts of anti-CD3. Furthermore, these results underline earlier functional data showing that the enzymatic ac-

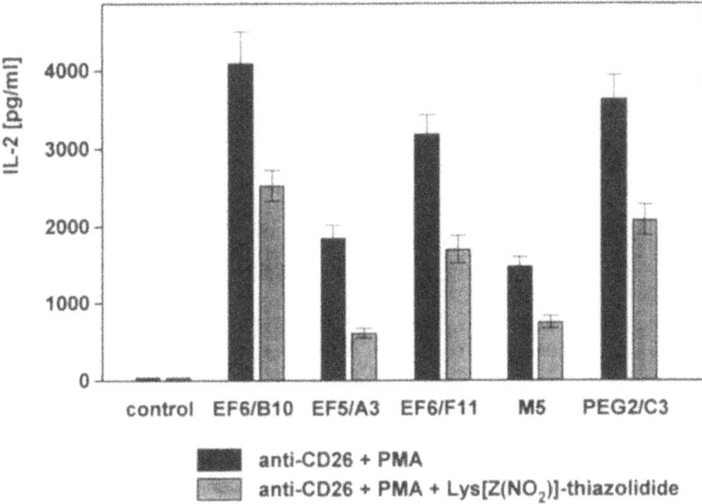

Figure 4. Influence of anti-CD26 antibodies and DP IV inhibitors on IL-2 production of purified T cells. Anti-CD26/PMA-stimulated T cells (10^6 cells/ml) were incubated with the reversible DP IV inhibitor Lys[Z(NO_2)]-thiazolidide (10^{-5} M). After 48 hours supernatants were harvested and the IL-2 concentrations of these supernatants were determined by ELISA (R & D Systems, Minneapolis, MN).

tivity of DP IV or at least the catalytic domain is important for mediating the CD26 effect on DNA synthesis and cytokine production [25,28,31].

3. ACKNOWLEDGMENTS

This work was supported by Deutsche Forschungsgemeinschaft SFB 387.

4. REFERENCES

1. Fleischer, B. (1994) CD26: a surface protease involved in T-cell activation. *J. Immunol.* **15**, 180–184.
2. Morimoto, C. & Schlossman, S. F. (1994) A key costimulatory molecule on CD4 memory T cells. *The Immunologist* **2**, 4–7.
3. Bühling, F., Junker, U., Reinhold, D., Neubert, K., Jäger, L. & Ansorge, S. (1995) Functional role of CD26 on human B lymphocytes. *Immunol. Lett.* **45**, 47–51.
4. Bühling, F., Kunz, D., Reinhold, D., Ulmer, A. J., Ernst, M., Flad, H.-D. & Ansorge, S. (1994) Expression and functional role of dipeptidyl peptidase IV on human natural killer cells. *Nat. Immun.* **13**, 270–279.
5. Yaron, A. & Naider, F. (1993) Proline-dependent structural and biological properties of peptides and proteins. *Crit. Rev. Biochem. Mol. Biol.* **28**, 31–81.
6. Ulmer, A. J., Mattern, T., Feller, A. C., Heymann, E. & Flad, H.-D. (1990) CD26 antigen is a surface Dipeptidyl peptidase IV (DPPIV) as characterised by monoclonal antibodies clone TII-19-4-7 and 4EL1C1. *Scand. J. Immunol.* **31**, 429–435.
7. Hegen, M., Niedobitek, C., Klein, C. E., Stein, H. & Fleischer, B. (1990) The T cell triggering molecule Tp103 is associated with Dipeptidyl aminopeptidase IV activity. *J. Immunol.* **144**, 2908–2914.
8. Mattern, T., Flad, H.-D., Feller, A. C., Heymann, E. & Ulmer, A. J. (1989) Anti-DPP-IV and anti-Ta1 but not IOT 15 should be clustered in CDw26. In: Leukocyte Typing IV. Knapp, W., Dörken, B., Gilks, W. R., Rieber, E. P., Schmidt, R. E., Stein, H. & Von Dem Borne, A. E. G. K. (Eds.), 416.

9. Kameoka, J., Tanaka, T., Nojima, Y., Schlossman, S. F. & Morimoto, C. (1993) Direct association of adenosine deaminase with a T cell activation antigen, CD26. *Science* **261**, 466–469.
10. Morimoto, C., Torimoto, Y., Levinson, G., Rudd, C. E., Schrieber, M., Dang, N. H., Letvin, N. L. & Schlossman, S. F. (1989) 1F7, a novel cell surface molecule, involved in helper function of CD4 cells. *J. Immunol.* **143**, 3430–3439.
11. Fleischer, B., Sturm, E., De Vries, J. E. & Spits, H. (1988) Triggering of cytotoxic T lymphocytes and NK cells via the Tp103 pathway is dependent on the expression of the T cell receptor/CD3 complex. *J. Immunol.* **141**, 1103–1107.
12. Hegen, M., Camerini, D. & Fleischer, B. (1993) Function of dipeptidyl peptidase IV (CD26, Tp103) in transfected human T cells. *Cell. Immunol.* **146**, 249–260.
13. Ansorge, S., Schön, E. & Kunz, D. (1991) Membrane-bound peptidases of lymphocytes: Functional implications. *Biomed. Biochem. Acta* **50**, 799–807.
14. Schön, E., Demuth, U. H., Barth, A. & Ansorge, S. (1984) Dipeptidyl peptidase IV of human lymphocytes. Evidence for specific hydrolysis of glycylprolin in T-lymphocytes. *Biochem. J.* **223**, 255–258.
15. Schön, E., Eichmann, E., Grunow, R., Jahn, S., Kiessig, S. T., Volk, H. D. & Ansorge, S. (1986) Dipeptidyl peptidase IV in human T lymphocytes. An approach to the role of a membrane peptidase in the immune system. *Biomed. Biochem. Acta* **45**, 1523–1528.
16. Dang, N. H., Torimoto, Y., Deusch, K., Schlossman, S. F. & Morimoto, C. (1990) Comitogenic effect of solid-phase immobilized anti-1F7 on human CD4 T cell activation via CD3 and CD2 pathways. *J. Immunol.* **144**, 4092–4100.
17. Dang, N. H., Torimoto, Y., Sugita, K., Daley, J. F., Show, P., Porado, C., Schlossman, S. F. & Morimoto, C (1990) Cell surface modulation of CD26 by anti-1F7 monoclonal antibody. Analysis of surface expression and human T cell activation. *J. Immunol.* **145**, 3963.
18. Dang, N. H., Torimoto, Y., Shimamura, K., Tanaka, T., Daley, J. F., Schlossman, S. F. & Morimoto, C. (1991) 1F7 (CD26): A marker of thymic maturation involved in the differential regulation of the CD3 and CD2 pathways of human thymocyte activation. *J. Immunol.* **147**, 2825–2832.
19. Schön, E., Born, I., Demuth, H.-U., Faust, J., Neubert, K., Steinmetzer, T., Barth, A. & Ansorge, S. (1991) Dipeptidyl peptidase IV in the immune system. Effects of specific enzyme inhibitors on activity of dipeptidyl peptidase IV and proliferation of human lymphocytes. *Biol. Chem. Hoppe Seyler* **372**, 305–311.
20. Schön, E., Demuth, H.-U., Eichmann, E., Horst, H.-J., Körner, I.-J., Kopp, J., Mattern, T., Neubert, K., Noll, F., Ulmer, J. A., Barth, A. & Ansorge, S. (1989) Dipeptidyl peptidase IV in human T lymphocytes Impaired induction of interleukin 2 and gamma interferon due to specific inhibition of dipeptidyl peptidase IV. *Scand. J. Immunol.* **29**, 127–132.
21. Schön, E., Jahn, S., Kiessig, S. T., Demuth, H.-U., Neubert, K., Barth, A., von Baehr, R. & Ansorge, S. (1987) The role of dipeptidyl peptidase IV in human T lymphocyte activation. inhibitors and antibodies against dipeptidyl peptidase IV suppress lymphocyte proliferation and immunoglobulin synthesis in vitro. *Eur. J. Immunol.* **17**, 1821–1826.
22. Torimoto, Y., Dang, N. H., Vivier, E., Tanaka, T., Schlossman, S. F. & Morimoto, C (1991) Coassociation of CD26 (dipeptidyl peptidase IV) with CD45 on the surface of human T lymphocytes *J. Immunol.* **147**, 2514–2517.
23. Torimoto, Y., Dang, N. H., Tanaka, T., Prado, C., Schlossman, S. F & Morimoto, C. (1992) Biochemical characterization of CD26 (dipeptidyl peptidase IV): Functional comparison of distinct epitopes recognized by various anti-CD26 monoclonal antibodies. *Mol. Immunol.* **29**, 183–192.
24. Schön, E., Eichmann, E., Jahn, S., Kopp, J., Volk, H. D. & Ansorge, S. (1988) On the role of dipeptidyl peptidase IV (DP IV) in thew immune system. Relations to the cell cycle and lymphokine production of activated lymphocytes. *Biol. Zent. bl.* **107**, 141–149.
25. Reinhold, D., Bank, U., Bühling, F., Neubert, K., Mattern, T., Ulmer, A. J., Flad, H.-D. & Ansorge, S. (1993) Dipeptidyl peptidase IV (CD26) on human lymphocytes. Synthetic inhibitors of and antibodies against Dipeptidyl peptidase IV suppress the proliferation of pokeweed mitogen-stimulated peripheral blood mononuclear cells and IL-2 and IL-6 production. *Immunobiol.* **188**, 403–414.
26. Reinhold, D., Bank, U., Bühling, F., Kähne, T., Kunz, D., Faust, J., Neubert, K. & Ansorge, S. (1994) Inhibitors of Dipeptidyl peptidase IV (DP IV, CD26) specifically auppress proliferation and modulate cytokine production of strongly CD26 expressing U937 cells. *Immunobiology* **192**, 121–136.
27. Fleischer, B. (1987) A novel pathway of human T cell activation via a 103 kD T cell activation antigen. *J. Immunol.* **138**, 1346–1350.
28. Ansorge, S., Bühling, F., Hoffmann, T., Kähne, T., Neubert, K. & Reinhold, D. (1995) DP IV/CD26 on human lymphocytes: Functional roles in cell growth and cytokine regulation. In: Dipeptidyl Peptidase IV (CD26) in metabolism and the immune response. Fleischer, B. (Eds.)R.G. Landes Company, Austin, (10)163–184.

29. Kähne, T., Kröning, H., Thiel, U., Ulmer, A. J., Flad, H.-D. & Ansorge, S. (1996) Alterations in structure and cellular localization of molecular forms of DP IV/CD26 during T cell activation. *Cell. Immunol.* **170**, 63–70.
30. Kähne, T. & Ansorge, S. (1994) Non-radioaktive labelling and immunoprecipitation analysis of leukocyte surface proteins using different methods of protein biotinylation. *J. Immunol. Methods* **168**, 209–218.
31. Kähne, T., Neubert, K. & Ansorge, S. (1995) Enzymatic activity of DP IV/CD26 is involved in PMA-induced hyperphosphorylation of $p56^{lck}$. *Immunol. Lett.* **46**, 189–193.

20

NEW FLUOROGENIC DIPEPTIDYL PEPTIDASE IV/CD26 SUBSTRATES AND INHIBITORS

Susan Lorey,[1] Jürgen Faust,[1] Uta Hermanns,[1] Frank Bühling,[2] Siegfried Ansorge,[2] and Klaus Neubert[1]

[1]Department of Biochemistry and Biotechnology
Institute of Biochemistry
Martin-Luther-University Halle-Wittenberg
Kurt-Mothes-Str. 3, D-06099 Halle, Germany
[2]Institute of Experimental Internal Medicine
Center of Internal Medicine
Otto-von-Guericke-University Magdeburg
Leipziger Str. 44, D-39106, Germany

1. INTRODUCTION

Dipeptidyl peptidase IV (DP IV, CD26) is a membrane bound serine protease which ubiquitously occurs in epithelial tissues of mammalian organisms, above all in kidney, liver and intestine[1,2] but also on the surface of immune cells[3]. DP IV is involved not only in activation of B- and T-lymphocytes[4,5] and NK-cells[6] with peptidase activity but also in the regulation of growth process, DNA-synthesis and cytokine production[7]. Therefore it is important for the signal transduction during the immune response. DP IV catalyses the cleavage of N-terminal dipeptides from oligo- and polypeptides where the penultimate residue is proline or -with less efficiency- alanine[8,9]. To investigate the physiological role of DP IV in further detail, specific high sensitive substrates or inhibitors would be useful tools. Here we describe the synthesis and enzymatic characterisation of sensitive fluorogenic rhodamine 110 (R110)-based substrates and fluoresceine- or biotin-labelled inhibitors and their use for the sensitive detection of DP IV on the surface of T-cell lines.

2. SUBSTRATES

2.1. Synthesis of the Substrates

The DP IV substrates of the type $(Xaa\text{-}Pro)_2$-R110 with Xaa = Gly, Ala, Abu, Leu, Ser, Lys and Phe were synthesized by mixed anhydride method using Boc-Pro-OH for the

Cellular Peptidases in Immune Functions and Diseases, edited by Ansorge and Langner
Plenum Press, New York, 1997

P1 position and Boc-Xaa-OH for the P2 position. The mono-substituted Xaa-Pro-R110 derivatives were obtained by coupling Boc-Xaa-Pro-OH with rhodamine 110 using 1-(3-dimethylamino-propyl)-3-ethylcarbodiimid (EDC). All compounds were purified by chromatographic methods (MPLC for protected, HPLC and RP 8 mini-columns for unprotected derivatives).

2.2. Results and Discussion

The rhodamine-based substrates of the types Xaa-Pro-R110 and $(Xaa\text{-}Pro)_2$-R110 differ in size and hydrophobicity. Therefore it was of interest to observe the kinetic character of their hydrolysis by isolated DP IV in comparison with the cleavage by cell surface DP IV.

The kinetic experiments with the isolated enzyme were performed according to Leytus et al.[10] at an enzyme concentration where less than 5% of the substrates were cleaved within measuring time. Under these conditions regarding the bis-dipeptide rhodamines the mono-substituted compounds are formed nearly exclusively. All synthesized substrates with exception of $(Phe\text{-}Pro)_2$-R110 are processed by DP IV. Most of the bis-substituted and all mono-substituted compounds are hydrolysed by a classical Michaelis Menten kinetic, only $(Abu\text{-}Pro)_2$-R110 and $(Leu\text{-}Pro)_2$-R110 are cleaved according to the model of substrate inhibition, where a second substrate molecule interacts with the enzyme forming an uncleavable SES-complex. The hydrolysis of the mono- and bis-substituted R110-derivatives by the isolated DP IV shows kinetic characteristics comparable to the cleavage of the well known dipeptide-4-nitroanilides[9]. Despite the size of the rhodamine molecule the kinetic constants of the hydrolysis of the R110- and the 4-nitroanilide derivatives are in the same range.

The kinetic constants shown in Table 1 represent the decrease of the hydrolysis in dependence on the increasing size of the amino acid in P2 position. Ala-Pro-R110 and the bis-substituted compounds where Xaa = Ala or Abu are the best substrates of the isolated DP IV with k_{cat}/K_m values between 2.90 and $4.28 \cdot 10^6$ $[M^{-1} \cdot s^{-1}]$. Possible interactions of polar (OH-, NH_2-) or hydrophobic (alkyl-) groups could explain a less efficient enzymatic cleavage of the bis-substituted compounds with Xaa = Ser, Lys and Leu. Despite the small size of the amino acid in P2 Position, $(Gly\text{-}Pro)_2$-R110 is cleaved with less efficiency.

Table 1. Kinetic constants of the mono- and bis-substituted rhodamine substrates

Compound	K_m $[10^{-5}$ M]	k_{cat} $[s^{-1}]$	k_{cat}/K_m $[M^{-1}$ $s^{-1}]$	K_i $[10^{-4}$ M]
Gly-Pro-R110	4.59 ± 0.71	72.02 ± 2.59	$1.59 \cdot 10^6$	
Ala-Pro-R110	2.50 ± 0.11	72.39 ± 3.46	$2.90 \cdot 10^6$	
Phe-Pro-R110	4.65 ± 0.58	54.72 ± 4.26	$1.18 \cdot 10^6$	
$(Gly\text{-}Pro)_2$-R110	9.91 ± 1.57	118.15 ± 18.70	$1.20 \cdot 10^6$	
$(Ala\text{-}Pro)_2$-R110	1.47 ± 0.19	63.19 ± 11.40	$4.28 \cdot 10^6$	
$(Abu\text{-}Pro)_2$-R110	2.55 ± 0.07	84.84 ± 2.11	$3.32 \cdot 10^6$	5.10 ± 1.00
$(Leu\text{-}Pro)_2$-R110	2.42 ± 0.23	58.36 ± 1.45	$2.42 \cdot 10^6$	4.69 ± 1.82
$(Ser\text{-}Pro)_2$-R110	5.89 ± 1.29	67.14 ± 5.78	$2.39 \cdot 10^6$	
$(Lys\text{-}Pro)_2$-R110	3.81 ± 0.20	75.02 ± 5.32	$1.30 \cdot 10^6$	
$(Phe\text{-}Pro)_2$-R110		no hydrolysis		

All measurements were performed at substrate concentrations of 10^{-7} - 10^{-4} M in 0.1 M Tris buffer, pH 7.6, I = 0.125 M, at 494 nm and 30 °C. DP IV was used at a concentration of $2.28 \cdot 10^{-10}$ M in case of the bis-substituted compounds and between 0.89 and $3.60 \cdot 10^{-9}$ M in case of the mono-substituted compounds. The initial velocities were recorded over 120 sec. The kinetic constants were determined by at least 2 independent measurements.

$(Phe\text{-}Pro)_2$-R110 is not hydrolysed and not bound by the enzyme due to a special conformation with a sandwich-like interaction between aromatic moieties of both phenylalanine side chains[11].

Measurements on DP IV-rich T-cells (U937) indicated $(Ala\text{-}Pro)_2$-R110 and $(Gly\text{-}Pro)_2$-R110 as the best substrates. $(Xaa\text{-}Pro)_2$-R110 with Xaa = Abu, Leu, Ser and Lys are cleft with less efficiency although the k_{cat}/K_m value of $(Abu\text{-}Pro)_2$-R110 at the isolated enzyme is comparable with the best of the R110-based substrates. Therefore amino acids greater than alanine in P2 position seem to slow down the hydrolysis by cell surface DP IV possibly due to steric hindrance by neighbouring proteins.

3. INHIBITORS

3.1. Synthesis

A series of fluorogenic DP IV inhibitors derived from Lys-pyrrolidides[12] and Lys-2-cyano-pyrrolidides[13] were synthesized by coupling of fluoresceine and biotin (complexation with fluorescent-labelled streptavidin after enzyme inhibition) with the side chain of lysine using the carbodiimide method. Furthermore the Lys-pyrrolidides were modified by insertion of 6-aminocaproic acid as a spacer between the lysine and the marker molecule.

3.2. Results and Discussion

The Lys-pyrrolidides and Lys-2-cyano-pyrrolidides could be characterized as very effective competitive inhibitors of the isolated DP IV with K_i values in the range of 10^{-7} M in case of the pyrrolidides and in the range of 10^{-8} M in case of the 2-cyano-pyrrolidides (Table 2). The activity of the isolated enzyme is inhibited by Lys-pyrrolidides and Lys-2-cyano-pyrrolidides (10^{-5} M) by 95% and 98%, respectively. The insertion of a C6-spacer between the lysine and the fluorophoric group does not affect the inhibitory activity of the compounds on the isolated DP IV.

At intact cells only the pyrrolidides (10^{-5} M) induced an inhibition of DP IV by 65%, the 2-cyano-pyrrolidides (10^{-5} M) with only 27% inhibition were uneffective inhibitors. In case of the pyrrolidides the insertion of an aminocaproic spacer led to a more effective inhibition by 87% whereas the spacer-free pyrrolidides inhibit cell surface DP IV

Table 2. Kinetic constants of the inhibition of isolated DP IV by Lys-pyrrolidides and Lys-2-cyano-pyrrolidides

Compound	K_i [10^{-7} M]
Lys(biotinyl)-pyrrolidide	1.98 ± 0.56
Lys(fluoresceinyl)-pyrrolidide	1.86 ± 0.12
Lys(biotinyl-capronyl)-pyrrolidide	1.53 ± 0.09
Lys(fluoresceinyl-capronyl)-pyrrolidide	2.08 ± 0.04
Lys(biotinyl)-2-cyano-pyrrolidide	0.58 ± 0.02
Lys(fluoresceinyl)-2-cyano-pyrrolidide	0.64 ± 0.03

All measurements were performed in 0.1 M Tris buffer, pH 7.6, I = 0.125 at five inhibitor and three substrate concentrations and at an enzyme concentration of $3.60 \cdot 10^{-9}$ M. The inhibitor as well as the substrate Gly-Pro-4-nitroanilide were dissolved in water. The free 4-nitroaniline was measured at 390 nm and 30°C over 120 sec.

by only 65%, due to the more flexible molecule structure which improved the steric accessibility of the catalytic site of the cell surface DP IV.

4. ACKNOWLEDGMENT

This work was supported by the Deutsche Forschungsgemeinschaft, SFB 387, Project A 5.

5. REFERENCES

1. Kenny, J., Booth, A.G. and Macnair R.D.C.: Peptidases of the kidney microvillus membrane. Acta biol. med. germ. 36 (1977) 1575–1585
2. Hertel, S., Gossrau, R. Hanski, C. and Reutter, W.: Dipeptidylpetidase IV in organs. Comparison of immunochemistry and activity histochemistry. Histochemistry 89 (1988) 151–167
3. Mentlein, R., Heymann, E., Scholz, W., Feller, A. and Flad, H.-D.: Dipeptidyl peptidase IV as a new surface marker for a subpopulation of human T-lymphocytes. Cell. Immunol. 89 (1984) 11–19
4. Bühling, F., Junker, U., Reinhold, D., Neubert, K., Jäger, L. and Ansorge, S.: Functional role of CD26 on human B lymphocytes. Immunol. Lett. 45 (1995) 47–51.
5. Reinhold, D., Bank, U., Bühling, F., Neubert, K., Mattern, T., Ulmer, A.J., Flad, H.-D. and Ansorge S.: Dipeptidyl peptidase IV (CD26) on human lymphocytes. Synthetic inhibitors of and antibodies against dipeptidyl peptidase IV suppress the proliferation of pokeweed mitogen-stimulated peripheral blood mononuclear cells, and IL-2 and IL-6 production. Immunobiol. 188 (1993) 403–414
6. Bühling, F., Kunz, D., Reinhold, D., Ulmer, A.J., Ernst, M., Flad, H.D. and Ansorge, S.: Expression and functional role of dipeptidyl peptidase IV (CD26) on human natural killer cells. Nat. Immun. 13 (1994) 270–279
7. Ansorge, S., Bühling, F., Hoffmann, T., Kähne, T., Neubert, K. and Reinhold, D.: DP IV/CD26 on human lymphocytes: Functional roles in cell growth and cytokine regulation. In: Dipeptidyl peptidase IV (CD26) in metabolism and the immune response (Fleischer, B., Ed.) R.G. Landes Company Texas. Springer Heidelberg, (1995) pp. 163–184
8. Rahfeld, J., Schutkowski, M., Faust, J., Neubert, K., Barth, A. and Heins J.: Extended investigations of the substrate specificity of dipeptidyl peptidase IV from pig kidney. Biol. Chem. Hoppe-Seyler 372 (1991) 313–318
9. Heins, J., Welker, P., Schönlein, C., Born, I., Hartrodt, B., Neubert, K., Tsuru, D. and Barth, A.: Mechanism of proline-specific proteinases: (I) Substrate specifity of dipeptidyl peptidase IV from pig kidney and proline-specific endopeptidase from *Flavobacterium meningosepticum*. Biochim. et Biophys. Acta 954 (1988) 161–169
10. Leytus, S.P., Melhado, L.L. and Mangel, F.: Rhodamine-based compounds as fluorogenic substrates for serine proteinases. Biochem. J. 209 (1983) 299–307
11. Mrestani-Klaus, C., Brandt, W., Faust, J., Hermanns, U., Lorey, S. and Neubert, K.: Structural studies of rhodamine 110 peptide derivatives representing a new class of fluorogenic substrates for dipeptidylpeptidase IV (DP IV). Peptides 1996, Proc. 24th Europ. Peptide Symp., Edinburgh 1996, in press
12. Schön, E., Born, I., Demuth, H.-U., Faust, J., Neubert, K., Steinmetzer, T., Barth, A. and Ansorge, S.: Dipeptidylpeptidase IV in the immune system. Effects of specific enzyme inhibitors on activity of dipeptidylpeptidase IV and proliferation of human lymphocytes. Biol. Chem. Hoppe-Seyler 372 (1991) 305–311
13. Li, J., Wilk, E. and Wilk, S.: Aminoacylpyrrolidine-2-nitriles: potent and stable inhibitors of dipeptidyl peptidase IV (CD26). Arch. of Biochem. and Biophys. 323 (1995) 148–154

21

MOLECULAR ANALYSES OF HUMAN AND RAT DIPEPTIDYL PEPTIDASE IV

C. A. Abbott,* M. D. Gorrell, M. T. Levy, and G. W. McCaughan

A.W Morrow Gastroenterology and Liver Centre
Centenary Institute for Cancer Medicine and Cell Biology
Royal Prince Alfred Hospital
Missenden Road, Sydney, 2050, Australia

1. INTRODUCTION

Dipeptidyl peptidase IV (DPP IV) is a highly glycosylated 110 kDa multifunctional protease. DPP IV has a broad tissue distribution but is mainly expressed in epithelial cells, endothelial cells and lymphocytes in both rat and human tissues[1-3]. DPP IV on the human T cell surface binds or coassociates with adenosine deaminase (ADA)[4,5], collagen[6-8] and CD45[9] and has a role in T cell costimulation[7,10]. Our group has investigated the role of DPP IV in hepatocyte function. DPP IV levels are altered during liver regeneration[11], human cirrhosis[12] and liver tumorigenesis[13]. It is still unclear which DPP IV functions are involved in hepatocyte growth and regulation. DPP IV is a type II protein, its N terminus is anchored to the cell surface through a hydrophobic domain with a short, six amino acid cytoplasmic tail. Thus, most of the protein is extracellular. Recent work in our group has concentrated on understanding the DPP IV glycoprotein at the molecular level.

2. MOLECULAR ANALYSIS OF THE HUMAN DPP IV GENE

2.1. The Human DPP IV Gene

Our group has cloned the human DPP IV gene and determined the exon/intron boundaries [14]. The DPP IV gene is composed of 26 small exons ranging in size from 45 bp to 1.4 kb. Mapping of the genomic clones demonstrated that the DPP IV gene was approximately 90 kb. The structure of the mouse DPP IV gene is identical to the human gene [15]. In addition we have precisely localised the DPP IV gene to chromosome 2q24.3 using fluorescent *in situ* hybridi-

* E-mail: c.abbott@centenary.usyd.edu.AU

Cellular Peptidases in Immune Functions and Diseases, edited by Ansorge and Langner
Plenum Press, New York, 1997

sation[14]. Other workers have localised DPP IV to 2q23 using the same technique[16]. Essentially 2q24.3 and 2q23 are indistinguishable as localisations are dependent upon banding method and degree of signal amplification. The murine DPP IV gene has been mapped to the equivalent region [2C2–2D] by *in situ* hybridisation[15].

Mutations in the DPP IV gene or variations in protein expression have not yet been directly associated with a single disease. Tanaka *et al* have shown that there are large variations in the levels of human plasma DPP IV *in vivo*[17]. Variations in DPP IV levels may be due to genotypic differences in DPP IV alleles. It is intriguing to speculate on whether the DPP IV gene has a direct role in any disease; or whether the changes in its expression relate to its involvement in immune responses or extracellular matrix remodelling.

In rats a mutant DPP IV phenotype has been detected. Japanese Fischer 344- rats[18, 19] and German Fischer 344 rats[20] are deficient in DPP IV enzyme activity. A single substitution of Gly^{633} to Arg near the catalytic serine is found in cDNA's isolated from both substrains[21, 22]. Identifying markers to screen for differences in the human DPP IV gene between individuals may help to isolate a mutant human phenotype. Restriction fragment length polymorphism (RFLP) studies revealed a highly polymorphic *Eco RI* band in the introns surrounding exons 8 to 10. This RFLP showed Mendelian inheritance in five families[23].

RFLP's as genetic markers rely on Southern blotting for their detection which is time consuming and expensive. It also requires DNA of high quality and analysis of archival specimens is difficult. The identification of a polyallelic marker linked to the DPP IV gene that could be typed by means of polymerase chain reaction (PCR) would be a useful diagnostic tool. The lambda genomic clones isolated for DPP IV were screened for CA repeats with an end-labelled CA repeat probe, poly[dC-dA]-poly[dG-dT]. Four CA repeats were identified in the introns between exons 12 and 23. Preliminary data indicates that one of these repeats contains 21 repeating units, is polymorphic and would be suitable for linkage analysis.

2.2. The Evolution of DPP IV and the Prolyl Oligopeptidase Family

DPP IV is a member of the prolyl oligopeptidase family[24, 25]. The catalytic triad of members of the prolyl oligopeptidase family has the unique order: Ser, Asp, His. This order is the reverse of that found in other serine proteases (e.g. trypsin and subtilisin families). The catalytic triad of the prolyl oligopeptidase family is superimposable with that of the lipase/esterase family[26]. Members of the prolyl oligopeptidase and lipase/esterase families share no other homology apart from the three catalytic residues. These two families appear to have arisen from divergent evolution from an α/ hydrolase ancestor, where distinct secondary and tertiary structures may be retained despite complete differences in sequence [15, 27, 28].

The gene structures for two other prolyl oligopeptidases are known: rat acylaminoacyl peptidase [29] and *Caenorhabditis elegans*[30]. The human DPP IV gene has a unique arrangement of exons around the serine recognition site compared to rat acylaminoacyl peptidase and serine protease genes[31–33]. In the human DPP IV gene the serine recognition site (GWSYGG) is divided between two exons (GW and SYGG) whereas in other proteases it is contained within one exon (Figure 1). This unexpected gene arrangement also occurs within the mouse DPP IV gene [15] and the *C. elegans* aminopeptidase gene[30]. This implies that the DPP IV ancestor diverged from the rest of the prolyl oligopeptidase family early in evolution. More genes of this family will need to be analysed to determine whether the serine recognition site was originally split in the α/ hydrolase ancestral gene or whether this occurred later in evolution.

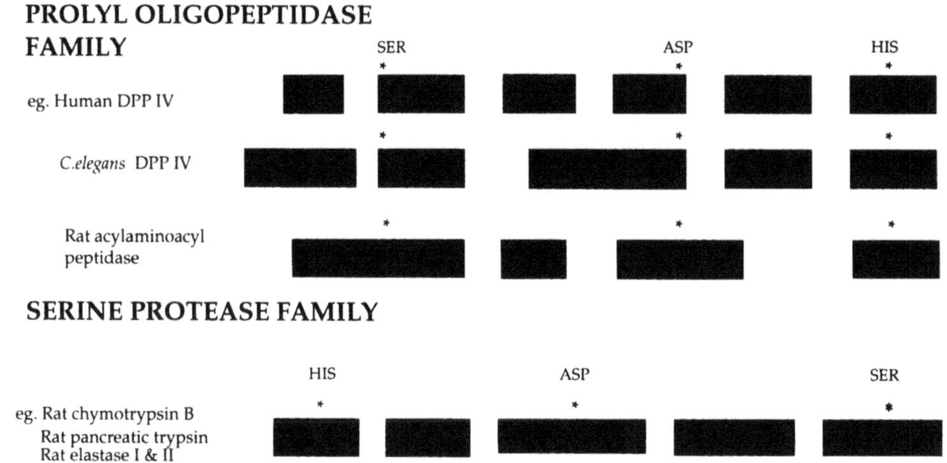

Figure 1. A schematic diagram comparing the arrangement of exons around the three members (Ser, Asp and His) of the catalytic triad in members of the prolyl oligopeptidase and serine protease families.

The human DPP IV gene spans 90 kb. In contrast, the *C. elegans* aminopeptidase gene[30] spans only 6.5 kb and rat acylaminoacyl peptidase spans 10 kb[29]. The dramatic increase in human DPP IV gene size implies the rearrangement and insertion of genetic material. This may have led to the development of many new and specialised functions, other than the original enzyme activity, for mammalian DPP IV.

2.3. The 5′ Flanking Region of Human DPP IV

The 5′ flanking region of the human DPP IV gene contains neither a TATA box nor a CAAT box, but a 300 bp region extremely rich in G-C (72%) contains potential binding sites for several transcription factors[14]. The G-C rich nature of the DPP IV promoter is typical of housekeeping genes. In contrast, most 5′ flanking potential transcription sites found in the promoter sequence are typical of promoters that regulate the expression of proto-oncogenes. Despite the house keeping nature of the promoter it has been demonstrated that the 5′ flanking sequence of DPP IV is capable of initiating transcription in a tissue specific manner[34]. Further studies into the promoter region of DPP IV will be required in order to fully understand the tissue specific control mechanisms. DPP IV appears to have a role in reproduction and progesterone stimulates the expression of DPP IV[35, 36]. Moreover, the cytokines tumour necrosis factor-α and IL-1 upregulated expression of DPP IV on human luteinising cells[37]. It would be interesting to study whether regions of the promoter are directly sensitive to progesterone or these cytokines.

2.4. Analysis of the Human and Rat DPP IV mRNA Transcripts

Transcription of the human DPP IV gene results in two different mRNAs sized at 4.2 and 2.8 kb[14]. These are both detected at high levels in the placenta and kidney and at moderate levels in the lung and liver. Only low levels of the 4.2 kb transcript are detected in mRNA from skeletal muscle, heart, brain and pancreas. This is the first evidence of human DPP IV expression in skeletal muscle, heart and brain. No evidence for alternative

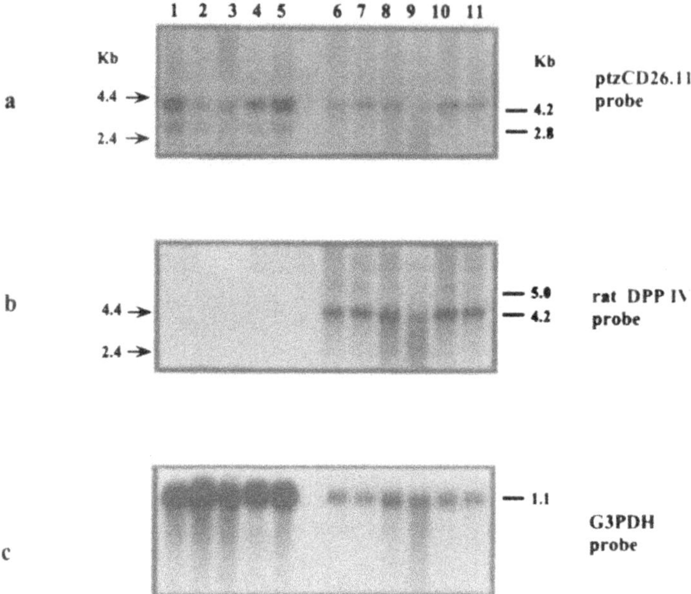

Figure 2. Northern blots of human liver (20 μg) and rat kidney (10 μg) RNA. Lanes 1–5 contain human liver RNA and lanes 6–11 contain rat kidney RNA. The Northern was hybridised with: (a) *Pst I* fragments of human DPP IV cDNA; (b) the rat DPP IV insert[38]; (c) the control probe human glyceraldehyde-3 phosphate dehydrogenase (G3PDH).

splicing of the human DPP IV gene is seen and the two mRNAs probably result from alternative polyadenylation mechanisms.

A Northern blot was performed to compare directly the size of major and minor, human and rat DPP IV mRNA transcripts. Figure 2 demonstrates clearly that the predominant 4.2 kb transcript expressed in human tissue is the same predominant transcript expressed in rat tissue. In all human tissues studied the 4.2 kb message was the predominant transcript seen. Figures 2a and 2b also show that the rat and human tissues express a different sized minor transcript. In human tissue a smaller 2.8 kb transcript was detected while in rat tissue a larger 5 kb transcript was detected.

3. THE PRODUCTION OF STABLE MUTANT DPP IV TRANSFECTANTS BY CHO CELLS FOR USE IN DPP IV FUNCTIONAL STUDIES

Functional aspects of the role and ligands of DPP IV in the numerous tissues and cells in which it is expressed still remain elusive. One approach to the molecular analysis of membrane-bound molecules is to produce large amounts of the extracellular portion of the molecule, i.e. soluble protein using an expression system. Soluble recombinant human DPP IV glycoprotein would be a useful tool to further analyse the function of DPP IV. Ultimately, production of soluble DPP IV glycoprotein may facilitate the elucidation of the tertiary structure of DPP IV.

Our group has developed a system for expressing soluble recombinant human DPP IV (srhDPP IV) at high levels. Human DPP IV has nine potential N-glycosylation sites[39, 40]. About thirty percent of DPP IV is carbohydrate. In order to study all DPP IV functions it was considered important to produce a recombinant protein possessing mammalian glycosylation. A relatively new mammalian gene expression system developed at Celltech (Celltech Ltd., Slough, United Kingdom) was adopted to express DPP IV. This system has been very useful in producing high levels of soluble glycosylated forms of CD4, CD45 and CD48 [41–43]. The function of DPP IV which has been the most extensively characterised is the enzyme activity. Despite this, the contribution and role of the DPP IV enzyme activity in immune function remains controversial [17, 44]. For this reason our group decided to also produce enzyme negative soluble recombinant human DPP IV (E-srhDPP IV) for use in functional studies.

3.1. Preparation of Constructs for the Expression of srhDPP IV, E-srhDPP IV and Cell Surface Human and Rat DPP IV

To create srhDPP IV, a consensus signal peptidase cleavage site[45] was engineered into the transmembrane domain of human DPP IV. A similar mutation of Leu^{28} to Ala had been successfully used to create soluble rat DPP IV[46]. Site directed mutagenesis was used on pTZ-CD26.11cDNA to introduce a 2 bp substitution that led to the mutation of Leu^{28} to Ala. Site-directed mutagenesis by overlap extension using PCR[47] was used to direct a Ser^{630} to Ala substitution in the srhDPP IV (pTZ-CD26.M1) construct. This expressed a protein that was not only enzyme negative but also soluble. In addition the full length human and rat cDNA's for DPP IV were cloned into pEE14 for surface expression of human and rat DPP IV.

3.2. Transfection, Selection and Expression of srhDPP IV, E-srhDPP IV and Cell Surface Rat and Human DPP IV

The constructs were transfected into CHO cells using LipofectAMINE™ (Gibco BRL). Transfected cell supernatants were screened by enzyme activity[48]. Supernatants of enzyme negative mutants were tested by dot and Western blot with 1F7 mAb. A cell line expressing high levels of srhDPP IV, 1B7, was cloned. By Western Blot with 1F7 mAb supernatant from mAb 1B7 was shown to contain a 110 kDa molecule equivalent in size to natural DPP IV derived from human liver and activated T cells in (Figure 3). 1B7 supernatant was also positive by Western blot with the 13 other human DPP IV mAbs tested. The concentration of DPP IV in static culture was estimated to be 9 µg/ml. Supernatant from 5–2.74, expressing E-srhCD26 produced an intense band at 110 kDa on a Western Blot with 1F7 mAb (Data not shown). Cells transfected with the full length rat and human DPP IV constructs were shown by flow cytometry with a panel of mAbs to express DPP IV on their surface[49].

3.3. Production of srhDPP IV Protein in the CELLMAX™

For larger scale expression, the stable clone 1B7 secreting srhDPP IV was placed into a CELLMAX™ QUAD (Cellco). The concentration of DPP IV in the supernatant collected from the CELLMAX was approximately 40 µg/ml. Thus the CELLMAX enhanced expression about five fold over static culture. This supernatant was purified using ADA affinity chromatography [50].

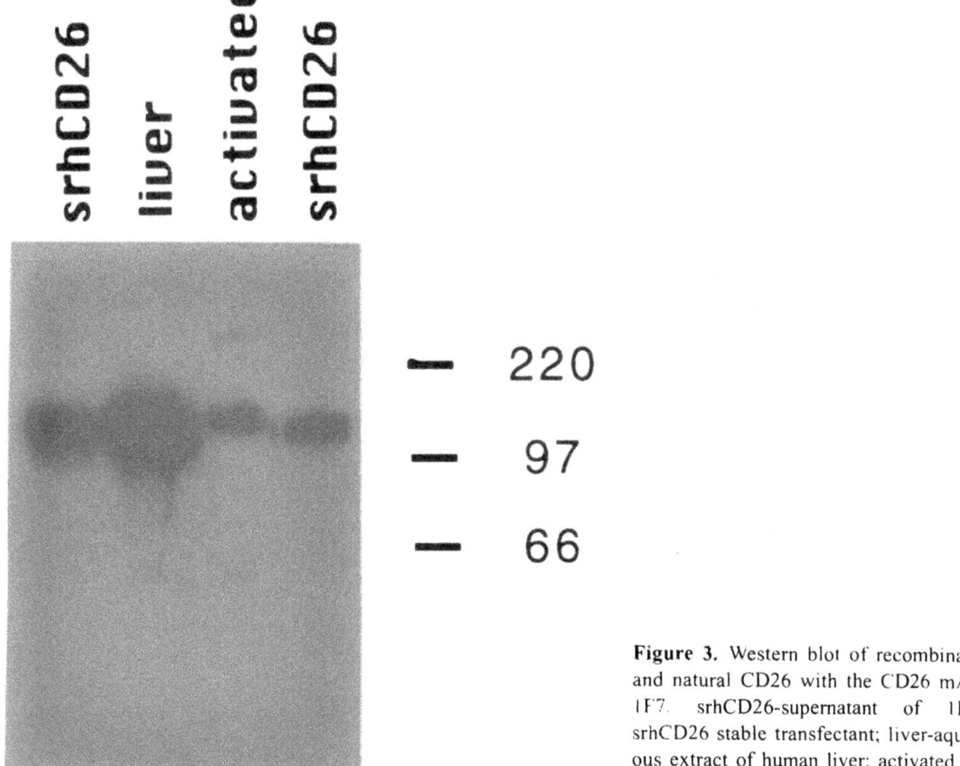

Figure 3. Western blot of recombinant and natural CD26 with the CD26 mAb 1F7. srhCD26-supernatant of 1B7 srhCD26 stable transfectant; liver-aqueous extract of human liver; activated T-activated human T cell sonicate.

4. ADA BINDING TO RECOMBINANT RAT AND HUMAN DPP IV

The binding of calf ADA to the surface of DPP IV positive CHO cells inhibited binding by the DPP IV mAbs TA5.9, 5F8, 134–2C2, S23 and 22C3[49]. ADA binding did not inhibit the binding of EF6/B10, M5, EF5/A3, EF6/F11, PEG2/C3, Ts145, 1F7, CB.1, 236.3 202–36, L272, BA5, 2A6 and Ta1 to these cells. Similarly, when the DPP IV mAbs TA5.9, 5F8, 134–2C2, S23 and 22C3 were bound to the surface of DPP IV positive CHO cells they inhibited ADA binding. The mAbs EF6/B10, M5, EF/A3, 1F7, 202–36, L272, 2A6, BA5 and Ta1 did not inhibit ADA binding[49]. These results confirm a previous study of TA5.9, 134–2C2, and 5F8[51] and in addition show mAbs S23 and 22C3 also recognise the ADA binding epitope. E-srh DPP IV also bound ADA, confirming that ADA binding does not require the catalytic triad of DPP IV[51]. The rat DPP IV positive CHO cells did not bind to ADA.

5. CONCLUSIONS

DPP IV is a large multifunctional glycoprotein, binding to the extracellular matrix and adenosine deaminase. The human DPP IV gene located at 2q24.3 shows both RFLP and CA repeat polymorphisms. Levels of DPP IV enzyme activity are altered in the sera of patients during some autoimmune diseases, many liver and renal diseases, and during various cancers. These polymorphic markers may be used to study the relationship of the DPP IV gene to human disease. The elucidation of the structure of the DPP IV gene gave new insights into the evolution of DPP IV and the prolyl oligopeptidase family but as yet no new insights into the many functions of DPP IV. Further understanding of the DPP IV molecule will rely on pinpointing the structural elements which are required for each of its functions. The pEE14/CHO cell approach is very appropriate for producing intact and mutant forms of DPP IV that possess glycosylation and all or selected biological activities of the natural form. These properties are necessary for producing DPP IV molecules which can be used for *in vivo* and *ex vivo* studies.

6. ACKNOWLEDGMENTS

This work would not have been possible without the gifts of human and rat cDNAs from Dr Morimoto and Dr Doyle respectively. We would like to especially acknowledge Dr Ingrid De Meester (University of Antwerp, Belgium) for helping purify srhCD26 and the gift of mAb TA5.9. We would also like to thank Prof. Ansorge and colleagues M. Taeger, H Kroening and T. Kaehne for generously giving us mAbs. Other mAbs utilised during this work were gifts from Drs: Morimoto, Plana, Old, Hixson, Prof. Fleischer, Prof. Ueda and contributors to the 6th Human Leucocyte Differentation Workshop[49].

7. REFERENCES

1. McCaughan G W, Wickson J E, Creswick P F and Gorrell M D, Identification of the bile canalicular cell surface molecule GP110 as the ectopeptidase dipeptidyl peptidase IV: an analysis by tissue distribution, purification and N-terminal amino acid sequence. Hepatology, 1990. 11(4): 534–44.
2. Gorrell M D, Wickson J and McCaughan G W, Expression of the rat CD26 antigen (dipeptidyl peptidase IV) on subpopulations of rat lymphocytes. Cell Immunol, 1991. 134(1): 205–15.
3. Hegen M, Niedobitek G, Klein C E, Stein H and Fleischer B, The T cell triggering molecule Tp103 is associated with dipeptidyl aminopeptidase IV activity. J Immunol, 1990. 144(8). 2908–14.
4. Morrison M E, Vijayasaradhi S, Engelstein D, Albino A P and Houghton A N, A marker for neoplastic progression of human melanocytes is a cell surface ectopeptidase. J Exp Med, 1993. 177(4): 1135–43.
5. Kameoka J, Tanaka T, Nojima Y, Schlossman S F and Morimoto C, Direct association of adenosine deaminase with a T cell activation antigen, CD26. Science, 1993. 261(5120): 466–9.
6. Bauvois B, A collagen-binding glycoprotein on the surface of mouse fibroblasts is identified as dipeptidyl peptidase IV. Biochem J, 1988. 252(3): 723–31.
7. Dang N H, Torimoto Y, Schlossman S F and Morimoto C, Human CD4 helper T cell activation: functional involvement of two distinct collagen receptors, 1F7 and VLA integrin family. J Exp Med, 1990. 172(2): 649–52.
8. Hanski C, Huhle T, Gossrau R and Reutter W, Direct evidence for the binding of rat liver DPP IV to collagen in vitro. Exp Cell Res, 1988. 178(1): 64–72.
9. Torimoto Y, Dang N H, Vivier E, Tanaka T, Schlossman S F and Morimoto C, Coassociation of CD26 (dipeptidyl peptidase IV) with CD45 on the surface of human T lymphocytes. J Immunol, 1991. 147(8): 2514–7.
10. Dang N H, Torimoto Y, Deusch K, Schlossman S F and Morimoto C, Comitogenic effect of solid-phase immobilized anti-1F7 on human CD4 T cell activation via CD3 and CD2 pathways. J Immunol, 1990. 144(11): 4092–100.
11. Abbott C A, Gorrell M D, Kobayashi Y, Kawasaki T, Liddle C, Bishop G A and McCaughan G W, Increased serum levels of dipeptidyl peptidase IV (CD26) in rats undergoing liver regeneration. Int Hepatology Commun, 1995. 4(3): 165–74.

12. Matsumoto Y, Bishop G A and McCaughan G W, Altered zonal expression of the CD26 antigen (dipeptidyl peptidase IV) in human cirrhotic liver. Hepatology, 1992. 15(6): 1048–53.
13. McCaughan G W, Siah C L, Abbott C, Wickson J, Ballesteros M and Bishop G A, Dipeptidyl peptidase IV is down-regulated in rat hepatoma cells at the mRNA level. J Gastroenterol Hepatol, 1993. 8(2): 142–5.
14. Abbott C, Baker E, Sutherland G R and McCaughan G W, Genomic organisation, exact localization, and tissue expression of the human CD26 (dipeptidyl peptidase IV) gene. Immunogenetics, 1994. 40(5): 331–8.
15. Bernard A M, Mattei M G, Pierres M and Marguet D, Structure of the mouse dipeptidyl peptidase IV (CD26) gene. Biochemistry, 1994. 33: 15204–14.
16. Mathew S, Morrison M E, Murty V V, Houghton A N and Chaganti R S, Assignment of the DPP4 gene encoding adenosine deaminase binding protein (CD26/dipeptidylpeptidase IV) to 2q23. Genomics, 1994. 22(1): 211–2.
17. Tanaka T, Duke-Cohan J S, Kameoka J, Yaron A, Lee I, Schlossman S F and Morimoto C, Enhancement of antigen-induced T-cell proliferation by soluble CD26/dipeptidyl peptidase IV. Proc Natl Acad Sci U S A, 1994. 91(8): 3082–6.
18. Watanabe Y, Kojima T and Fujimoto Y, Deficiency of membrane-bound dipeptidyl aminopeptidase IV in a certain rat strain. Experientia, 1987. 43(4): 400–1.
19. Tiruppathi C, Miyamoto Y, Ganapathy V, Roesel R A, Whitford G M and Leibach F H, Hydrolysis and transport of proline-containing peptides in renal brush-border membrane vesicles from dipeptidyl peptidase IV-positive and dipeptidyl peptidase IV-negative rat strains. J Biol Chem, 1990. 265(3): 1476–83.
20. Thompson N L, Hixson D C, Callanan H, Panzica M, Flanagan D, Faris R A, Hong W J, Hartel-Schenk S and Doyle D, A Fischer rat substrain deficient in dipeptidyl peptidase IV activity makes normal steady-state RNA levels and an altered protein. Use as a liver-cell transplantation model. Biochem J, 1991. 273(Pt 3): 497–502.
21. Tsuji E, Misumi Y, Fujiwara T, Takami N, Ogata S and Ikehara Y, An active-site mutation (Gly633—>Arg) of dipeptidyl peptidase IV causes its retention and rapid degradation in the endoplasmic reticulum. Biochemistry, 1992. 31(47): 11921–7.
22. Brandsch M, Ganapathy V and Leibach F, Chapter 6: Role of Dipeptidyl peptidase IV(DP IV) in intestinal and renal absorption of peptides, in Dipeptidyl peptidase IV(CD26) in metabolism and the immune response, B. Fleischer, Editor. 1995, R.G. Landes: Georgetown, Texas, USA. p. 111–29.
23. Abbott C A, Koorey D and McCaughan G W, An EcoRI polymorphism within the dipeptidyl peptidase IV (DPPIV) gene. Hum Mol Genet, 1993. 2(9): 1507.
24. Barrett A J and Rawlings N D, Oligopeptidases, and the emergence of the prolyl oligopeptidase family. Biol Chem Hoppe Seyler, 1992. 373(7): 353–60.
25. Polgar L and Szabo E, Prolyl endopeptidase and dipeptidyl peptidase IV are distantly related members of the same family of serine proteases. Biol Chem Hoppe Seyler, 1992. 373(7): 361–6.
26. Polgar L, Structural relationship between lipases and peptidases of the prolyl oligopeptidase family. FEBS Lett, 1992. 311(3): 281–4.
27. Ollis D L, Cheah E, Cygler M, Dijkstra B, Frolow F, Franken S M, Harel M, Remington S J, Silman I, Schrag J, *et al.*, The alpha/beta hydrolase fold. Protein Eng, 1992. 5(3): 197–211.
28. Goossens F, De Meester I, Vanhoof G, Hendriks D, Vriend G and Scharpe S, The purification, characterization and analysis of primary and secondary-structure of prolyl oligopeptidase from human lymphocytes. Evidence that the enzyme belongs to the alpha/beta hydrolase fold family. Eur J Biochem, 1995. 233(2): 432–41.
29. Lin L W, Lee F J and Smith J A, Structural organization of the rat acyl-peptide hydrolase gene. Nucleic Acids Res, 1989. 17(11): 4397–400.
30. Wilson R, Ainscough R, Anderson K, Baynes C, Berks M, Bonfield J, Burton J, Connell M, Copsey T, Cooper J, *et al.*, 2.2 Mb of contiguous nucleotide sequence from chromosome III of C. elegans. Nature, 1994. 368(6466): 32–8.
31. Bell G I, Quinto C, Quiroga M, Valenzuela P, Craik C S and Rutter W J, Isolation and sequence of a rat chymotrypsin B gene. J Biol Chem, 1984. 259(22): 14265–70.
32. Swift G H, Craik C S, Stary S J, Quinto C, Lahaie R G, Rutter W J and MacDonald R J, Structure of the two related elastase genes expressed in the rat pancreas. J Biol Chem, 1984. 259(22): 14271–8.
33. Craik C S, Choo Q L, Swift G H, Quinto C, MacDonald R J and Rutter W J, Structure of two related rat pancreatic trypsin genes. J Biol Chem, 1984. 259(22): 14255–64.
34. Bohm S K, Gum J R, Jr., Erickson R H, Hicks J W and Kim Y S, Human dipeptidyl peptidase IV gene promoter: tissue-specific regulation from a TATA-less GC-rich sequence characteristic of a housekeeping gene promoter. Biochem J, 1995. 311(Pt 3): 835–43.

35. Liu W J and Hansen P J, Progesterone-induced secretion of dipeptidyl peptidase-IV (cluster differentiation antigen-26) by the uterine endometrium of the ewe and cow that costimulates lymphocyte proliferation. Endocrinology, 1995. 136(2): 779–87.
36. Ohta N, Takahashi T, Mori T, Park M K, Kawashima S, Takahashi K and Kobayashi H, Hormonal modulation of prolyl endopeptidase and dipeptidyl peptidase IV activities in the mouse uterus and ovary. Acta Endocrinol Copenh, 1992. 127(3): 262–6.
37. Fujiwara H, Fukuoka M, Yasuda K, Ueda M, Imai K, Goto Y, Suginami H, Kanzaki H, Maeda M and Mori T, Cytokines stimulate dipeptidyl peptidase-IV expression on human luteinizing granulosa cells. J Clin Endocrinol Metab, 1994. 79(4): 1007–11.
38. Hong W and Doyle D, cDNA cloning for a bile canaliculus domain-specific membrane glycoprotein of rat hepatocytes. Proc Natl Acad Sci U S A, 1987. 84(22): 7962–6.
39. Misumi Y, Hayashi Y, Arakawa F and Ikehara Y, Molecular cloning and sequence analysis of human dipeptidyl peptidase IV, a serine proteinase on the cell surface. Biochim Biophys Acta, 1992. 1131(3): 333–6.
40. Tanaka T, Camerini D, Seed B, Torimoto Y, Dang N H, Kameoka J, Dahlberg H N, Schlossman S F and Morimoto C, Cloning and functional expression of the T cell activation antigen CD26. J Immunol, 1992. 149(2): 481–6.
41. Davis S J, Ward H A, Puklavec M J, Willis A C, Williams A F and Barclay A N, High level expression in Chinese hamster ovary cells of soluble forms of CD4 T lymphocyte glycoprotein including glycosylation variants. J Biol Chem, 1990. 265(18): 10410–8.
42. McCall M N, Shotton D M and Barclay A N, Expression of soluble isoforms of rat CD45. Analysis by electron microscopy and use in epitope mapping of anti-CD45R monoclonal antibodies. Immunology, 1992. 76(2): 310–7.
43. Van der Merwe P A, McNamee P N, Davies E A, Barclay A N and Davis S J, Topology of the CD2-CD48 cell-adhesion molecule complex: implications for antigen recognition by T cells. Curr Biol, 1995. 5(1): 74–84.
44. Steeg C, Hartwig U and Fleischer B, Unchanged signaling capacity of mutant CD26/dipeptidylpeptidase IV molecules devoid of enzymatic activity. Cell Immunol, 1995. 164(2): 311–5.
45. von Heijne G, Signal sequences. The limits of variation. J Mol Biol, 1985. 184(1): 99–105.
46. Weisz O A, Machamer C E and Hubbard A L, Rat liver dipeptidylpeptidase IV contains competing apical and basolateral targeting information. J Biol Chem, 1992. 267(31): 22282–8.
47. Ho S N, Hunt H D, Horton R M, Pullen J K and Pease L R, Site-directed mutagenesis by overlap extension using the polymerase chain reaction. Gene, 1989. 77(1): 51–9.
48. Nagatsu T, Hino M and Fuyamada H, New chromogenic substrates for X-prolyl dipeptidyl-aminopeptidase. Anal Biochem, 1976. 74: 466–76.
49. Gorrell M D, Levy M T, Abbott C A, Washington E A and McCaughan G W, CD26: ADA binding, immunoblotting and species cross reactivity examined using recombinant proteins, in Leucocyte Typing VI, T. Kishimoto, Editor. 1997, Garland Publishing Inc.: New York.
50. De Meester I, Vanhoof G, Lambeir A and Scharpe S, Use of immobilized adenosine deaminase(EC 3.5.4.4) for the rapid purifaication of native human CD26/dipeptidyl peptidase IV(EC 3.4.14.5). J Immunol Methods, 1996. 189: 99–105.
51. De Meester I, Vanham G, Kestens L, Vanhoof G, Bosmans E, Gigase P and Scharpe S, Binding of adenosine deaminase to the lymphocyte surface via CD26. Eur J Immunol, 1994. 24(3): 566–70.

A MOLECULAR MODEL OF THE ACTIVE SITE OF DIPEPTIDYL PEPTIDASE IV

Explanation of the Substrate Specificity and Interaction with Inhibitors

Wolfgang Brandt

Fachbereich Biochemie/Biotechnologie
Martin-Luther-Universität Halle-Wittenberg
Kurt-Mothes Str. 3, D-06099 Halle, Germany

1. INTRODUCTION

The ectoenzyme dipeptidyl peptidase IV (DPP IV) is an integral plasma membrane glycoprotein abundantly expressed on a variety of cell surfaces.[1] It is a serine peptidase of broad medical and biochemical significance. It plays a role in the degradation and post translational processing of bioactive peptides such as substance P, ß-casomorphins, growth hormone-releasing hormone and promelittin.[2–9] DPP IV has been identified as CD26.[10] Recently, it could be demonstrated that it is not only a surface differentiation marker involved in the transduction of mitogenic signals in thymocytes and T lymphocytes in human but also a cofactor in AIDS expression.[10–13]

The substrate specificity has been well characterised. This enzyme sequentially removes a dipeptide of the N-terminus of a peptide chain whenever a proline or alanine residue is present in the penultimate position (scissile residue, P_1- position using the nomenclature of *Schechter and Berger*[14]) of the substrate. Peptides with Pro at the P'_1-position (C-terminal of the scissile residue) are not hydrolysed. Any native amino acid can occupy the P_2-position (N-terminal from the scissile residue) as long as the N-terminus is unprotected and the N-terminal amino function is protonated. Interesting is the stereospecificity of DPP IV.[15,16] In the case of proline substrates (Pro in P_1-position), DPP IV has an absolute requirement for the S-configuration of both the penultimate and the N-terminal amino acid residues. In the case of alanine substrates (Ala in P_1-position), the configuration of the N-terminal residue may be R or S, but the penultimate residue must have S-configuration for enzymatic activity. Furthermore, the inhibition abilities of DPP IV by a number of dipeptides and amino acid pyrrolidides were determined by kinetic measurements (see Table I).[17] These kinetic data clearly demonstrate not only the importance of

Cellular Peptidases in Immune Functions and Diseases, edited by Ansorge and Langner
Plenum Press, New York, 1997

position P_1 but also the influence of the amino acid residue in P_2 for the recognition by DPP IV.

Recently, gentechnique experiments on the human DPP IV have been reported by *Ogata et al.*[18] They were able to identify the active serine from the 67 serine residues occurring in this enzyme by radioactive labelling with [^{3}H]diisopropyl-fluorophosphate. From these studies it is now known that the active serine is situated within the sequence Ile^{627}-Trp^{628}-Gly^{629}-Trp^{630}-Ser^{631}-Tyr^{632}-Gly^{633}. Interestingly, there is a correlation with the sequences found in other serine-proteases (Gly-X-Ser-X-Gly).[18] The importance of these five amino acid residues has been examined by site directed mutagenesis.[18] From these studies it seems to be clear that the Trp^{630} is of no or less importance. Furthermore, the substitution of Tyr^{632} by a phenylalanine residue did not reduce the DPP IV activity, however, a replacement by Gly or Leu led to a dramatic loss of the activity. The conclusion that may be drawn is that the aromatic ring of the residue in position 632 is of high importance in the recognition of substrates and inhibitors, respectively. Furthermore, the substitution of Gly^{629} to alanine or arginine leads to the complete loss of the catalytic activity of DPP IV. Further informations with respect to the active site of DPP IV were obtained by *David et al.*[19] They identified Ser^{624}, Asp^{702} and His^{734} of the mouse DPP IV to be involved in the catalytic triad residues.

In a recent study we derived a model of the recognition conformation of DPP IV substrates from theoretical conformational analyses of peptides.[20] This model was able to explain some findings of the substrate specificity of DPP IV.

By means of molecular modelling of a part of the active site of DPP IV under consideration of all experimental facts we want to contribute to the understanding of the properties of this enzyme.

2. METHODS

All structure manipulations, the conformational analyses and energy minimisations were performed with the molecular modelling program SYBYL 5.5 and 6.0 on an Evans & Sutherland workstation 10 and a Silicon Graphics Workstation VGXT.[21]

All conformations were minimised using the TRIPOS force field including electrostatic interactions based on Gasteiger partial charge distributions by using a dielectric constant of $\varepsilon=4$ with a distance dependent function.[22,23]

The energy minimum was considered to be reached when the energy change was not larger than 0.0001 kcal/mol using the Powell minimiser included in SYBYL/MAXIMIN2. In all cases the N-terminus was calculated in its protonated, positively charged state.

3. RESULTS

3.1. Development of a Model of the Active Site of DPP IV

The proline specificity of several serine proteases is a phenomenon that is not clarified in detail until now.

We postulate a specific interaction of the pyrrolidine ring with an aromatic ring of the enzyme. This assumption is supported by several findings. Thus, this type of interactions occurs in a number of enzyme inhibitor complexes as well as for instance between a phenylalanine side chain and proline in peptides containing the sequence Phe-Pro.[24] Furthermore, the site directed mutagenesis experiments of *Ogata et al.* demonstrated the high importance of Tyr^{632} in the recognition of substrates and the activity of DPP IV.[18] The importance of tyrosine in the recognition of a proline residue seems to be very likely since in all the three known X-prolyl dipeptidyl amino peptidases (DPP IV, yeast dipeptidyl aminopeptidase B (62) and X-prolyl dipeptidyl aminopeptidase from *Lactococcus lactis* [25]) the consensus sequence surrounding the active site serine is Gly-X-**Ser-Tyr**-X-Gly.

Taking all these results together it seems to be obvious that the proline specificity of the mentioned enzymes is mainly caused by the recognition of proline by the phenolic tyrosine side chain following in the sequence after the active serine residue.

On the basis of our model of the recognition conformation of substrates of DPP IV (C_7-conformation of the P_1 residue) we considered at first several energy minimum conformations of the sequence Ser^{631}-Tyr^{632} and examined their ability in interacting with the substrates.

In the next step we considered the lengthening of the sequence in N-terminal direction, that is, the conformation of the Trp^{628}-Gly^{629}-Trp^{630} residues should be derived. The resulting low energy conformation of this chain is shown in Fig. 1.

This conformation correlates with all experimental findings such as the outside orientation of the Trp^{630} side chain from the active site and the recognition of the P_2- amino acid residues by the side chain of Trp^{628}. Mutations of Gly^{629} would influence immediately the docking of substrates into the active site. Furthermore, it is worth mentioning that hydrogen bonds are formed between the carbonyl function of the scissile peptide bond of the substrates and the NH- groups of Trp^{630} and Gly^{629} (see Fig. 1). These hydrogen bonds are of importance to stabilise the correct conformation of the substrates in such a manner that the peptide bond to be cleaved is in more or less perpendicular orientation to the oxygen atom of the active serine residue. Earlier assumptions[26] that these interactions may play an essential role in stabilising the oxo-anion formed in the peptide bond hydrolysing reaction step is unlikely (for a detailed discussion of the possible mechanism in the DPP IV catalysis see [27]).

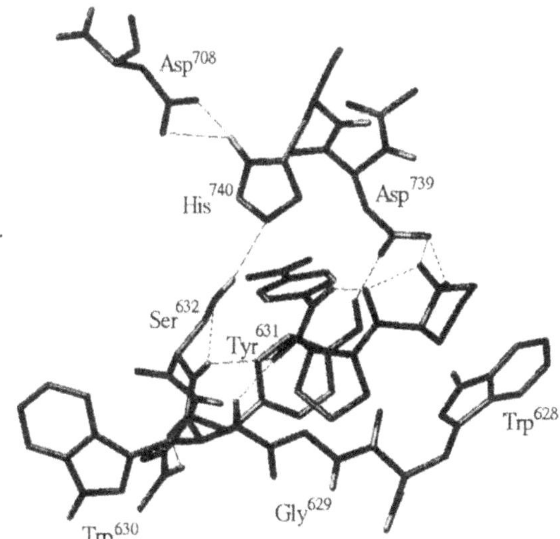

Figure 1. Model of the active site of DPP IV with docked (force field optimized) substrate H'Pro-Pro-pNA.

The next question to solve was: How to complete the catalytic triad? It has been suggested by Misumi *et al.*[28] that the Asp^{708} and His^{740} of the human DPP IV or Asp^{702} and His^{734} of the mouse DPP IV, respectively, are the essential residues comprising the triad.

General investigations of the active sites of serine proteases served as the basis for the development of the structure of the recognition site of DPP IV. The steric arrangements of the catalytic triads or tetrads, respectively, of all serine proteases of which the X-ray structure is stored in the Brookhaven Protein Data Base have been analysed.

Only in the case of subtilisin (Ser^{125}) the backbone dihedral angles (Φ = - 61,7°, Ψ = -27.8°) are in close correspondence to the dihedral angles (Φ = - 37.0°, Ψ = -41.8°) determined for Ser^{631} in our model. In the cases of chymotrypsin, kallikrein and tonin the corresponding dihedral angles are about -47° and 135°. The correlation of the backbone dihedrals of subtilisin to our model favours to use the arrangement of this catalytic triad for further development of the model. We placed the imidazole ring of His^{64} of subtilisin (becomes His^{740} of DPP IV) in a spatial position in the model by matching the backbone of Ser^{125} (from subtilisin) to the Ser^{631} of DPP IV. The fixed position of the imidazole side chain of His^{740} served then to determine the probable conformation of this residue as well as of the preceding Asp^{739}. For this purpose we searched for alignments of the sequence Asp^{739}-His^{740}-Gly^{741}-Ile^{742}. The corresponding sequence was found in apo-lactate dehydrogenase (Asp^{294}-His^{295}-Gly^{296}-Ile^{297}). We took the Asp-His sequence and matched the imidazole ring with the already determined spatial position in the DPP IV active site model. The resulting arrangement of these residues with regard to the other chain and to the enclosed substrate is represented in Fig. 1.

There are some very interesting findings worthy to discuss. The aspartic acid side chain (Asp^{739}) forms a hydrogen bond with the phenolic hydroxyl group of Tyr^{632}. Such a type of interaction does occur also in the active sites or recognition areas of several lipases and might be important for the stabilisation of the active site of DPP IV. Furthermore, the Asp side chain forms a salt bridge with the positively charged N-terminus of the substrates (and inhibitors) of DPP IV just in a conformation (Ψ_1 ~ 100°) predicted for the substrates. Even if we used the lowest energy substrate conformation with Ψ_1 = 180° this dihedral angle changes during an energy optimisation to a more or less perpendicular conformation as postulated earlier.[20]

Table I. Inhibitory activities (K_i in [mol/l]) of inhibitors of DPP IV (see also [26])

Name	Compound	pK_i	Name	Compound	pK_i
DI1	Gly-Pro	2.796	DI15	Phe-Pyd	5.638
DI2	Gly-Pyd	3.854	DI16	Asp-Pro	3.699
DI3	Ala-Pro	3.699	DI17	Asp-Pyd	5.252
DI4	Ala-Pyd	4.824	DI18	Sar-Pyd	3.468
DI5	Abu-Pyd	5.347	DI19	Arg-Pro	4.377
DI6	Val-Pro	5.000	DI20	β-Ala-Pro	1.769
DI7	Val-Pyd	6.319	DI21	Ala-Ala	3.387
DI8	Nva-Pyd	5.620	DI22	Ile-Ala	5.000
DI9	Leu-Pro	4.155	DI23	Pro-Gly	2.046
DI10	Leu-Pyd	5.456	DI24	Gly-Phe	2.495
DI11	Ile-Pro	5.161	DI25	Leu-Phe	3.854
DI12	Ile-Pyd	6.620	DI26	Ile-Val	4.398
DI13	Pro-Pyd	5.027	DI27	Pro-Hyp	0.155
DI14	Phe-Pro	4.328			

3.2. Conclusions and Application of the Model to Explain Experimental Results

The inhibitors of DPP IV listed in Table I were placed in the pocket of the model recognition site and an energy optimisation was performed by keeping the backbone atoms of the model fixed. The interaction energies between the inhibitors and the model were calculated by subtraction of the energy of the most stable conformations of each isolated inhibitor and the model without inhibitor from the energy of the complex (see Fig. 2).

We did not expect complete correlation between calculated interaction energies and the experimental determined pK_i-values but considering the amino acid pyrrolidids (Pyd) only the very satisfying correlation ($r^2 = 0.86$) clearly demonstrates that the interaction of the side chains of the amino acid residues in P_2-position with the tryptophan side chain of the DPP IV model reflects the experimental findings very well. There is favoured an interaction between hydrophobic (or positively charged) side chains and the aromatic side chain of Trp^{628} (see Fig. 3). This is indicated by the pK_i values as well as by the calculated interaction energies.

When comparing the pK_i-values of the Xaa-Pyd inhibitors with the corresponding Xaa-Pro compounds it seems to be unintelligible in the first view why the proline derivatives show decreased affinity in each case. The C-terminal carboxyl group forms two hydrogen bonds with the receptor model, that is, with the NH groups of Trp^{630} and Gly^{629} or in the case of the alanyl (P_1-position) derivatives with the NH groups of Tyr^{628} and Gly^{629}, respectively.

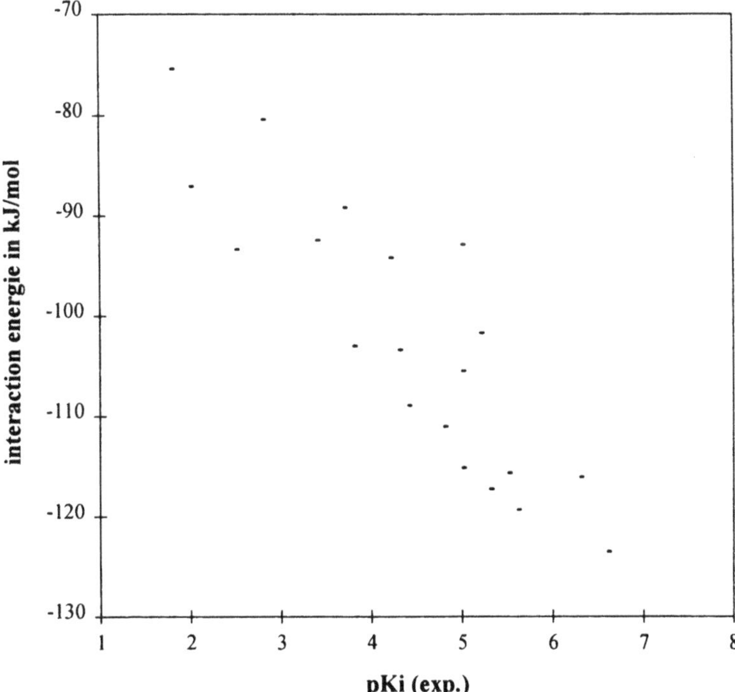

Figure 2. Graphical representation of the correlation between calculated interaction energies of inhibitors listed in Table 1 and experimenal pK_i-values.

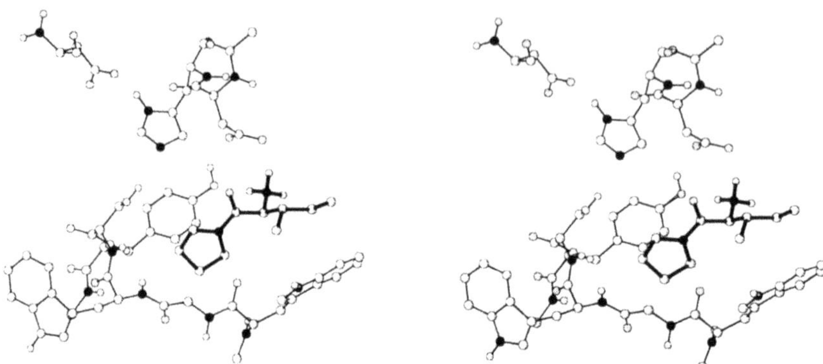

Figure 3. Model of the active site of DPP IV showing the interaction with H_2Ile-Pyd (DI12) the best inhibitor from Table I.

These interactions should lead to higher pK_i values which is in contradiction to the experimental findings. However, if we consider the most stable conformations of the isolated inhibitors a *cis*-conformation is favoured which includes a salt bridge between the charged N- and C-terminals. This conformation does not correspond with the recognition conformation and is approximately more than 4 kcal/mol more stable than the "correct" *trans*-conformation. This energy effort has to be overcome by docking at the DPP IV binding site and therefore reduces the energy gain mainly responsible for pK_i (except of entropic and additional solvation contributions in comparison to the pyrrolidids). The correlation of the calculated interaction energies with the pK_i-values strongly supports this hypothesis.

A further interesting experimental finding can be explained by the model. It has been found that Ala-anti-5-MeOxa-pNA ([4S,5R]-5-methyl-oxazolidine-4-carboxylic acid) is a substrate of DPP IV whereas the syn analogue is not (see [26]). The results of the energy optimisation of both complexes are represented in Figures 4 and 5 for comparison.

In the case of Xaa-Gly derivatives the interaction between a glycine and the aromatic side chain of Tyr is very weak. Furthermore, because of these very weak interac-

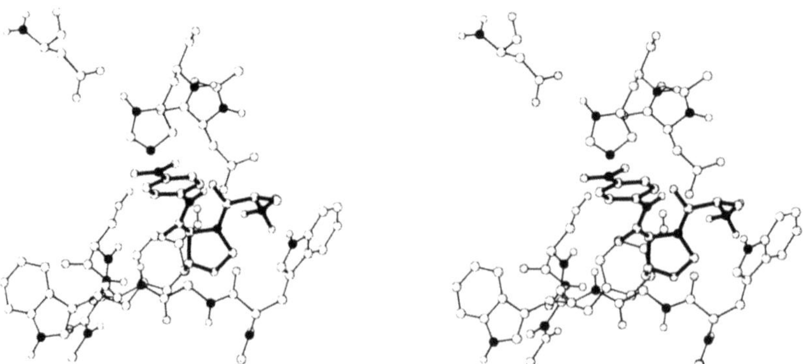

Figure 4. Interaction of the substrate H^+Ala-anti-5-MeOxa-pNA with the model of the active site of DPP IV.

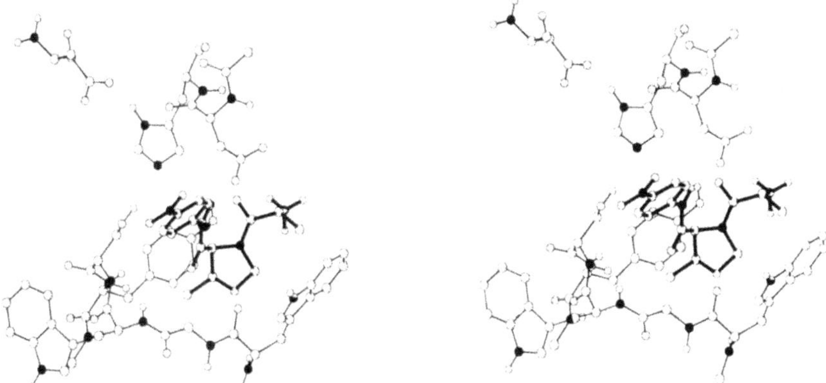

Figure 5. Interaction of the non-substrate H^+Ala-syn-5-MeOxa-pNA with the model of the active site of DPP IV. In the case of the Ala-anti-5-MeOxa-pNA the distance between the carbonyl carbon atom of the bond to be cleaved and the serine oxygen atom is 3.3 Å whereas the corresponding distance of the syn analogue is 4.3 Å. This difference is caused by steric repulsions between the syn oriented methyl group and the Ser and Tyr backbone atoms whereas in the case of the anti-derivative the methyl group is directed in a sterically accessible space and supports the recognition by the tyrosine side chain. The relatively high distance of 4.3 Å between the carbonyl carbon atom of the scissile bond for Ala-syn-5-MeOxa-pNA and the serine oxygen atom might be the reason for the impossibility to enter into a reaction.

tions and a missing anchor the correct positioning inside the active pocket is not guaranteed. This might be the reason for the very low k_{cat}/K_m-values of these compounds.

All other amino acid residues cannot occupy position S_1 to become substrates of DPP IV. Either the side chains are too bulky to fit into the pocket in a correct manner or they possess no hydrophobic side chains for an optimal interaction with the tyrosine side chain. The only native amino acid residue that could be a candidate for a weak substrate seems to be valine in position S_1.

The favour of proline in position P_1 with regard to alanine is caused by the optimal fit into the pocket of the active site and the higher interaction energy with tyrosine connected with optimal positioning of the scissile bond. In the case of proline the correct conformation is favoured also in the isolated state whereas the alanine residue possesses higher conformational degrees of freedom. Probably this is also the reason why the rate limiting step is in the case of the Xaa-Pro-pNA derivatives the deacylation but in the Xaa-Ala-pNA substrates the acylation.

The experimental finding that proline in position P_1' prevents the hydrolysis of molecules by DPP IV even if all other conditions for a substrate are fulfilled has been already explained and can now be validated by the model in detail.[20] In the case of an imino function at the scissile bond the formation of a C_7-conformation including a hydrogen bond to the carbonyl oxygen atom of the residue in position P_2 is impossible. As the consequence, another conformation (described in [20]) in this area is preferred and therefore, these molecules cannot adopt the recognition conformation demanded and graphically demonstrated in Fig. 1.

An explanation of the substrate specificity of DPP IV with regard to stereoselectivity of amino acid residues at the N-terminus (P_2) in dependence on the residues in position P_1 (alanine or proline, respectively) has already been discussed in [20] where we could show that it is impossible to adopt the correct recognition conformation for compounds possessing proline in P_1-position and a R-configured amino acid residue in position P_2 because of steric hindrance between the side chain of the N-terminal residue and the C_δ methylene group of proline. In the case of alanine in position P_1 this steric hindrance is not so strong (N-H group instead of the CH_2 group) as in the former and, therefore, such molecules can be slowly recognised and hydrolysed by DPP IV. This hypothesis can now be confirmed by the model.

On the basis of the presented model of the active site not only a multitude of experimental facts with regard to the substrate specificity and inhibition of DPP IV can be explained but also a number of proposals for the synthesis of new highly active inhibitors have been derived.

REFERENCES

1. Hartel, S., Gossrau, R., Hanski, C. & Reutter, W. (1988) Histochemistry **89**, 151–161
2. Nausch, I., Mentlein, R. & Heymann, E. (1990) Biol. Chem. Hoppe-Seyler **371**, 1113–1118
3. Nausch, I. & Heymann, E. (1985) J. Neurochem. **44**, 1354–1357
4. Ahmad, S., Wang, L. & Ward, P. E. (1992) J. Pharmacol. Exp. Ther. **260**, 1257–1256
5. Hartrodt, B., Fischer, G., Schulz, H.& Barth, A. (1982) Pharmazie **37**, 165–169
6. Kikuchi, M.; Fukuyama, K. & Epstein, W. L. (1988) Arch. Biochem. Biophys. **266**, 369–76
7. Kreil, G., Umbach, M., Brantl, V. & Teschemacher, H. (1983) Life Sciences **33**, 137–140
8. Frohman, L.A., Downs, T.R., Heimer, E.P. & Felix, A.M. (1989) J. Clin. Invest. **83**, 1533–1540
9. Kreil, G., Haiml, L. & Suchanek, G. (1980) Eur. J. Biochem. **111**, 49–58
10. Callebaut, C., Krust. B., Jacotot, E. & Hovanessian, A.G. (1993) Science **262**, 2045–2050
11. Fox, D.A., Hussey, R.E., Fitzgerald, K.A., Acuto, O., Poole, C., Palley, L., Daley, J.F., Schlossman, S.F. & Reinherz, E.L. (1984) J. Immunol. **133** 1250–1256
12. Fleischer, B. (1987) J. Immunol. **138**, 1346–1350
13. Reinhold, D., Bank, U., Bühling, F., Neubert, K., Mattern, T., Ulmer, A. J., Flad H.-D., & Ansorge, S. (1993) Immunobiol. **188**, 403–414
14. Schechter, I. & Berger, A. (1967) Biochem. Biophys. Res. Commun. **27**, 157–162
15. Rahfeld, J., Schutkowski, M., Faust, J., Neubert, K., Barth, A. & Heins. J. (1991) Biol. Chem. Hoppe-Seyler **372**, 313–318
16. Heins, J., Welker, P., Schönlein, C.,Born, I., Hartrodt, B., Neubert, K.,Tsuru, D. & Barth, A. (1988) Biochim. Biophys. Acta **954**, 161–169
17. Born, I., Faust, J., Heins, J., Barth, A. & Neubert, K. (1994) Eur. J. Cell Biol. Supl.**40**, 23
18. Ogata, S., Misumi, Y., Tsuji, E., Takami, N., Oda, K. & Ikehara, Y. (1992) Biochemistry **31**, 2582–2587
19. David, F., Bernard, A.-M., Pierres, M. & Marguet, D. (1993) J. Biol. Chem. **268**, 17247–17252
20. Brandt, W., Lehmann, T., Hofmann, T., Schowen, R.L. & Barth, A. (1992) J. Comp.-Aided Mol. Design **6**, 159–174
21. Tripos Associates Inc., 1699 S. Hanley Road, Suite 303, St. Louis, MO 631–44
22. Clark, M., Cramer III, R.D. & Van Opdenbosch, N. (1989) J. Comp. Chem. **10**, 982–1012
23. Gasteiger, J. & Marsili, M. (1980) Tetrahedron **36**, 3219–3238
24. Mrestani-Klaus, C., Brandt, W., Schmidt, R., Schiller, P.W.., Neubert K. (1996) Arch. Pharm. Pharm. Med. Chem., **3**, 133–142
25. Chich, J.-F., Chapot-Chartier, M.-P., Ribadeau-Dumas, B. & Gripon, J.-C. (1992) FEBS **314**, 139–142.
26. Brandt, W., Lehmann, T., Thondorf, I., Born, I., Schutkowski, M., Rahfeld, J., Neubert, K., Barth A. (1995) Int. J. Peptide Protein Res., 46, 494–507
27. Brandt, W., Ludwig, O., Thondorf, I., Barth, A. (1996) Europ. J. Biochem., **236**, 109–114
28. Misumi, Y., Hayashi, Y., Arakawa, F. & Ikehara, Y. (1992) Biochim. Biophys. Acta **1131**, 333–336

23

THE LEVEL OF CD26 DETERMINES THE RATE OF HIV ENTRY IN A CD4$^+$ T-CELL LINE

Christian Callebaut,* Etienne Jacotot, Julià Blanco, Bernard Krust, and Ara G. Hovanessian

Unité de Virologie et Immunologie Cellulaire
Unité Associée CNRS 1157
Département Retrovirus-SIDA
Institut Pasteur
28, rue du Dr Roux, 75724 Paris cedex 15, France

ABSTRACT

We have reported that CD26 could serve as a cofactor of CD4 in HIV entry. Recently, more evidence has been provided for the implication of CD26 in HIV entry, replication and cytopathic effect. Along with, we have demonstrated that the level of CD26 may determine the rate of HIV-envelope induced-apoptosis. The role of CD26 in HIV entry was further investigated using CEM T-cell line. Clones were established by transfection, expressing different levels of CD26. Entry, infection and cytopathic effect were monitored in several independent clones, and were found to be delayed in clones CD26-Low and CD26-SuperHigh compared to clones CD26-High. The delay was most significant in clones CD26-AntiSense, without any apparent cytopathic effect. These results demonstrate that relatively enhanced levels of CD26 contribute to an increased virus infection. Furthermore, they illustrate that CD26-SuperHigh clones manifest a phenotype similar to CD26-Low clones. This point out the critical role of CD26 in the rate of HIV entry and its cytopathic effect, two events which are initiated by the interaction of HIV envelope glycoproteins with cell-surface CD4.

INTRODUCTION

CD26, a T-cell activation antigen, is a proteolytic enzyme (dipeptidyl-peptidase IV, DPP IV) with a wide tissue distribution and a unique specificity. Recent developments in-

* Corresponding author. Tel: 33-1-45-65-88-97; Fax: 33-1-40-61-30-12; E-mail: chrcall@pasteur.fr.

Cellular Peptidases in Immune Functions and Diseases, edited by Ansorge and Langner
Plenum Press, New York, 1997

dicate that CD26 is a multifunctional molecule that may have important functions in the immune system, espacially in T-lymphocytes [1]. We have previously reported that CD26 through a potential interaction with the V3 loop of HIV envelope glycoproteins, could serve as a cofactor of CD4 in entry of lymphotropic HIV-1 Lai and HIV-2 EHO isolates [2, 3]. In contrast, using other experimental approches, several groups have rapidly contested our results; their criticisms with our response have been published [4–8]. However, recently more evidence has been provided by two other groups. In one hand, they show the implication of CD26 in HIV entry and its cytopathic effect, and this has been correlated with the structure of the V3 loop [9, 10]. On the other hand, the second complementarity determining region CDRH2 of the heavy chain of an anti-V3 loop neutralizing antibody was found to contain a short stretch of amino acids homologous to the sequence GWSYG of CD26 containing the catalytic serine-630. And a V3 loop peptide was shown to bind a synthetic CD26 peptide containing a catalytic sequence GWSYG involved in the DPP IV activity [11]. Along with, we have further demonstrated that the level of CD26 may determine the rate of HIV-envelope mediated cell to cell membrane fusion and the induction of cell killing by apoptosis [12, 13]. The results discuted here are consistent with the involvement of CD26 in HIV entry, infection and its cytopathic effect.

CELLULAR RECEPTORS FOR HIV ENVELOPE GLYCOPROTEINS

The HIV envelope glycoproteins, the gp120/gp41 complex, play a major role in HIV infection [14]. For HIV particles, this complex is essential for binding of particles to cell membrane receptors to initiate viral/cell membrane fusion, thus allow viral entry, whereas in infected cells, the expression of this complex on the cell membranes initiates the cytopathic effect which is manifested by cell killing via apoptosis, with or without the formation of syncytia [15]. The binding of the gp120/gp41 complex to the CD4 receptor is essential but not sufficient for the initiation of HIV entry and its cytopathic effect [16]. By analogy with certain other enveloped viruses, receptor binding by HIV may be followed by cleavage of the V3 loop by an exogenous proteinase [17, 18]. Accordingly, several studies have emphasized the implication of other factors, reviewed in [19]. We have ourselves proposed the implication of CD26 in HIV entry [2]. More recently, convincing evidence has been provided by several laboratories to show that the chemokine receptors, such as CXCR4 or "fusin" and CCR-5, serve as species specific cofactors for the entry of HIV-1 isolates [20, 21].

CD26 AND VIRAL INFECTION

The role of CD26 in the mechanism of HIV entry was further investigated using CEM CD4$^+$ T-cell line, which are permissive to HIV infection and express very low but reproducibly detectable levels of CD26. CEM clones were established by transfection of CEM clone 13 [15], in order to express different levels of CD26 and anti-sense mRNA of CD26. The CEM cell clones obtained are the following: CEM CD26-Low expressing endogenous low level of CD26, CEM CD26-High expressing an enhanced level of recombinant CD26, CEM CD26-SuperHigh overexpressing recombinant CD26 and CEM CD26-AntiSense without expression of CD26. All of the CEM cell clones expressed similar level of CD4 and other cell surface molecules implicated in HIV infection, like CD95,

CD45, CD44, CD7, LFA-1 and "fusin" mRNA. Three independent clones of each cell type (Low, High, SuperHigh and AntiSense) were infected with HIV Lai. The kinetics of virus infection was monitored by the concentration of major viral core protein p24 in the culture supernatant. The entry of HIV was found to be delayed in clones CD26-Low and CD26-SuperHigh compared to clones CD26-High. Interestingly, the delay was most significant in clones CD26-AntiSense, in which the virus production was also found to be very low. HIV-infected CD26-AntiSense clones did not show any apparent cytopathic effect. These results demonstrate that relatively enhanced levels of CD26 contribute to an increased virus infection (see Table 1). Furthermore, they illustrate that CD26-SuperHigh clones manifest a phenotype similar to that observed in CD26-Low clones. Thus suggesting that overexpression of CD26 may interfere with the balanced distribution of cell surface components and disturb their proper functioning in the virus to cell membrane fusion process, to allow HIV entry. This is an important observation, pointing out how critical is the optimal level of CD26 expression for its function. Whatever is the case, relatively enhanced levels of CD26 contribute to a significant increase in the rate of infection. This latter should be the consequence of slow rate of entry in clones CD26-Low and CD26-SuperHigh. Indeed, monitoring HIV proviral DNA levels at the third day of infection, revealed a significant difference in these clones which was correlated with the kinetics of virus production (see Table 1). The proviral DNA level was also monitored early after virus addition (90 min) and was shown to be several fold lower in CD26-Low compared to CD26-High clones. In an another set of experiments, virus infection was followed by the presence of HIV envelope glycoproteins (env) in cell extracts. Once again there was a significant delay in the synthesis of viral envelope glycoproteins in CD26-Low and CD26-SuperHigh clones. In CD26-High clones, a significant degree of cell death by apoptosis occured at day 7, 24 hours following the peak of the accumulation of the HIV envelope glycoproteins. In contrast, cell death did not occur in CD26-Low, CD26-SuperHigh and CD26-AntiSense clones, in which the synthesis of viral envelope glycoproteins was delayed (see Table 1). The mechanism by which CD26 is responsible for the occurence of apoptosis remains to be shown [22]. It is possible to suggest that the interaction of the gp120/gp41 complex directly or indirectly leads to a modified CD26-signaling, which might then be responsible for the initiation of apoptosis. Taken together, our results indicate that CD26 is implicated in the events associated with the proper functioning of the gp120/gp41 complex by assisting viral entry and its cytopathic effect. In other words, CD26 is acting as a cofactor of CD4 in HIV infection, and we have indirect evidence for the specific interaction of gp120 with CD26 [23]. Our results also point out that over-expression of CD26 in a given cell line could account for negative results.

Table 1. Different parameters of HIV-1 Lai infection in CEM-CD26 clones

CD26 Level	HIV-DNA 90 min	HIV-DNA 3 days	p24 5 days	env 5 days	Syncytia 6 days	Apoptosis 7 days
Low	-/+	+	+	+	+	+
High	++++		++++	++++	++++	++++
SuperHigh	ND	+	-/+	+	+	+
Anti-sens	ND	ND	-/+	ND	–	–

++++: 100%; +++: 75%, ++: 50%; +: 25%; +/-: 10%; -: 0%; ND: not determined.

IMPLICATION OF COFACTORS IN THE FUNCTIONING OF THE gp120/gp41 COMPLEX

HIV-1 gp120/gp41 complex initiates the fusion processes between virus and cell membranes to allow HIV entry, and between cell to cell membranes to generate the formation of multinucleated cells or syncytia [14]. Independent of syncytium formation, cell membrane expressed gp120/gp41 complex is responsible for the occurence of apoptosis in HIV-infected cell cultures[15]. In all of these functions, the cell-surface CD4 molecule is essential but not sufficient[16]. Accordingly, throughout the years, several potential cofactors of CD4 have been proposed. By biochemical approaches, different cell surface proteins have been reported to interact with the V3 loop of gp120. However, in most cases the relationship between the interaction and a putative role in HIV infection has not been determined. The implication of CD26 in HIV entry and its cytopathic effect has to be correlated with decrease of immune function during the disease. In view of all these observations, it is conceivable that several cell-surface antigens may coordinate the complex machinery of membrane fusion process in which the HIV gp120/gp41 envelope complex plays a key role. The requirement for some of the individual components might depend on the cell line studied and also might vary between the virus to cell or cell to cell fusion processes [24]. Whatever is the case, the number of cofactors implicated and their precise role in the HIV-mediated membrane fusion process remains still to be determined. Recently, evidence has been provided to show that the large family of seven-transmembrane-spanning cell-surface proteins, such as CXCR4 and CCR5, serve as species specific cofactors for the entry of T-cell- and macrophage-tropic HIV-1 isolates, respectively [20, 21]. Nevertheless, it will be essential to determine the implication of CD26 in relation to these cofactors, in the HIV-envelope mediated membrane fusion mechanism. Since chemokines can control HIV infection, and are potential substrates for the DPP-IV activity of CD26, it could be essential to determine the relationship between CD26 and chemokine receptors functions.

CD26 IN THE IMMUNE SYSTEM

Review of the current knowledge, show CD26 as a multifunctional molecule with important functions in the immune system, particulary in T lymphocytes [1]. CD26 can regulate immune responses by allowing antigen-specific T cells to exert a maximal response to their specific antigen [25].Besides its catalytic activity, CD26 is the ADA binding protein (ADAbp) [26]. In T cells, ADA and CD26 colocalized on the surface of T cells. Results from several laboratories strongly indicates that ADA binding to CD26 produces a costimulatory response in T cell activation events [27, 28]. Accordingly, it has been described that inherited deficiency of ADA leads to the syndrome of Severe Combined Immuno Deficiency [29]. ADA-ADAbp decrease have also correlated to natural immunomodulation: it has been reported a reduced DPP IV activity of PBMC and an impairement of cellular immunity during pregnancy [30]. Rich et al. have also reported a common immune function changes during pregnancy and HIV infection: T helper function are altered in pregnancy and return to baseline postpartum in uninfected but not HIV-infected women [31]. These defects could explain the different functional states of the immune system observed during the progression of AIDS [32]. Concerning HIV infection, we and others have shown a selective decrease in $CD4^+/CD26^+$ T-lymphocytes [33], which is in agreement with the potential involvement of CD26 as a cofactor of CD4 in HIV pathogenesis.

REFERENCES

1. Fleischer, B, *CD26: a surface protease involved in T-cell activation.* Immunol Today, 1994. **15**(4): p. 180–4.
2. Callebaut, C, *et al.*, *T cell activation antigen, CD26, as a cofactor for entry of HIV in CD4+ cells.* Science, 1993. **262**(5142): p. 2045–50.
3. Callebaut, C, *et al.*, *Entry of HIV into CD4+ cells requires the T-cell activation antigen CD26/DPP IV,* in *VIIIè Colloque des Cent Gardes,* M. Girard and B. Dodet, Editors. 1993b, In Retrovirus of Human AIDS and related diseases: p. 187–193.
4. Broder, CC, *et al.*, *CD26 antigen and HIV fusion? [letter; comment].* Science, 1994. **264**(5162): p. 1156–9.
5. Patience, C, *et al.*, *CD26 antigen and HIV fusion? [letter; comment].* Science, 1994. **264**(5162): p. 1159–60.
6. Camerini, D, Planelles, V, and Chen, IS, *CD26 antigen and HIV fusion? [letter; comment].* Science, 1994. **264**(5162): p. 1160–1.
7. Alizon, M and Dragic, T, *CD26 antigen and HIV fusion? [letter; comment].* Science, 1994. **264**(5162): p. 1161–2.
8. Callebaut, C, *et al.*, *CD26 antigen and HIV fusion? [Response].* Science, 1994. **264**(20 May 1994): p. 1156–1165.
9. Oravecz, T, *et al.*, *CD26 expression correlates with entry, replication and cytopathicity of monocytotropic HIV-1 strains in a T-cell line [see comments].* Nat Med, 1995. **1**(9): p. 919–26.
10. Callebaut, C and Hovanessian, AG, *CD26 and HIV infection.* Res.Virol., 1996. **147**: p. 67–69.
11. Chin, LT, *et al.*, *Molecular characterization of a human anti-HIV 1 monoclonal antibody revealed a CD26-related motif in CDR2.* Immunology Letters, 1995. **44**: p. 25–30.
12. Callebaut, C, *et al.*, *The role of CD26 in HIV infection : viral entry and its cytopathic effect,* in *IXè Colloque des Cent Gardes,* M. Girard and B. Dodet, Editors. 1994b, In Retrovirus of Human AIDS and related diseases: p. 141–149.
13. Jacotot, E, *et al.*, *HIV envelope-glycoproteins induced cell-killing by apoptosis is enhanced with increased expression of CD26 in CD4$^+$ T cells.* *Virology*, 1996. **223**: p. 318–330.
14. Moore, JP, *et al.*, *The HIV-cell fusion reaction,* in *Viral Fusion Mechanisms,* J. Bentz, Editor. 1993, p. 233–289.
15. Laurent-Crawford, AG, *et al.*, *Membrane expression of HIV envelope glycoproteins triggers apoptosis in CD4 cells.* Aids Res Hum Retroviruses, 1993. **9**(8): p. 761–73.
16. Maddon, PJ, *et al.*, *The T4 gene encodes the AIDS virus receptor and is expressed in the immune system and the brain.* Cell, 1986. **47**: p. 333–348.
17. Sattentau, QJ and Moore, JP, *Conformational changes induced in the human immunodeficiency virus envelope glycoprotein by soluble CD4 binding.* J Exp Med, 1991. **174**(2): p. 407–15.
18. Schulz, TF, *et al.*, *Effect of mutations in the V3 loop of HIV-1 gp120 on infectivity and susceptibility to proteolytic cleavage.* Aids Res Hum Retroviruses, 1993. **9**(2): p. 159–66.
19. Dragic, T, Picard, L, and Alizon, M, *Proteinase-resistant factors in human erythrocyte membranes mediate CD4-dependent fusion with cells expressing human immunodeficiency virus type 1 envelope glycoproteins.* J Virol, 1995. **69**(2): p. 1013–8.
20. Feng, Y, *et al.*, *HIV-1 entry cofactor: functional cDNA cloning of a seven-transmembrane, G protein-coupled receptor [see comments].* Science, 1996. **272**(5263): p. 872–7.
21. Dragic, T, *et al.*, *HIV-1 entry into CD4+ cells is mediated by the chemokine receptor CC-CKR-5.* Nature, 1996. **381**: p. 667–673.
22. Jacotot, E, *et al.*, *CD26 as positive regulator of HIV envelope-glycoproteins induced apoptosis in CD4$^+$ T cells,* in *Cellular Peptidases in Immune Functions and Diseases,* S. Ansorge and J. Langner, Editors. Plenum: 1997, p. 207–216.
23. Valenzuela, A, *et al.*, *HIV-1 envelope gp120 and viral particles block adenosine deaminase binding to human CD26,* in *Cellular Peptidases in Immune Functions and Diseases,* S. Ansorge and J. Langner, Editors. Plenum: 1997, p. 185–192.
24. Pantaleo, G, *et al.*, *Dissociation between syncytia formation and HIV spreading. Suppression of syncytia formation does not necessarily reflect inhibition of HIV infection.* Eur. J. Immunol., 1991. **21**: p. 1771–1774.
25. Tanaka, T, *et al.*, *Enhancement of antigen-induced T-cell proliferation by soluble CD26/dipeptidyl peptidase IV.* Proc Natl Acad Sci U S A, 1994. **91**(8): p. 3082–6.
26. Kameoka, J, *et al.*, *Direct association of adenosine deaminase with a T cell activation antigen, CD26.* Science, 1993. **261**(5120): p. 466–9.

27. Martin, M, *et al.*, *Expression of ecto-adenosine deaminase and CD26 in human T cells triggered by the TCR-CD3 complex. Possible role of adenosine deaminase as costimulatory molecule.* J Immunol, 1995. **155**(10): p. 4630–43
28. Dong, RP, *et al.*, *Characterization of adenosine deaminase binding to human CD26 on T cells and its biologic role in immune response.* J Immunol, 1996. **156**(4): p. 1349–55.
29. Hershfield, MS and Mitchell, BS, *Immunodeficiency diseases caused by adenosine deaminase deficiency and purinenucleoside phosphorylase deficiency,* in *The Metabolic and Molecular Bases of Inherited Disease,* S. C.R., *et al.*, Editors. 1995, p. 1745–1768.
30. Mentlein, R, *et al.*, *Influence of pregnancy on dipeptidyl peptidase IV activity (CD 26 leukocyte differentiation antigen) of circulating lymphocytes.* Eur J Clin Chem Clin Biochem, 1991. **29**(8): p. 477–80.
31. Rich, KC, *et al.*, *CD4+ lymphocytes in perinatal human immunodeficiency virus (HIV) infection: evidence for pregnancy-induced immune depression in uninfected and HIV-infected women.* J Infect Dis, 1995. **172**(5): p. 1221–7.
32. Graziosi, C, *et al.*, *HIV-1 infection in the lymphoid organs.* Aids, 1993. **7**(2): p. S53–8.
33. Gougeon, M-L, *et al.*, *Selective loss of CD4+/CD26+ T cell subset during HIV infection. Res. Immunol,* 1996. **147**, 5–8.

HIV-1 ENVELOPE gp120 AND VIRAL PARTICLES BLOCK ADENOSINE DEAMINASE BINDING TO HUMAN CD26

Agustín Valenzuela,[1] Julià Blanco,[2] Christian Callebaut,[2] Etienne Jacotot,[2] Carmen Lluis,[1] Ara G. Hovanessian,[2] and Rafael Franco[1*]

[1]Departament de Bioquímica i Biologia Molecular
Facultad de Química
Universidad de Barcelona
Marti i Franques 1, 08028 Barcelona, Spain
[2]Unité de Virologie et d'Immunologie Cellulaire
UA CNRS 1157, Institut Pasteur
28 rue du Dr. Roux, 75724 Paris Cedex 15, France

1. ABSTRACT

CD26, known to be the adenosine deaminase (ADA) binding protein, has been implicated in HIV infection. In human B and T cell lines we show that, irrespective of CD4 expression, ^{125}I-labeled ADA binding to CD26 is inhibited by recombinant soluble HIV-1 envelope glycoprotein gp120 and by HIV-1 infectious particles. Overlapping synthetic peptides covering the entire gp120 sequence were tested to map the region in gp120 responsible for ADA binding inhibition. Only peptides in the C3 region significantly inhibited the binding of ADA to CD26. These results indicate that a specific function of gp120 is the inhibition of ADA binding to CD26 in both CD4$^+$ and CD4$^-$ cells. Since the interaction ecto-ADA/CD26 is required for the activation of T cells, it remains possible that HIV particle-mediated blockade of ecto-ADA/CD26 interaction may have significant consequences in the pathogenesis of AIDS disease.

2. INTRODUCTION

Adenosine deaminase (ADA; adenosine aminohydrolase, E.C. 3.5.4.4), an enzyme involved in purine metabolism, catalyzes the hydrolytic deamination of adenosine or 2'-

* Author for correspondence. Fax 343-4021219. E-mail: r.franco@sun.bq.ub.es.

Cellular Peptidases in Immune Functions and Diseases, edited by Ansorge and Langner
Plenum Press, New York, 1997

deoxyadenosine to inosine or 2′-deoxyinosine and ammonia. Although this enzyme is present in all mammalian tissues, it appears to have a major role in the development and function of lymphoid cells. Thus, inherited deficiency of adenosine deaminase leads to the syndrome of severe combined immunodeficiency, which is characterized by abnormal development of lymphoid tissue, lymphopenia and impairment of cell-mediated (T-cell) and humoral (B-cell) immunity (1). In lymphocytes, as in a variety of cell types, the location of the enzyme is mainly cytosolic. However by fluorescence labeling we found that ADA is present on the surface of a high proportion of B lymphocytes and macrophages, and also in some T lymphocytes from peripheral blood (2). In lymphocytes ectoADA binds to the T-cell activation antigen CD26 (3) also known as the serine protease dipeptidyl peptidase IV (DPP IV). The use of specific antibodies against different epitopes in CD26 have suggested a costimulatory role for CD26 in cells triggered via the T-cell receptor (TCR) (for references see (4)), and this event appears to be associated with increased activity of CD4-associated $p56^{lck}$ tyrosine kinase and enhanced phosphorylation of the ζ chain of TCR/CD3 complex (5, 6). One of the functions of ecto-ADA might be the protection of cells against the inhibitory effect of adenosine in the culture medium (7). Furthermore, we have recently shown that the interaction of ADA and CD26 on the cell surface provides a costimulatory signal in the activation events mediated by the TCR/CD3 complex (8). Therefore, the multifunctional role of ecto-ADA and CD26, as enzymes and as costimulatory proteins, may be essential for a proper funtioning of the immune system.

Among other cell surface molecules CD26 has been shown to influence HIV-1 infection (9, 10). Independently of the controversy about the role of CD26 in the entry of T cell-tropic HIV-1 viruses into lymphocytes (9–12), a selective decline of CD26-expressing T cells was found to occur in HIV-1 infected individuals (13–16). Oravecz et al., (9) have found a correlation between cell surface expression of CD26 and high efficiency infection of the human PM1 T-cell line by macrophage-tropic viruses and their cytopathic effect. Recent evidences have also linked the expression of CD26 with the T cell-tropic HIV-1 gp120/gp41-induced cell-killing by apoptosis (17). The implication of CD26 in the HIV entry and its cytopathic effect appears to be related to the V3 loop region in the gp120 (9, 18). In this regard, a V3 loop peptide was shown to bind synthetic CD26 peptide containing the catalytic consensus sequence implicated in the DPP IV activity of CD26 (19). However, there is still no evidence of a direct interaction between gp120 and CD26. In this report we show that, irrespective of CD4 expression, three recombinant gp120 preparations inhibit the binding of ADA to CD26. Also HIV-1 viral particles block ADA binding to human CD26. This HIV-induced disruption of the ADA/CD26 interaction might have functional consequences in lymphocytes and in other cell types where the cell-surface ADA/CD26 complex is present.

3. MATERIALS AND METHODS

3.1. Antibodies

MAb OKT4A (Ortho Diagnostics Systems) blocks the binding of gp120 to CD4 and thus HIV infection. Three different mAbs against the gp120 of the HIV-1 Lai/IIIB isolate were used: mAb N11–20 which is directed against the V3 loop (ANRS/Pasteur-Merieux); mAb 110–1 which recognizes an epitope in the sequence 489–511; and mAb 110–4 which recognizes an epitope located between amino acids 303 and 323 corresponding to V3 loop

(20). The anti-human CD26 mAb TA5.9-CC1–4C8 (a generous gift from Dr. E. Bosmans, Eurogenetics, Tessenderlo, Belgium) was used to block the ADA binding site of CD26 (21).

3.2. ^{125}I-Labeled ADA Binding to CD26

Aliquots of 0.5x10^6 cells were preincubated for 15 minutes at 37 °C in the presence of different reagents (unlabeled ADA, antibodies, gp120, peptides or HIV preparations) and then incubated with the desired amounts of ^{125}I-labeled ADA (Bolton-Hunter reagent) in a final volume of 200 µl. Incubations were performed in non supplemented RPMI 1640 medium, except for HIV experiments, where supplemented medium was used. After incubation at 37°C for 90 minutes ^{125}I-labeled ADA bound to cells was also monitored by polyacrylamide gel electrophoresis (SDS-PAGE). Following SDS-PAGE, gels were fixed, dried and radioactivity associated with the 43 kDa ^{125}I-labeled ADA band was either quantified in a Phosphorimager (Molecular Dynamics, Sunnyvale, CA, USA) or visualized by autoradiography using Agfa-Curix-Rp2 films.

4. RESULTS

4.1. Effect of gp120 from HIV-1 on ^{125}I-Labeled ADA Binding to CD26 in Human Cells

^{125}I-labeled ADA binding to surface CD26, detected by SDS-PAGE and subsequent autoradiography, was found to be inhibited significantly by 100 nM of three preparations of recombinant soluble gp120, corresponding to HIV-1 isolates IIIB (or Lai, construct made by Dr. I. Jones), MN (Dr. N. Haigwood, Chiron Corporation) and SF2 (Dr. C. Bruck). There was up to 78% inhibition in Jurkat cells, which express low CD26 levels, and up to 60% in Jurkat #11 (a generous gift from C. Morimoto), which express high CD26 levels, and SKW6.4 cells, which are CD4$^-$ and express high CD26 levels. In all cells analyzed inhibition of ^{125}I-labeled ADA binding by TA5.9 mAb, which binds to the ADA binding site of CD26, was similar to that exerted by unlabeled 4 µM ADA (75–90%), thus confirming that under our experimental conditions, CD26 is the only ^{125}I-labeled ADA receptor on the surface of these cells. The inhibition of ^{125}I-labeled ADA binding to Jurkat cells by gp120 was not affected significantly by the mAb OKT4A, known to block the binding of gp120 to the CD4 receptor (23). This observation confirmed that the binding of gp120 to CD4 is not a necessary event for the interaction of gp120 with CD26. It could be argued that gp120-mediated inhibition of ^{125}I-labeled ADA binding to CD26 might be the consequence of a direct interaction between gp120 and ADA. This was however unlikely because no interaction between purified ADA and gp120 was detected using the surface plasmon resonance technique, which is useful for detecting molecular interactions (BIAcore^R).

4.2. Characterization of the gp120 Region Responsible for the Inhibition of ^{125}I-Labeled ADA Binding to CD26

In order to determine which region of gp120 was responsible for the competence in ^{125}I-labeled ADA binding, a series of peptides of 14–22 amino acids in length and covering the complete gp120 IIIB sequence was assayed for their ability to inhibit the binding of ^{125}I-labeled ADA to CD26. In these assays, to avoid possible interferences by the high-affinity interaction that is established between gp120 and human CD4, the CD4$^-$ human cell line SKW6.4 was selected. The peptides that significantly inhibited ^{125}I-labeled ADA binding to CD26, measured by cell harvesting and radioactivity counting (See Methods), were those corresponding to the C3 region, located between V3 and V4 loops, and that corresponds to the amino acid residues 335 to 375 of gp120 IIIB. These peptides, which markedly inhibited ^{125}I-labeled ADA binding (>70%), and those covering the V3 loop region were also assayed in experiments of ^{125}I-labeled ADA binding analyzed by SDS-PAGE (See Methods). Once again, the two peptides covering the domain of gp120 comprised between V3 and V4 loops, as well as the recombinant gp120 protein, were able to reduce markedly the binding of ^{125}I-labeled ADA to CD26, whereas peptides from the V3 loop did not exert any significant effect.

4.3. HIV-1 Particles Inhibit ^{125}I-Labeled ADA Binding to CD26 on Jurkat Cells

SDS-PAGE analysis demonstrated that preincubation of Jurkat cells with viral stocks inhibited specific ^{125}I-labeled ADA binding. This inhibition is due to the presence of HIV particles rather than to that of free gp120 molecules since both HIV-1 Lai stock and pellet inhibited to a similar extent whereas the supernatant after ultracentrifugation was not able to induce any significant binding inhibition. Moreover, the inhibitory effect of HIV-1 Lai on ^{125}I-labeled ADA binding was abolished when virus preparation was inactivated by heating at 56°C for 30 min, thus indicating that the inhibition required a functional virus particle. The potency of HIV-1 particles as blockers of ADA binding to CD26 was very high as indicated by an IC_{50} value of less than 1 nM. The possibility of a HIV-induced down-regulation of CD26 was discarded since similar levels of cell surface CD26 were found, by FACS analysis, in cells before and after incubation with HIV. The 64% inhibition of the specific binding of ^{125}I-labeled ADA to CD26 induced by HIV-1 Lai particles was not affected by preincubating cells with the anti-CD4 mAb OKT4A (1 μg/ml), which completely blocked HIV-1 Lai infection. In contrast both anti-gp120 mAbs tested (against V3 loop and against C5 region) were able to reverse the inhibitory effect of HIV-1 Lai particles upon ^{125}I-labeled ADA binding to CD26 (Figure 1), irrespective of their ability to neutralize HIV-1 Lai infection: mAb 110–4 against the V3 loop is a neutralizing antibody and, accordingly, it blocked virus infection; mAb 110–1 directed against an epitope at the C-terminus of gp120 is not a neutralizing antibody, and, accordingly, it did not affect HIV infection. These results confirm that HIV inhibits the binding of ADA to CD26 through its envelope glycoprotein gp120, and that this may occur whether or not HIV binds to its principal CD4 receptor molecule.

5. DISCUSSION

The results presented in this paper demonstrate that recombinant soluble HIV-1 glycoprotein gp120 inhibits ^{125}I-labeled ADA binding to CD26 in CD4$^+$ and CD4$^-$ human cell lines. In fact, three different preparations of recombinant gp120 were able to displace ^{125}I-labeled ADA binding to CD26 in all human cells assayed thus suggesting the occurrence of an interaction between gp120 and cell surface CD26. A lack of binding of gp120 to soluble CD26 has been reported (12, 24); these results may be due to non-optimal conforma-

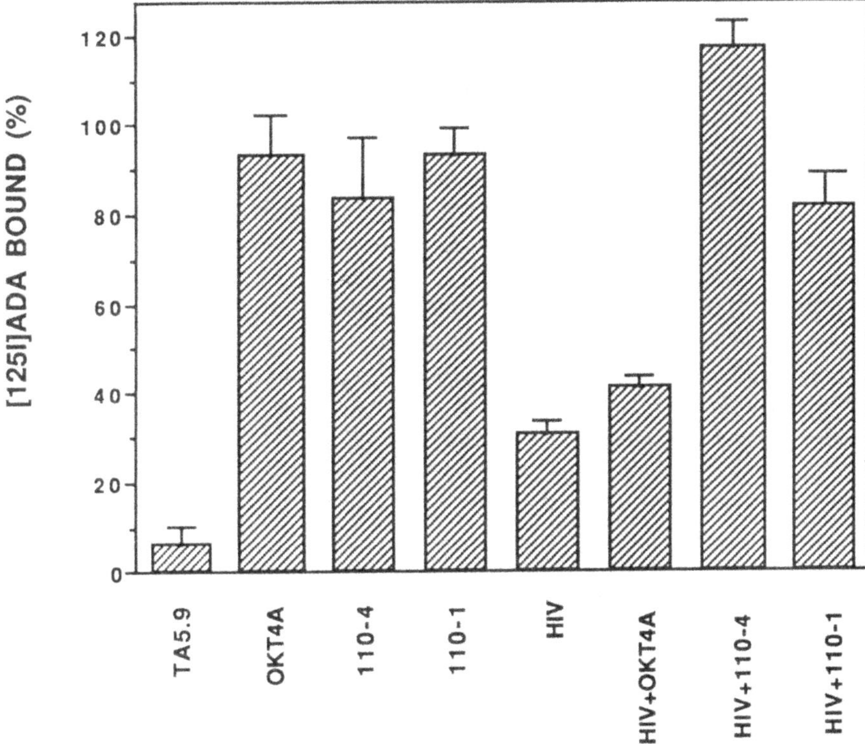

Figure 1. HIV-1 Lai Inhibition of specific ^{125}I-labeled ADA binding to CD26 is independent of HIV infection in Jurkat cells. Different mAbs were assayed for their ability to reverse the inhibiton of 15 μM ^{125}I-labeled ADA binding induced by viral particles. Final concentrations in the different samples were: 1 μg/ml (TA5.9 mAb), 1 μg/ml (OKT4A mAb) and 1/200 dilution of the ascitic fluid (110–1 and 110–4 mAbs). Data given in percentage of specific binding come from quantitation of the radioactivity after SDS/PAGE (mean ± SEM from triplicates). TA5.9 mAb is directed against the ADA binding site in CD26; for this reason it blocks ADA binding to CD26 by more than 90%. 110–4, a mAb against gp120, blocks HIV-1 infection whereas 110–4 does not. OKT4A, a mAb against CD4 blocks HIV-1 infection. ADA was not able to block HIV-1 infection.

tion of the soluble CD26 or to the participation of other cell surface molecules in a complex interaction.

Mapping of the gp120 domain responsible for the inhibition of ADA binding to CD26 was performed in the CD4$^-$ B cell line SKW6.4, using synthetic peptides covering the entire gp120 protein sequence. Peptides corresponding to the C3 region of gp120 were able to markedly inhibit the binding of ^{125}I-labeled ADA to CD26, thus suggesting that this region located between the V3 and V4 loop may play an important role in the interaction of gp120 with CD26. This does not rule out the involvement of additional regions in HIV-1 envelope glycoprotein for the interaction between gp120 and CD26. Indeed, a V3 loop peptide corresponding to the HIV-1 MN isolate has been shown to bind a synthetic peptide containing the consensus catalytic domain of CD26 (19). In our experiments however, synthetic peptides corresponding to the V3 loop of gp120 did not affect the binding of ADA to CD26. Therefore, in spite of their potential capacity to bind CD26, V3 loop peptides would not trigger major conformational changes in the CD26 structure to interfere with the specific binding of ADA. Taken together these observations suggest that two specific interactions may occur between gp120 and CD26, one between the catalytic domain of CD26 and the V3 loop of gp120, and a second contact between C3 region in gp120 and a region in CD26 which is closely associated with the ADA binding site. It is interesting to note that the ADA binding site is not located in the vicinity of the catalytic site of CD26, since ADA binding does not affect the dipeptidyl peptidase activity of CD26 (25). The finding that mAbs specific for two domains in gp120 (V3 loop and C5 region) abolished the inhibitory effect of HIV on ADA binding to CD26, is consistent with the involvement of different domains of gp120 for a multiple interaction bewteen gp120 and CD26. Moore et al., (26) have probed the structure of the gp120 with a panel of mAbs and have found that C3 domain is not accessible in the soluble gp120 molecule. If this is also the case for the gp120 complexed to gp41 on HIV particles, it is then possible that the interaction of the V3 loop of gp120 with the catalytic domain of CD26, and/or with other cell surface components such as heparan sulfate proteoglycans (27), may induce conformational changes making the C3 domain accessible for interaction with CD26. On the other hand, the observation that virus particles were 10 to 100 fold more efficient than soluble gp120 to inhibit the binding of ADA to CD26, is consistent with a higher accessibility of the C3 domain in the gp120 presented by the HIV-1 particles for interactions with CD26.

The lack of involvement of CD4 in the HIV-1- or gp120-induced inhibition of ADA binding to CD26 has been demonstrated by using CD4$^-$ cells, and by blocking gp120/CD4 interaction using mAb OKT4A. Thus, the binding of gp120 to CD4, and consequently the potential post-CD4-binding conformational modifications in the structure of gp120, may not be essential for the interaction between gp120 and CD26. Consistent with the fact that mAb OKT4A prevents viral entry but does not affect the interaction of HIV-1 with the ADA binding site of CD26, saturation of this binding site with excess ADA has no apparent effect on HIV-1 infection. Therefore, the ADA binding site in CD26 is not directly implicated in HIV infection. This does not rule out the involvement of other domains of CD26 in HIV-1 entry process, in which chemokine receptors play a central role (28, 29). In view of this, and since many chemokines are potential substrates of the DPPIV activity of CD26, it is plausible to consider that CD26 may have a role in HIV-1 infection through its interaction with certain chemokines, as it has been recently proposed (30).

Dong et al., (7) have shown that adenosine inhibits T cell proliferation and IL-2 production induced by various stimuli in CD26 negative cells, which is a consequence of the lack of ecto-ADA expression. In contrast, cells expressing CD26 and thus cell-surface ADA, are resistant to the inhibitory effects of adenosine. In view of these findings, it has been suggested that ADA bound to cell surface CD26 could be responsible for the reduction of the local concentration of adenosine (7). Besides its role in the metabolism of adenosine, Martin et al. (8) have demonstrated that surface ADA in peripheral blood T cells, is a costimulatory molecule since the interaction ADA/CD26 is essential for activation via the T-cell receptor-CD3 complex. Blockade of the ADA/CD26 interaction induced by HIV gp120 may result in a defective ADA costimulatory effect and a modification of signaling through CD26, thus perturbing T cell activation. Consequently, a new pathophysiological mechanism of HIV is to block the interaction between ecto-ADA and CD26 established on the plasma membrane of lymphocytes and other cell types. The inhibition of ADA binding to CD26 by soluble gp120 and HIV particles is consistent with the elevated levels of serum ADA reported in AIDS patients (31, 32) and might have negative consequences in T cell activation events (7, 8). The possibility of ecto–ADA having a specific role in processes such as the ontogenesis and maturation of the cells of the immune system would give insight about the selective damage that lymphoid tissues and cells suffer in the severe combined immunodeficiency caused by a congenital ADA–defect. Therefore HIV disruption of ADA binding to cells might affect a variety of signalling events that could alter the maturation and activation of lymphocytes thus contributing to the development of AIDS.

6. ACKNOWLEDGMENTS

Supported by a joint (Echevarne Fundation and Spanish Ministry of Education) PETRI Grant (PTR92/0047) and by Grants from *Fondo de Investigaciones Sanitarias de la Seguridad Social* (no. 87/1389 and 91/0272), from CIRIT-CICYT (QFN93/4423), from Institut Pasteur and "*Agence Nationale de la Recherche sur le SIDA*" (ANRS). AV is the recipient of a grant from CIRIT, *Generalitat de Catalunya*, JB of a post-doctoral fellowship from Spanish *Ministerio de Educacion y Ciencia*. CC is the recipient of a fellowship from the French *Ministère de l'Enseignement Superieur et de la Recherche*. EJ is supported by *Fondation Mondiale Recherche et Prévention SIDA* France. We thank Bernard Krust for advice and critical review of this manuscript and Sara Sarrio for the technical help in the analysis of molecular interactions by the biosensors technology. We acknowledge the Medical Research Council AIDS Directed Programme Reagent Project (Potters Bar, U.K.) and Drs. I. Jones, N. Haigwood and C. Bruck for kindly providing the gp120-related reagents.

7. REFERENCES

1. N. M. Kredich, M. S. Hershfield, in *The Metabolic Basis of Inherited Disease* S. C.R., B. A.L., S. W.S., V. D., Eds. 1989) pp. 1045–1075.
2. J. M. Aran, D. Colomer, E. Matutes, J. L. Vives-Corrons, R. Franco, *J. Histochem. Cytochem.* **39**, 1001–1008 (1991).
3. J. Kameoka, T. Tanaka, Y. Nojima, S. F. Schlossman, C. Morimoto, *Science.* **261.**, 466–469 (1993).
4. B. Fleischer, *Immunol. Today* **15**, 180–184 (1994).
5. Y. Torimoto, et al., *J Immunol* **147**, 2514–7 (1991).
6. H. W. Mittrücker, C. Steeg, B. Malissen, B. Fleischer, *Eur. J. Immunol.* **25**, 295–297 (1995).
7. R.-P. Dong, et al., *J.Immunology.* **156.**, 1349–1355. (1996).
8. M. Martin, J. Huguet, J. J. Centelles, R. Franco, *J.Immunol.* **155.**, 0001–0014 (1995).
9. T. Oravecz, et al., *Nature Medicine.* **1.**, 919–926 (1995).
10. C. Callebaut, E. Jacotot, B. Krust, A. G. Hovanessian, *Science* **262**, 2045–2052 (1993).
11. C. Callebaut, E. Jacotot, B. Krust, A. Hovanessian and C. Broder, et al., *Science* **264**, 1156–1165 (1994).
12. C. Morimoto, et al., *Proc. Natl. Acad. Sci. USA* **91**, 9960–9964 (1994).
13. A. De Pasquale, L. Ginaldi, P. Limoncelli, D. Quaglino, *Acta Haemat.* **81**, 19–21 (1989).
14. M. Valle Blazquez, et al., *J.Immunology.* **149**, 3073–3077 (1992).
15. G. Vanham, et al., *J. AIDS* **6**, 749–757 (1993).
16. M.-L. Gougeon, et al., *Res. Immunol.* **147**, 5–8 (1996).
17. E. Jacotot, et al., *Virology* **223**, 318–330 (1996).
18. C. Callebaut, A. G. Hovanessian, *Res.Virol.* **147**, 67–69 (1996).
19. L. T. Chin, et al., *Immunology Letters* **44**, 25–30 (1995).
20. P. S. Linsley, J. A. Ledbetter, E. Kinney-Thomas, S. L. Hu, *J. Virol.* **62**, 3695–3702 (1988).
21. J. Blanco, et al., *Exp. Cell Res.* **225**, 102–111 (1996).
22. A. G. Laurent-Crawford and A. G. Hovanessian, *J. Gen. Virol.* **74**, 2619–2628 (1993).
23. Q. J. Sattentau, R. A. Weiss, *Cell* **52**, 631–633 (1988).
24. Y. H. Wang, A. H. Davies, I. M. Jones, *Virology* **208**, 142–146 (1995).
25. I. De Meester, et al., *Eur J Immunol* **24**, 566–70 (1994).
26. J. P. Moore, Q. J. Sattentau, R. Wyatt, J. Sodroski, *J Virol* **68**, 469–84 (1994).
27. G. Roderiquez, et al., *J.Virol.* **69.**, 2233–2239 (1995).
28. T. Dragic, et al., *Nature* **381**, 667–673 (1996).

29. Y. Feng, C. C. Broder, P. E. Kennedy, E. A. Berger, *Science* **272**, 872–876 (1996).
30. T. Oravecz, M. Pall, M. A. Norcross, *J. Immunol.* **157**, 1329–1332 (1996).
31. H. C. Lane, et al., *N Engl J Med* **313**, 79–84 (1985).
32. J. L. Murray, et al., *Blood* **65**, 1318–24 (1985).

FURTHER CHARACTERIZATION OF DPP IV-β, A NOVEL CELL SURFACE EXPRESSED PROTEIN WITH DIPEPTIDYL PEPTIDASE ACTIVITY

Julià Blanco,* Etienne Jacotot, Christian Callebaut, Bernard Krust, and Ara G. Hovanessian

Unité de Virologie et Immunologie Cellulaire
UA CNRS1157, Institut Pasteur
28 rue Dr. Roux, 75724, Paris Cedex 15, France

1. ABSTRACT

By using a CD26 negative human lymphoblastoid cell line (C8166), here we describe the characterization of a cell-surface protein which manifests CD26-like dipeptidyl peptidase IV (DPP IV) activity. This protein, referred to as DPP IV-β, shows a higher K_M value for Gly-Pro-pNA than CD26 (0,31 mM compared to 0,11 mM, respectively). In addition, DPP IV-β was found not to bind ^{125}I-labeled adenosine deaminase (a property of human CD26). Gel filtration experiments using extracts from C8166 and MOLT4 (a CD26 positive human T cell line) cells, revealed that the apparent molecular mass of DPP IV-β is 82 kDa, whereas that of CD26 is 110 kDa. In order to conveniently differentiate both activities, a new family of inhibitors, that selectively blocks peptidase activity associated to CD26, has been developed.

2. INTRODUCTION

The DPP IV activity belongs to the family of dipeptidyl aminopeptidases, which includes four activities (DPP I, DPP II, DPP III, and DPP IV), associated to least to four different serine proteases[1]. These activities are characterized by their ability to release N-terminal dipeptides from a variety of small proteins. Dipeptidyl peptidases show different substrate specificity and cellular location, suggesting different functions of each activity in peptide

* Author for correspondence. Fax: 33-1-40613012; E-mail: jblanco@pasteur.fr.

Cellular Peptidases in Immune Functions and Diseases, edited by Ansorge and Langner
Plenum Press, New York, 1997

processing. DPP I and DPP II are lysosomal enzymes with an acidic optimal pH, DPP III is a less known cytosolic enzyme, probably involved in intracellular peptide degradation[2].

Among dipeptidyl peptidase activities, DPPIV is the only activity known to be located on the outer side of plama membrane, thus acting as an actoenzyme. In man, DPP IV activity has been associated to a 105 kDa cell surface protein, known to be identical to the adenosine deaminase binding protein (ADAbp), and to the T cell activation antigen CD26[3–5]. The deduced amino acid sequence of DPP IV/CD26 in man and other different species[6–10], revealed the presence of a conserved sequence Gly-Trp-Ser-Tyr-Gly towards the COOH-terminal end which corresponded to the consensus sequence Gly-Xaa-Ser-Xaa-Gly representing the active site of serine proteases[9]. CD26/DPP IV is constitutively expressed on many cell types, showing high levels of expression in brush border epithelia from kidney, intestine and placenta[11]. The promoter of the human CD26 gene contains the sequence characteristic of a housekeeping gene, but has the capacity to regulate the expression of CD26 mRNA in a tissue specific fashion[12]. This is the case of the peripheral blood T lymphocytes, where the expression of CD26 is highly enhanced upon T-cell activation. In T cells, CD26 is physically associated with the membrane-linked tyrosine phosphatase CD45[13], and modulation of CD26 with an anti-CD26 monoclonal antibody or with its natural ligand, adenosine deaminase, leads to a comitogenic effect on anti CD3 T cell activation[14,15]. In the immune system, CD26 has been also reported to play two controversed roles in HIV infection, on one hand CD26 may act as a cofactor for viral entry into permissive cells[16,17], and on the other hand CD26 could be the cellular target of HIV-1 tat protein[18,19].

Recently two novel proteins with DPP IV activity have been described. Duke-Cohan et al[20,21] have reported a 175 kDa membrane bound form, which is expressed by different epithelial and lymphoblastoid cell lines, and can be secreted from activated peripheral blood T cells. Our laboratory has recently reported the existence of a CD26-like cell-surface 82 kDa protein which has been failed to be detected before[22]. Characterization of some new properties of this CD26-like DPP IV activity which we refer to as DPP IV-β are described here.

3. RESULTS

3.1. C8166 Cells Exhibit a DPP IV Activity Different from CD26

The expression of CD26 in different $CD4^+$ T cell lines was investigated by immunoblotting using mAb 1F7, specific for human CD26, and by determining DPP IV activity on the cell surface of intact cells. Among several $CD4^+$ cell lines we tested Jurkat parental cells (clone P32), Jurkat cells transfected to express enhanced levels of recombinant CD26 (clone J37, both Jurkat clones were kindly provided by Dr. C. Morimoto, Boston), C8166 and its clonal derivative M8166 cells, and MOLT4 cells. By immunoblotting analysis, both of the Jurkat cell lines and MOLT4 cells were found to express the 110 kDa CD26 protein, and as it was expected, the level of CD26 was much higher in the transfected Jurkat cell line J37 (see Table I). However, the expression of CD26 was not detectable in C8166 and M8166 cells, even after overexposure of the autoradiograph. This lack of CD26 protein was in accord with the previously reported lack of CD26 mRNA expression in these cells investigated by the reverse-transcriptase/polymerase-chain-reaction technique (RT/PCR)[23].

In contrast to the lack of CD26 expression, when we determined the DPPIV activity (by measuring GPpNA cleavage) on the cell surface of C8166 or M8166 cells we found activity levels similar to those found in parental Jurkat or MOLT4 cells (Table I), suggest-

Table I. Lack of correlation between CD26 expression and DPP IV activity in C8166 cells

	Jurkat P32	Jurkat J37	C8166	M8166	MOLT4
CD26 (1F7 reactivity)	+	+++	−	−	++
DPP IV activity	+	+++	+	+	++

ing to us the presence of a DPP IV activity on the surface of these CD26⁻ cells. In order to confirm this hypothesis, the expression of different cell-surface peptidases by CD26⁻ C8166 cells was studied. These experiments were carried out in the absence or presence of 25 μM Lys-[Z(NO_2)]-pyrrolidide (kindly provided by Dr. A. Barth, Halle, Germany), a potent specific inhibitor of the DPP IV activity[24,25]. Peptidase activities were detected as follows: DPP IV by the cleavage of Gly-Pro- and Ala-Pro-pNA, DPP I by the cleavage of Gly-Arg-pNA, DPP II by the cleavage of Lys-Ala- and Ala-Ala-pNA, proline endopeptidase by the cleavage of Succinyl-Ala-Pro-pNA, and Ala-, Pro- and Arg-peptidases by the cleavage of Ala-, Pro- and Arg-pNA. It should be noted that, the DPP II substrates Lys-Ala- and Ala-Ala-pNA can also be cleaved by DPP IV but at a much lower efficiency[26].

Determination of p-nitroanilide release from DPP IV substrates clearly showed the presence of a CD26-like peptidase detectable on the cell surface of intact cells using classical substrates, GPpNA or APpNA, for DPP IV activity. As the cleavage of these dipeptides was inhibited by more than 85 % at 25 μM of Lys-[Z(NO_2)]-pyrrolidide, it is possible to suggest that C8166 cells express a cell-surface protein which has DPP IV activity. We called this protein DPP IV-β. The specificity of Lys-[Z(NO_2)]-pyrrolidide upon DPP IV activities was demonstrated by the lack of its action on other cell-surface peptidases such as Ala-, Pro-, Arg- and Gly-Arg-peptidase. Lys-Ala-pNA which is a substrate of DPP II can also be cleaved by DPP IV. However the degree of cleavage of Lys-Ala-pNA by CD26 is at a much lower efficiency than that of Gly-Pro- or Ala-Pro-pNA. This was clearly demonstrated when compared to an immunoprecipitated CD26 preparation. Interestingly, the degree of cleavage of Lys-Ala-NH-Np compared to Gly-Pro- or Ala-Pro-pNA was also lower by the cell-surface peptidase activity on C8166 cells. Thus suggesting that DPP IV-β and CD26 might manifest similar catalytic properties. Consistent with this, the cleavage of Lys-Ala-pNA was inhibited only partially by Lys-[Z(NO_2)]-pyrrolidide in both cases.

Furthermore, as we have previously reported[22], the determination of optimum pH for this DPP IV-β activity on the C8166 cell surface showed a optimum value of 7.5–8, in strong discrepancy with DPPII and close to optimum pH of CD26. We have also characterized this novel DPPIV activity by determining the IC_{50} of different general protease and DPPIV specific inhibitors. General protease inhibitors (Aprotinin, EDTA and PMSF) do not show important effect on DPP IV-β or CD26 activities, only PEFABLOC (4-(2-aminoethyl)-benzenesulphonyl fluoridehydrochloride) showed in both cases an inhibitory effect at 1 mM. Among the specific DPPIV inhibitors used (Lys-[Z(NO_2)]-pyrrolidide, Lys-[Z(NO_2)]-thiazolidide and Ile-pyrrolidide), no major differences were observed in the inhibitory effect on CD26 and DPP IV-β activities.

3.2. DPP IV and DPP IV-β Activities Are Associated to Different Proteins

Since DPP IV activity associated to both CD26 and DPP IV-β, are very similar, we studied other characteristics that could differentiate both proteins, as the ability to bind adenosine deaminase (ADA), a well described characteristic of human CD26. ADA binding was assessed by determining the binding of ^{125}I labelled bovine ADA to the surface of a CD26$^+$ cell line (MOLT4) and to the CD26$^-$ C8166 cells. After incubation with labelled ADA, cells were washed and ADA bound determined by SDS-PAGE and quantitative autoradiography. ADA binds specifically to CD26 expressed on the surface of MOLT4 cells, as demonstrated by the almost complete ADA binding inhibition induced by the mAb TA5.9 (kindly provided by Dr. Bosmans, Belgium) directed against the ADA binding site of CD26[19]. ADA fails to bind to C8166 cells, demonstrating that DPP IV-β expressed on the cell surface of these cells lacks the ability to bind ADA.

We determined also the molecular mass of DPP IV-β by gel filtration chromatography loading nuclei free 1% Triton X-100 extracts from C8166 cells onto a Superose 6 column (1.6x50 cm). Peptidase activities were monitored by the cleavage of Gly-Pro-pNA at pH 7.5 (in these conditions DPPIV and DPPII activities are detected), and by the cleavage of Ala-Ala-pNA at pH 5.5 (only DPPII activity is detected). Two active peaks were detected using Gly-Pro-pNA as substrate with molecular weights of 82 and 70 kDa, respectively. Ala-Ala-pNA cleavage demostrated that the 70 kDa peak corresponded to DPPII, thus showing that DPPIV activity of C8166 cells, DPP IV-β is associated to a 82 kDa protein.

Since we used MOLT4 cells in several studies to compare CD26 and DPP IV-β activities, we were interested to determine whether these cells express DPP IV-β in addition to CD26. We used the above described chromatographic method to separate CD26, DPP IV-β and DPPII using MOLT4 extracts. Activities were monitored as above, and in addition fractions were immunoprecipitated with the anti CD26 mAb 1F7, which recognizes CD26 but not DPP IV-β. Using Gly-Pro-pNA as substrate we found also two active peaks, corresponding to a molecular mass of 110 kDa and 82 kDa. Immunoprecipitation with 1F7 confirmed that CD26 corresponds to the 110 kDa protein, being the 82 kDa protein most probably DPP IV-β. Due to the low expression of DPPII in MOLT4 cells, compared to C8166 cells, this activity is weakly observed at pH 7.5, but by using using Ala-Ala-pNA as substrate at pH 5.5, this activity was observed at the expected molecular weight (70 kDa).

3.3. Kinetic Differences between CD26 Peptidase Activity and DPP IV-β

In order to further characterize DPP IV-β and to demonstrate that the 82 kDa peak found in MOLT4 cell extract corresponded to the same DPP IV-β protein found in C8166 cells, we have determined the K_M for Gly-Pro-pNA of different partially purified preparations of both enzymes, and we used a new family of specific inhibitors that inhibit selectively CD26 activity.

Michaelis constants of both enzymes for Gly-Pro-pNA were determined in PBS containing 0.1% Triton X-100 using Gly-Pro-pNA concentrations ranging between 0,001 and 10 mM. K_M value of DPP IV-β purified from C8166 cells was found to be slightly higher than K_M of CD26 purified from MOLT4 cells (0.31 mM compared to 0.11 mM respectively). K_M value of the 82 kDa active peak present in MOLT4 cells was 0.27 mM in good agreement with the value found for DPP IV-β and significantly different from value observed for CD26. The same result was obtained using preparations of CD26 or DPP IV-β partially purified from two other human lymphoblastoide cell lines CEM and Jurkat (K_M values were 0.25–0.27 mM for DPP IV-β and 0.11–0.13 mM for CD26).

However the most convenient tool to differenciate both activities is a new family of DPPIV inhibitors developed in collaboration with Dr. C N'Guyen and Dr. M Wakselman (Université de Versailles, France). This family of irreversible inhibitors includes a cyclic pseudopeptide (Cyclo(NeLys(NaH)-Pro-m-aB-[$CH_2S^+Me_2$]-Gly-Gly 2 CF_3COO^-) that, when used at 1 μM, completely inhibit CD26 with only a weak effect on DPP IV-β activity. Inhibition studies using partially purified fractions of CD26 and DPP IV-β from C8166, MOLT4, CEM and Jurkat cells yield a IC_{50} value of 0.11–0.17 μM for CD26 peptidase activity and 2.6–2.9 μM for the 82 kDa peak of MOLT4, CEM and Jurkat cells, as well as for DPP IV-β from C8166 cells, thus confirming that this 82 kDa active peak is DPP IV-β. DPPIV activity present on the cell surface of these cells is therefore, a mixture of the well described peptidase activity of CD26 and the new MOLT4 cells is a mixture of CD26 and the new DPP IV-β. It is interesting to note that in all cell lines tested, as well as in resting PBMC, DPP IV-β activity was more important than CD26 peptidase activity, which represents only 18–35% of total activity.

3.4. Purification of DPP IV-β

With the purpose to prepare the large scale purification of DPP IV-β, a strategy to purify this activity has been developed. First we have biotinylated C8166 cell surface proteins by the method described by Cole et al[27]. Since biotinylation did not modify DPP IV-β activity, this step allowed us to follow DPP IV-β purification by monitoring its peptidase activity and by blotting active fractions, which were further revealed using streptavidin-horseradish peroxidase and chemoluminiscence detection.

Extracts from surface biotinylated C8166 cells were first loaded onto a MonoQ column (1x10 cm) equilibrated with Tris 20 mM, NaCl 50 mM, Triton X-100 0.1%, pH 7.4 buffer. Bound proteins were eluted with a linear NaCl gradient (50–500 mM). As expected, two active peaks were found: DPP II, that eluted at 150 mM NaCl and DPP IV-β, that eluted at 250 mM NaCl. Fractions containing this latter activity were loaded onto the above described Superose 6 column and eluted DPP IV-β activity was assayed for its binding to different lectins. Among the lectins tested (WGA, PHA-E, PHA-L, and concanavalin-A), wheat germ agglutinin (WGA) showed a high affinity binding to DPP IV-β and was then used in the last purification step. Active fractions eluted from lectin column, retaining 22% of the original activity, were analyzed by strptavidin-peroxidase blotting, showing a major band with a molecular mass of 80 kD, in agreement with the molecular mass of DPP IV-β observed in previous gel filtration experiments.

4. DISCUSSION

By the use of $CD26^-$ T lymphoblastoid C8166 cells, here we were able to demonstrate the existence of a CD26-like ectopeptidase manifesting a typical DPP IV activity, the DPP IV-β. Previously reported studies[22] showed that the profiles for subtrate and inhibitor molecules were almost indistinguishable for both CD26 and DPP IV-β. In addition, both proteins manifested optimum dipeptidyl peptidase activity at pH 7.5, and resisted proteinase K digestion. The only distinct property that we found, was the binding of ADA to CD26 but not to DPP IV-β. Among five CD26 mAbs, only mAb 4H12 stained C8166 cells, thus suggesting that it cross reacts with a cell surface antigen which shares a common epitope with CD26, it is most likely therefore, that this antigen is DPP IV-β. However, we have no direct evidence for the binding of this mAb to DPP IV-β.

Recently, CD26-like proteins have been described by several investigators in human cells and serum. Scanlan et al[28] have described a 95 kDa cell-surface antigen referred to as fibroblast activation protein α (FAPα), which shows 48% amino acid sequence identity to CD26. In fibroblasts, FAPα could exist in an oligomeric form with CD26 at the cell membrane. However, FAPα has been reported to manifest very little DPP IV activity[29]. Another CD26-like protein secreted by activated T lymphocytes, called DPPT-L, has been described by Duke-Cohan et al.[20,21] This latter is both a membrane bound and a soluble protein of 175 kDa which cross reacts with some mAbs against CD26, and also manifests DPP IV activity. Interestingly, this 175 kDa protein does not bind ADA. However, it is unlikely that DPP IV-β corresponds to this 175 kDa protein, since gel filtration experiments showed that DPP IV-β is a protein with a molecular mass of 82 kDa. In addition, MOLT4 cells, which express important levels of DPP IV-β, do not express the 175 kDa DPPT-L[21]. Furthermore, DPP IV-β as DPP IV/CD26, is not affected by EDTA even at 100 mM, whereas the DPP IV activity of the 175 kDa protein is inhibited by EDTA with an IC_{50} value of 8.9 mM[21]. No detectable DPP IV activity was observed in the supernatant of C8166 cells incubated several hours in NaCl/P_i containing 1% BSA. Thus DPP IV-β is membrane associated and does not appear to be liberated in the culture medium. Indeed, the cell-surface DPP IV activity in intact C8166 cells was not reduced in cells extensively washed with PBS containing 100 mM EDTA or after several hours of incubation of washed cells in PBS.

The DPP IV activity of CD26 has been suggested to be essential for determining the half-life and activity of some natural substrates, such as substance P, chorionic gonadotropin hormone, monomeric fibrin and glucose-dependent insulinotropic polypeptide[30,31]. Other N-terminus Xaa-Pro potential DPP IV substrates are some cytokines, such as IL-1β, IL-2, tumour necrosis factor β (TNF-β), granulocyte colony-stimulating factor (G-CSF) and the different chemokines that modulate HIV infection[32]. However, there is no evidence for the interaction of these cytokines with DPP IV under physiological conditions[11]. The role of the new DPP IV activities may be complementary to that of CD26. In this regard, the differences reported here between the affinity of DPPIV/CD26 and DPP IV-β for some small substrates (Gly-Pro-pNA) or inhibitors (Cyclo(NeLys(NaH)-Pro-m-aB-[$CH_2S^+Me_2$]-Gly-Gly 2 CF_3COO^-) probably reflect differences on the catalytic site of both enzymes. The effect of these differences on the natural substrate selectivity is a crucial point that can help to better understand the physiological role of DPP IV activity. An attractive possibility is to think that different DPP IV may be selective for different bioactive peptides, even if these activities can share some sustrates, as is suggested by the CD26 and DPPT-L T cell costimulatory role[20].

In view of the different functions of CD26 and its DPP IV activity involving different biochemical, immunological and viral processes, it remains essencial to characterize the DPP IV-β, determine its expression on different types of cells and define its role in comparison to CD26. Purification and complete elucidation of its sequence will provide the means to answer some of these questions.

5. REFERENCES

1. Alba, F. et al (1995) *Peptides*, **16**: 325–329
2. Smyth, M. and Cuinn G.O. (1994) *J. of Neurochem.*; **63**: 1439–1445
3. Morrison, M.E. et al (1993) *J. of Exp. Med.*; **177**: 1135–1143
4. Kameoka, J. et al (1993) *Science*; **261**: 166–469
5. De Meester, I. et al (1994) *Eur. J. of Immunol.*; **24**: 566–570
6. Darmoul, D. (1992) *J. Biol. Chem.*; **267**: 4824–4833
7. Tanaka, T. (1992) *J. Immunol.*; **149**: 481–486
8. Marguet, D. et al (1992) *J. Biol. Chem.*; **267**: 2200–2208
9. Ogata, S. et al (1992) *Biochemistry*; **31**: 2582–2587
10. Roberts, C.J. et al (1992) *J. Cell. Biol.*; **119**: 69–83
11. Fleischer, B. (1994) *Immunol. Today*; **15**: 180–184

12. Bohm, S.K. et al (1995), *Biochem. J.*; **311**: 835–843
13. Torimoto, Y. et al (991) *J. Immunol.*; **147**: 2514–2517
14. Dang, N.H. et al (1990) *J. Immunol.*; **145**: 3963–3971
15. Martin, M. et al (1995) *J. Immunol.*; **155**: 4630–4643
16. Callebaut, C. et al (1993) *Science*; **262**:2045–2050
17. Morimoto, C. et al (1994)*Proc. Natl. Acad. Sci. USA*; **91**: 9960–9964
18. Gutheil, W.G. et al (1994) *Proc. Natl. Acad. Sci. USA*; **91**: 6594–6598
19. Blanco, J. et al. (1996) *Exp. Cell Res.*; **225**: 102–111
20. Duke-Cohan, J. S., et al (1995) *J. Biol. Chem.*; **270**: 14107–14114
21. Duke-Cohan, J.S. et al (1996) *J. Immunol.*; **156**: 1714–1721
22. Jacotot, E. et al (1996) *Eur. J. Biochem.*, **239**: 248–258
23. Werner, A. et al (1994) *AIDS*; **9**: 1348–1349
24. Schön, E. et al (1991) *Biol. Chem. Hoppe Seyler*; **372**: 305–311
25. Born, J. (1994) *Eur. J. Cell. Biol.*;Supl. **40**: 23
26. Rahfeld, J. et al (1991) *Biol. Chem. Hoppe-Seyler*; **372**: 313–318
27. Cole, S.R. et al (1987) *Mol. Immunol.*; **24**: 699–705
28. Scanlan, M.J. (1994) *Proc. Natl. Acad. Sci. USA*; **91**: 5657–5661
29. Rettig, W.J. et al (1994) *Int. J. Cancer*; **58**: 385–392
30. Yaron, A. and Naider, F. (1993) *Critical Rev. in Biochem. and Mol. Biol.* ; **28**: 31–81
31. Kieffer, T. et al (1995) *Endocrinol.*; **136**: 3585–3596
32. Oravezc, T. et al (1996) *J. Immunol.*; **157**: 1329–1332

26

EXPRESSION OF DIPEPTIDYLPEPTIDASE IV (DPP IV/CD26) ACTIVITY ON HUMAN MYELOID AND B LINEAGE CELLS, AND CELL GROWTH SUPPRESSION BY THE INHIBITION OF DPP IV ACTIVITY

Anne Micouin* and Brigitte Bauvois

Unite 365 INSERM
Institut Curie
Section de Recherche
26 rue d'Ulm, 75231 Paris Cedex 05 France

1. BACKGROUND

Human CD26 antigen is a transmembrane glycoprotein with dipeptidyl peptidase IV (DPP IV) activity in its extracellular domain. Within the hematopoietic system, CD26/DPP IV has been thought to be exclusively expressed in activated T and NK cells [1]. However, by enzymatic assay, Bauvois et al. have characterized DPP IV activity (sensitive to diprotin A which is the specific inhibitor of DPP IV) present on the surface of human myeloblastic HL-60 and monoblastic U937 cell lines as well as on peripheral monocytes, granulocytes and macrophages [2]. Moreover, DPP IV activity increased strictly on cells undergoing macrophage maturation [2]. Reinhold et al. found that synthetic reversible inhibitors of DPP IV suppressed cell proliferation and modulated cytokine production in U937 cells expressing high levels of CD26 [3]. Monoclonal antibody (mAb) anti-Ta1 (purchased from Coulter, Hialeah, FL) and directed against CD26 of activated human T cells was recently shown to react with a few B cell lines including Daudi cells which are tumor cells derived from Burkitt lymphoma and EBV-transformed cells [4]. Another anti-CD26 mAb i.e. anti-TA5.9 (from Eurogenetics, Tessenderlo, Belgium) was shown to bind activated T cells as well as myeloid cells, Raji and U266 cells [5].

* Fax: 00.01.44.07.07.85; Phone: 00.01.42.34.67.13; E-mail: amicouin@curie.fr.

Cellular Peptidases in Immune Functions and Diseases, edited by Ansorge and Langner
Plenum Press, New York, 1997

2. CHARACTERIZATION OF DPP IV ACTIVITY IN MONOBLASTIC AND B CELL LINES

Several human cell lines were assessed for DPP IV activity : the myeloblastic U937 cell line, the plasma RPMI 8226, U266 and Eskol cell lines, and resting B cells isolated from normal blood. DPP IV activity was defined by the hydrolysis of Gly-Pro-paranitroanilide (pNA) substrate. Gly-Pro hydrolyzing activity was found in whole-cell homogenates from all cell populations tested (from 100 to 1300 pmol/30min/μg). Moreover, intact cell lines had the capacity to hydrolyze Gly-Pro-pNA in contrast to resting B cells, suggesting that hydrolyzing activity in intact B cells was mainly intracellularly associated.

Gly-Pro-pNA hydrolyzing activity was not inhibited by the inhibitors of metalloproteases (EDTA) and N-aminopeptidases (bestatin). The serine protease inhibitors diisopropylfluoridate (DFP) and diprotin A (Ile-Pro-Ile) which is known to be a specific inhibitor of DPP IV, prevented the hydrolysis of Gly-Pro-pNA (> 50 %) in U937, Eskol and B cells, leading to suggest that these cells expressed DPP IV activity. However, DFP and diprotin A were uneffective in U266 and RPMI 8226 lysates, indicating that Gly-Pro-pNA hydrolysing activity from these cells was not associated with DPP IV activity.

Assays to immunoprecipitate DPP IV activity from U937 cells, Eskol and B cells were conducted using the mAbs anti-Ta1 and anti-TA5.9. Both mAbs were capable of precipitating DPP IV activity from Eskol and B cell homogenates. Unexpectedly, DPP IV activity from U937 lysates was not provided by the anti-CD26 mAbs, suggesting that these cells express a different, from but related to, DPP IV protein.

3. EXPRESSION OF CD26 PROTEIN AT THE SURFACE OF U937 AND ESKOL CELLS

To detect surface expression of CD26 on U937, Eskol and B cells, we used anti-Ta1 and anti-TA5.9 mAbs that were above capable of immunoprecipitating DPP IV activity from Eskol and B cells. As determined by flow cytometry, CD26 was present at the surface of Eskol cells (Figure 1). Resting B cells were unreactive with these mAbs (Figure 1) and this result indeed correlated with the absence of cell surface associated DPP IV activ-

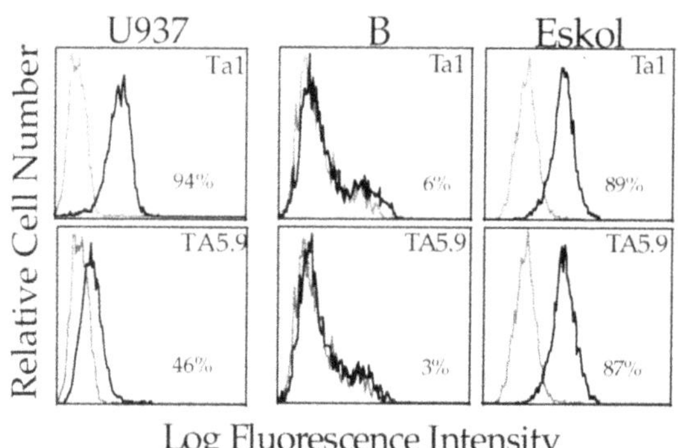

Figure 1. Expression of CD26 on the surface of U937, Eskol and B cells. Purified B cells, U937 and Eskol cells were incubated at 4°C with control isotype antibodies (mIgG1) (thin line) or with anti-Ta.1 or anti-TA5.9 mAbs (black line) and goat anti-mouse-FITC. Analysis was performed using a Becton-Dickinson cytofluorimeter.

ity. Surprisingly, both anti-Ta1 and anti-TA5.9 mAbs showed reactivity with U937 cells (Figure 1). Whether similar structures to CD26 protein are present on U937 cells remain to be demonstrated.

4. EFFECTS OF PRO-PRO DIPHENYL PHOSPHONATE ESTER ON DPP IV ACTIVITY AND GROWTH OF U937 AND ESKOL CELLS

4.1. Inhibition of DPP IV Activity by Pro-Pro Diphenyl Phosphonate Ester

Pro-Pro diphenyl phophonate ester (M1, Figure 2) was previously developed by A. M. Lambeir et al. [6]. M1 was shown to form a covalent adduct with the catalytic site of purified [6]. Due to its stability and the irreversible nature of the inhibition, it could be considered as a useful tool for studying the functional role of DPP IV activity. M1 specifically inhibited in a dose-dependent manner DPP IV activity from U937 and Eskol homogenates (Figure 2).

4.2. Influence of M1 on Growth of U937 and Eskol Cells

U937 and Eskol cells were cultured for up to three days in the absence or in the presence of increasing concentrations of M1 (from 10 to 300 μM). No necrotic cell death was noted as assessed by blue trypan exclusion. As shown in Figure 3, a growth inhibition of U937 and Eskol cells by M1 was already observed after 1 day of treatment within the range of concentrations of 100 to 300 μM, whereas cell surface DPP IV activity was already inhibited at 100 μM (data not shown).

5. CONCLUSIONS

Taken together, these data indicate that both myeloid and B lineage cells expressed a DPP IV activity. In monoblastic U937 cells as well as in monocytes and granulocytes (data not shown), DPP IV activity was however, not immunoprecipitated by monoclonal antibodies against CD26 (anti-Ta1 and anti-TA.5.9) strongly suggesting that myeloid cells

Pro-Pro-diphenyl phosphonate ester

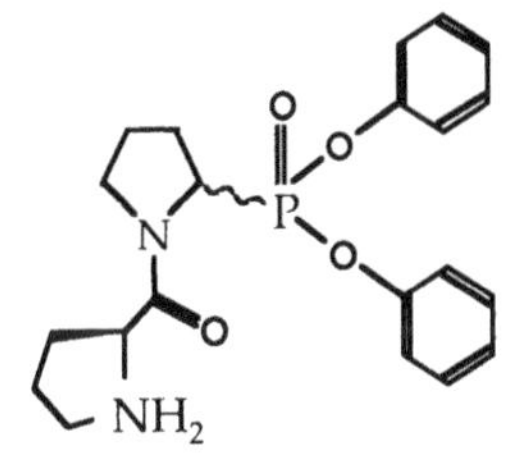

% DPP IV activity inhibition

M1	Eskol	U937
10 μM	8± 8%	39%
100μM	23 ± 15%	83%
1 mM	62 ± 6%	83%

Figure 2. Structure of M1 inhibitor, and effects on DPP IV activity of U937 and Eskol cell homogenates. Experiments were performed on whole-cell homogenates (in octyl-β-D-glucopyranoside) at 37°C for 30 min in the presence of 1 mg/ml Gly-Pro-pNA.

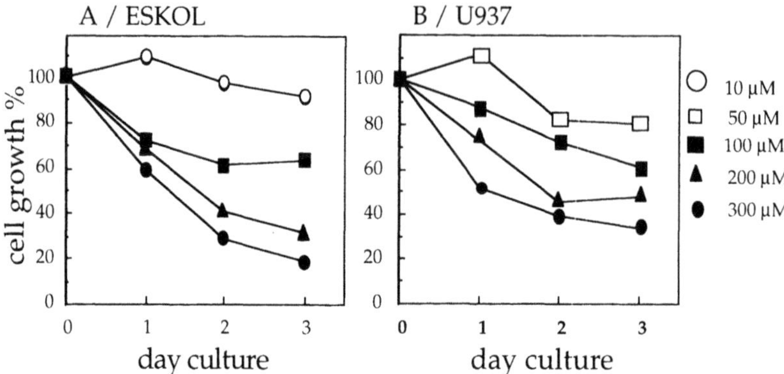

Figure 3. Effects of M1 on growth of Eskol and U937 cells. Eskol and U937 cells were cultured in the absence or in the presence of increasing concentration of Pro-Pro-diphenyl phosphonate ester (M1) (from 10 to 300 µM) for up to 3 days. After each day of incubation, cells were assayed for cell number and percent of cell growth was calculated.

possess a related to, but distinct from, standard DPP IV/CD26 protein. In agreement with previous studies [4, 5], B lineage cells express DPP IV/CD26 molecule.

Recent reports have shown that the addition of DPP IV inhibitors to peripheral blood mononuclear cells, T cells or monoblastic U937 cells, strongly suppressed cell proliferation, and modulated cytokine production [3, 7, 8]. Treatment of monoblastic or Eskol cells with a specific inhibitor of DPP IV i.e. Pro-Pro diphenyl phophonate ester [5] resulted in a time- and dose-dependent block of cell growth, suggesting that cell surface DPP IV (from Eskol cells) as well as DPP IV-like (from U937 cells) are involved in regulation of cell proliferation. Additional experiments are needed to evaluate if DPP IV (and its isoform) participates in other cellular responses including cytokine and immunoglobulin secretion.

ACKNOWLEDGMENTS

This work was suported by grants from Institut National de la Santé et de la Recherche Médicale (INSERM). The authors thank Dr I. de Meester for kindly providing M1 inhibitor and Dr E. Bosmans for the gift of the mAb anti-CD26 TA5.9.

6. REFERENCES

1. Fleischer B. (1994) CD 26: a surface protease involved in T-cell activation. *Immunology Today* **15** : 180–184.
2. Bauvois B, J. Sancéau J, Wietzebin J (1992) Human U937 cell surface peptidase activities: characterization and degradative effect on tumor necrosis factor-α. *Eur. J. Immunol.* **22** : 923–930.
3. Reinhold D, Bank Ubühling F, Kähne T, Kunt D, Faust J, Neubert K, Ansorge S. (1994) Inhibitors of dipeptidyl peptidase IV (DPP IV, CD 26) specifically supress proliferation and modulate cytokine production of strongly CD 26 expressing U 937 cells. *Immunobiol* **192** : 121–136.
4. Bühling F, Junker U, Reinhold D, Neubert K, Jäger L, Ansorge S. (1995) Functional role of CD26 on human B lymphocytes. *Immunol. Lett.* **45** : 47–51.
5. Shaw S, Gilks W. (1995). In : White Cell Differentiation Antigens, Schlossman SF, Boumsell L, Golks W, Harlan JM, kishimoto T, Morimoto C, Ritz J, Shaw S, Silverstein R, Springer T, Tedder TF, Todd RF eds. Leukocyte Typing V. Oxford University Press, Oxford-New York-Tokyo : pp 78.

6. Lambeir AM, Borloo M, De Meester I, Belyaev A, Augustyns Khendriks D, Scharpé S, Haemers A. (1996) Dipeptide-derived diphenyl phosphate ester : mechanism-based inhibitors of dipeptidyl peptidase IV. *Biochem. Biophys Acta.* **1290** : 76–82.
7. Schön E, Jahn S, Kiessig T, Demuth HU, Neubert K, Barth A, Von Baehr R, Ansorge S. (1987) The role of dipeptidylpeptidase IV in human T lymphocyte activation inhibitors and antibodies against dipeptidylpeptidase IV suppress lymphocyte proliferation and immunoglobulin synthesis in vitro. *Eur. J. Immunol.* **17**: 1821–1826.
8. Reinhold D, Bank DU, Bühling F, Neubert K, Mattern T, Ulmer AJ, Flad HD, Ansorge S. (1993) Dipeptidylpeptidase IV (CD26) on human lymphocytes. Synthetic inhibitors of and antibodies against dipeptidylpeptidase IV suppress the proliferation of pokeweed mitogen-stimulated peripheral blood mononuclear cells and IL-2 and IL-6 production. *Immunobiol.* **188:** 403–409.

27

CD26 AS A POSITIVE REGULATOR OF HIV ENVELOPE-GLYCOPROTEIN INDUCED APOPTOSIS IN $CD4^+$ T CELLS

E. Jacotot, C. Callebaut, J. Blanco, Y. Rivière, B. Krust, and
A. G. Hovanessian

Unité de Virologie et Immunologie Cellulaire, UA CNRS 1157
Institut Pasteur
28, rue du Dr. Roux, 75724 Paris cedex 15, France

SUMMARY

The membrane-expressed HIV-1 envelope glycoprotein complex, gp120/gp41, has been shown to be responsible for the initiation of cell killing by apoptosis in $CD4^+$ T cells. By using two experimental approaches we demonstrate that CD26, independent of its DPP IV activity, appears to be implicated in this function of the gp120/gp41 complex to initiate apoptosis.

INTRODUCTION

The human immunodeficiency virus (HIV) infects T lymphocytes, monocytes and macrophages by binding to its principal receptor, the CD4 molecule, through its envelope glycoprotein complex. The HIV-1 envelope gene codes for a precursor polyprotein gp160 which is processed by proteolytic cleavage to yield extracellular (gp120) and transmembrane (gp41) envelope glycoproteins. In spite of the multiple hypothesis on the indirect effects of viral infection in terms of pathogenesis (1,2) a clear link has been established in vivo between the degree of HIV replication, CD4+ T cells depletion, and disease progression towards AIDS (3,4). Consequently, it is plausible to consider that CD4+ T cell depletion is due, at least in part, to HIV mediated direct killing which is a property of the cell surface expressed HIV envelope glycoprotein complex, gp120/gp41 for HIV-1 (5,6). A number of in vitro experiments have shown that this cell death occurs in most cases through the mechanism of apoptosis (7–12) independent of syncytia formation (13–16). Apoptosis could be characterized by different techniques, such as by the fragmentation of the chromatin into the typical oligonucleosomal-length DNA ladder, by the presence of nucleosomal-histones in the nucleoplasm, and by electron microscopy (5,8).

Cellular Peptidases in Immune Functions and Diseases, edited by Ansorge and Langner
Plenum Press, New York, 1997

It has been demonstrated that in contrast to other activation molecules like CD38, HLA-DR, the expression of CD26 on T lymphocytes is decreased in HIV seropositive individuals during the progression toward AIDS , and this latter appears to be selective for the $CD4^+/CD26^+$ T cell subset (for references see 17). Furthermore, the decrease of CD26 expression on PBMC from seropositive individuals appears not to be correlated with the expression of Fas antigen (CD95/Apo-1; 18). Recently, CD26 has been implicated in HIV cytopathic effect in relation with the structure of the V3 loop (19). Here, we demonstrate that the level of CD26 in $CD4^+$ Jurkat cells could be a determining component for the rate of the gp120/gp41 complex-mediated initiation of apoptosis. Systematically, we observed a significant delay in the induction of apoptosis in Jurkat cell lines expressing low levels of CD26 compared to those expressing high levels of CD26. The delay in the induction of apoptosis was not observed in the low CD26 expressing cells, when apoptosis was triggered with anti-Fas monoclonal antibody in the different Jurkat cell lines. These observations indicate that the delay in the induction of apoptosis is not an intrinsic property of the low CD26 expressing cell lines, but it is a consequence of a lesser efficacy of events associated with the functioning of the gp120/gp41 complex when the expression of CD26 is low.

EXPRESSION OF CD26 IN JURKAT CELLS

CD26 as other members of the serine protease family contains the consensus sequence G-X-S-X-G (G for glycine, S for serine, X for any other amino acid) at its catalytic site, and point mutation of the serine residue to alanine causes inactivation of the DPP IV activity (20). In this study we used parental Jurkat cells (clone P32) or transfected with the plasmids containing human CD26 cDNA, either the wild type (clones J37 and T11) or mutated at the serine-630 codon (clones D28 and 3/13) under the control of a SRα promoter (21). Additional control Jurkat cells included in this study were JBK cells and the transfected clones JC31 and JC35 with a plasmid containing the SRα promoter without the CD26 cDNA.

The parental Jurkat cell lines, in general, are considered to be CD26 negative in comparison to cells expressing high levels of CD26. However, it should be emphasized that these cells do express low levels of CD26 which could be detected by fluorescence-activated cell sorting (FACS) analysis using different monoclonal antibodies specific to CD26, mAb BA5 and Ta1 (22). An observation which has been confirmed by the detection of CD26 mRNA (23; and unpublished data) and by an immunoblot assay using mAb 1F7 specific to CD26 (24). As expected the expression of CD26 was enhanced considerably in the recombinant CD26 expressing clones (22). The expression of others cell surface antigens, such as CD95, CD45RO, CD7, CD44 and the chain CD11a of LFA-1, were not significantly different between the different Jurkat cell lines (not shown).

THE LEVEL OF CD26 DETERMINES THE RATE OF APOPTOSIS

Cocultures of the chronically-infected cells and the different Jurkat cell lines, at a ratio of 1 to 10, were investigated for the occurence of apoptosis by assaying the presence of nucleosomal histones and the fragmentated nucleosomal-length DNA-ladder in the nucleoplasm fractions of such cocultured cells (Fig. 1). The histones were detected as soon as 18 hr post-coculturing with Jurkat cell lines expressing the recombinant forms of

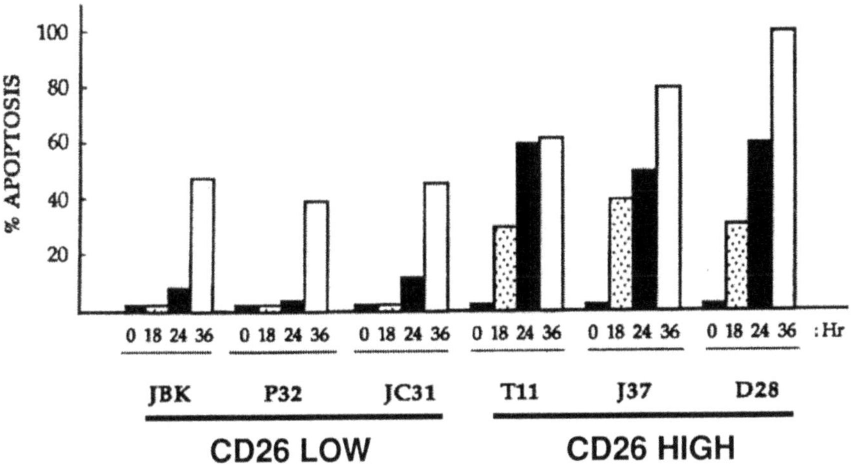

Figure 1. Enhanced expression of CD26 is correlated with an increased rate of HIV-mediated apoptosis. Coculturing of the different Jurkat cell lines with the chronically infected H9/IIIB cells was at a ratio of 10 to 1, with a final concentration of cells at 1 x 10^6 cells per ml of culture medium. Apoptosis was monitored at time 0 (to estimate the background in each cell line), and 18, 24 and 36 hr post-coculturing by assaying the nucleoplasm fractions for the presence of histones H2A, H2B, H3 and H4 to monitor apoptosis. The % apoptosis was estimated by quantitative densitometry of stained histones (as described in ref. 22).

CD26 and apoptosis reach 60% at 24hrs. In the coculture with the parental Jurkat cell line however, histones were not detectable before 36 hr , and even then, their level was found to be lower than that observed in the cocultures with the recombinant CD26 expressing cells. The presence of histones was indeed the consequence of the specific fragmentation of the chromatin associated with apoptosis, which was illustrated by the presence of typical LMW DNA ladder corresponding to the nucleosomal-length DNA fragments (22). In several independent experiments, apoptosis was delayed at least 24 hr in the cocultures with the low CD26 expressing Jurkat cell lines (JBK, P32, JC31) compared to Jurkat cell lines expressing enhanced levels of recombinant CD26, either wild-type (T11, J37) or mutated (D28). Furthermore, the DPP IV activity of CD26 was not required for its function in the occurence of apoptosis in such cocultured cells (Fig. 1; see also below).

The high CD26 expressing cell lines were generated by transfection of Jurkat cells with the plasmid pSRα, expressing either the wild-type or mutated form of CD26 (21). Consequently, it could be argued that transfection of Jurkat cells with the plasmid alone could trigger phenotypic modifications resulting in a more rapid induction of apoptosis. This possibility however is most unlikely, since the rate of apoptosis was found also to be delayed in Jurkat clones obtained by transfection with a similar plasmid containing the SRα promoter(Fig. 1). Clearly, no apparent differences were observed between the rate of apoptosis in the transfected clone JC31 compared to its corresponding untransfected Jurkat JBK cells.

CHARACTERIZATION OF APOPTOSIS IN COCULTURES OF H9/IIIB AND JURKAT CELLS

For further characterization of the occurence of apoptosis in cocultures of H9/IIIB with Jurkat cells, we investigated the effect of three reagents:

1. the reverse transcriptase inhibitor AZT which blocks new HIV infection as a consequence of cell/cell spreading of virus;
2. the monoclonal antibody mAb OKT4A which is directed against the gp120 binding site in the CD4 molecule;
3. $ZnSO_4$, an inhibitor of the nuclear endonuclease(s) responsible for fragmentation of cellular DNA during apoptosis.

In view of previous experiments, we used cocultures of H9/IIIB cells with the recombinant CD26 expressing cell line J37, that result in the generation of syncytia and the occurence of apoptosis clearly observed at 24 hr. AZT blocked completely cell/cell spreading of HIV, monitored by the lack of enhanced synthesis of viral proteins (assayed by immunoblotting using HIV-1 positive human serum; not shown). Interestingly, in the presence of AZT, no apparent modification was observed in the formation of syncytia and induction of apoptosis, thus suggesting that these events are due to the viral envelope gp120/gp41 complex already expressed on the surface of the chronically-infected cells and not due to new infection. Addition of mAb OKT4A blocked both syncytium formation and apoptosis, thus confirming that initiation of these events requires first the interaction of gp120 to the CD4 molecule. Finally, consistent with the role of the endonuclease in the apoptotic process, $ZnSO_4$ blocked apoptosis without any apparent effect on syncytium formation. This latter observation also illustrates once again that apoptosis is not a direct consequence of syncytium formation, as it has been pointed out before (9,13,14). In order to demonstrate that DNA fragmentation did occur in the uninfected Jurkat cells during the coculturing with the chronically-infected cells, the chromatin of Jurkat cells (clone J37) was ^{125}I-labeled by first labeling cells with ^{125}I-IUdR (a DNA precursor) before incubation with the chronically infected cells (22). These experiments illustrate that apoptosis occuring in cocultures of target Jurkat cells with chronically infected effector cells, indeed occured in the target cells. The close correlation between the intensity of nucleosomal histone bands with the amount of radioactivity and ^{125}I-labeled LMW DNA, provide further confirmation that the assay of histones is a sensitive, reliable and efficient experimental technique for monitoring apoptosis.

Several groups have demonstrated that in HIV infected subjects, $CD4^+$ T lymphocytes become primed for apoptosis upon further activation by mitogens or superantigens. This apoptosis is blocked by cycloheximide and cyclosporin A. In contrast, and as reported by others (16,25), we have found in the coculturing experimentel model that cycloheximide and cyclosporin A do not block HIV-induced apoptosis (26). These observations indicate that direct HIV-induced apoptosis is distinct from TCR-initiated cell death and from apoptosis of ex vivo PBMC from HIV infected individuals (1).

THE DPP IV ACTIVITY OF CD26 DOES NOT APPEAR TO BE IMPLICATED IN ITS FUNCTION IN THE HIV-ENVELOPE INITIATED APOPTOSIS

The results presented in Figure 1 demonstrated that increased rate of apoptosis is observed in cell lines expressing enhanced levels CD26, either wild-type (clones J37 and

T11) or catalytically inactive mutant form of CD26 (clone D28). This was further confirmed by investigating the rate of apoptosis in another clone expressing catalytically inactive mutant form of CD26 (clone 3/13). Once again, a significant degree of apoptosis was observed at 24 and 36 hr in this Jurkat clone expressing a high level of CD26, as was previously observed in other high CD26 expressing Jurkat clones (J37, T11, and D28), compared to that observed in low CD26 expressing Jurkat cell line P32 (22).

Lys-[Z(NO_2)]-pyrrolidide is a specific inhibitor of the DPP IV activity of CD26 with a 50% inhibitory concentration value around 1 μM (27). Addition of this inhibitor at 10 and 25 μM concentrations in the cocultures of H9/IIIB and J37 cells had no effect on syncytium formation and occurence of apoptosis (Fig. 2). This and the observations showing that cell lines expressing either the wild-type or the catalytically inactive mutant form of CD26, are equally effective in the coculture system to induce rapid syncytium formation and apoptosis, point out that the DPP IV activity of CD26 may not be essential for its function in HIV infection. Moreover, since the recently described DPPIV β (27; see also chapter from Blanco et al.) is also inhibited by Lys-[Z(NO_2)]-pyrrolidide, one can conclude that DPPIV β activity is not required for HIV-mediated apoptosis.

ANTI-Fas INDUCED APOPTOSIS IS INDEPENDENT OF CD26

Certain mAbs against the Fas cell-surface protein, also known as Apo-1 and CD95, have the capacity to kill cells by apoptosis (28). This property of mAb specific for the Fas antigen was investigated on the different Jurkat cell lines which were shown to express similar levels of cell-surface Fas (not shown). The different Jurkat cell lines were incubated for 16 hr in the presence of 100 ng/ml of the anti-Fas antibody and apoptosis was monitored by the assay of nucleosomal histones. The results indicated that all the different Jurkat cell lines, expressing low or high levels of CD26, are susceptible to anti-Fas antibody mediated apoptosis (22). The kinetics of anti-Fas induced apoptosis was further investigated in low and high CD26 expressing Jurkat cell lines and we have observed that the kinetics of FAS-mediated apoptosis are similar (22). Taken together, these observa-

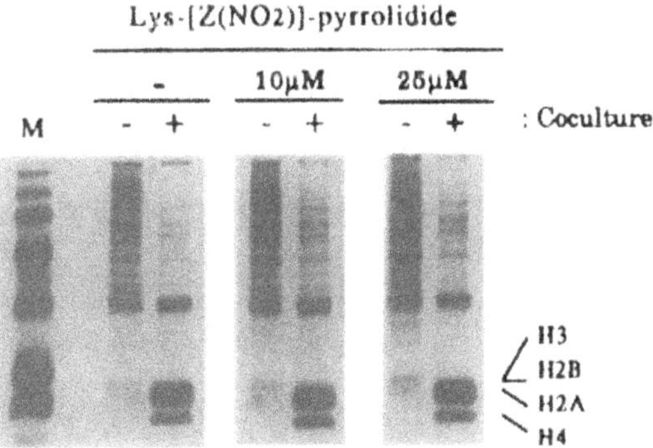

Figure 2. Inhibition of the DPP IV activity does not affect the occurence of HIV-mediated apoptosis. Coculturing (lanes +) of the Jurkat J37 cells expressing high levels of wild-type CD26 with the chronically infected H9/IIIB cells was at a ratio of 10 to 1, in the absence or presence of 10 and 25 μM of Lys-[Z(NO_2)]-pyrrolidide (panels -, 10μM and 25μM, respectively). Apoptosis was monitored at 36 hr post-coculturing by assaying the nucleoplasm fractions for the presence of histones H2A, H2B, H3 and H4. Lanes Coculture -, represent control Jurkat J37 cells incubated for 36 hr in the absence or presence of 10 and 25 μM of Lys-[Z(NO_2)]-pyrrolidide. Lane M shows the profile of molecular weight protein markers.

tions indicate that the delay in the occurence of apoptosis observed in our experimental models with the low endogenous CD26 expressing Jurkat cells, is not an intrinsic property of these cells to resist apoptosis. It is also of interest to note that the mechanism of HIV envelope-mediated apoptosis has recently been reported to be different from that of anti-Fas induced apoptosis (25).

THE gp120/gp41 INDUCED APOPTOSIS IN CD4$^+$ CELLS IS CORRELATED WITH THE ENHANCED EXPRESSION OF CD26

In order to evaluate the role of CD26 in the function of the gp120/gp41 complex to trigger apoptosis, the HIV-1 *env* gene was expressed in human CD4$^+$ Jurkat cell lines expressing low-endogenous CD26 (clones JC31 and P32) or high recombinant CD26, either the wild-type (clone J37) or mutated at its Ser-630 (Clone D28). For this purpose, Jurkat cell lines were infected with vaccinia recombinant viruses (VV) expressing the HIV-1 *env* gene, either the wild type (VVTG 9.1) to generate a functional gp120/gp41 complex or mutated at the cleavage site (VVTG 1139) to generate an uncleavable membrane-expressed precursor gp160. At 18 hr p.i., cytoplasmic extracts were analysed for the expression of the HIV envelope protein, whereas apoptosis was monitored in the nucleoplasm by two different techniques: by the fragmentation of the chromatin into typical oligonucleosomal-length DNA ladder (not shown; 22), and by the presence of nucleosomal-histones (Fig. 3). The level of gp160 and gp120 was found to be similar in the different Jurkat cell lines infected with VVTG 1139 and VVTG 9.1, respectively (Fig.3). As the recombinant HIV *env* gene was that of the HIV-1 Lai isolate with a syncytium inducing phenotype, syncytia were formed in both low and high CD26 expressing cell lines infected with VVTG 9.1 that produces a mature functional gp120/gp41 complex. However in these latter cultures, a significant degree of apoptosis occured only in Jurkat cell lines expressing enhanced levels of recombinant CD26, either wild-type or mutated (Fig. 3). Thus, expression of enhanced levels of CD26 in Jurkat CD4$^+$ cells appeared to favor the initiation of gp120/gp41 induced apoptosis, and the catalytic activity of CD26 was found not to be essential for this effect.

DISCUSSION

The mechanism responsible for the initiation of apoptosis in CD4$^+$ T lymphocytes by the membrane-expressed gp120/gp41 glycoproteins involves an array of complex molecular events initiated at the cell to cell interface. The first step is the interaction of the gp120/gp41 complex on the surface of infected cells with the CD4 molecule expressed by uninfected neighboring cells. This priming event induces a protein tyrosine kinase (PTK) signaling (25), along with the lateral interaction and the recruitement of membrane expressed proteins, including CD3, CD45 and CD26 (29). In view of the transductional capacities of these membrane antigens, it is plausible to consider that they could modulate or participate in the PTK signaling following the priming step. Consistent with this latter, the phosphorylation of the TCR ζ chain has been reported not to occur in the coculture-experimental model of apoptosis, despite of activation of $p56^{lck}$ and the related tyrosine kinase $p59^{fyn}$ (25). Our results indicate that the increased expression of CD26 is correlated with the rapidity of initiation of apoptosis induced by the gp120/gp41 complex in CD4+ cell lines. We propose that the interaction of the gp120/gp41 complex with cell surface

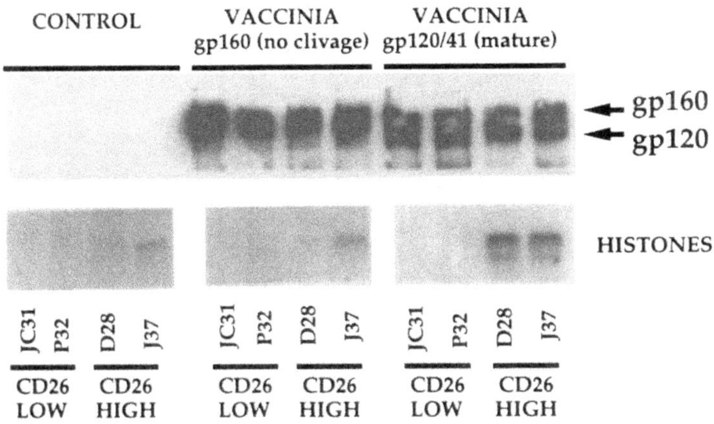

Figure 3. The gp120/gp41 induced apoptosis in Jurkat cells expressing high levels of recombinant CD26. The different Jurkat cell lines were the following: low-endogenous CD26 expressing cells (the control transfected cell line JC31 and the untransfected parental cells P32); high-recombinant CD26 expressing cell lines, either wild-type (clone J37) or mutated at its Ser-630 (clone D28). Cells were infected with VV recombinants expressing the HIV-1 Lai envelope glycoprotein: VVTG 1139 construct expresses gp160 that cannot be cleaved because of mutations in the cleavage site; VVTG 9.1 construct produces a gp160 precursor which is then cleaved into gp120 and gp41. Cells were infected with these VV recombinants at 50 pfu/cell. At 18 hr p.i., cell cultures were examined under a light microscope for syncytium formation (indicated as - or +) before preparation of cytoplasmic and nuclear extracts. Immunoblotting was carried out to monitor HIV *env* gene expression using cytoplasmic extracts and mAb 110/4 specific for the envelope gp120 (the position of gp160 and gp120 is as indicated on the right). Assay of apoptosis in nuclear extracts was carried out by monitoring the presence of nucleosomal histones (H2A, H2B, H3 and H4; a section of the protein-stained PAGE-SDS gel is shown). The control panel represents samples from cells which have not been infected with VV. Material corresponding to 0.5 x 10^6 were used for histone analysis.

CD4 is followed by a modified CD26 signaling, which might then be responsible for the initiation of apoptosis. Recently, Schlossman and colleagues have shown that the costimulatory activity of CD26 involves and requires its association with CD45 (see chapter from M. Hegen). In view of the fact that the cytoplasmic tail of CD26 contains only 6 amino acids (20), and the observations indicating that $p56^{lck}$ interacts directly with the cytoplasmic domain of CD45 (30) that antibody triggering CD26 up regulates $p56^{lck}$ and MAP kinase activities, increases $p56^{lck}$, $p59^{fyn}$, TCR ζ chain, ZAP-70, and phospholypase C $\gamma 1$ phosphorylation, and Ca^{2+} flux (see chapter from M. Hegen); one can postulate the existence of a functional association between CD4, CD45 and CD26. Thus the recruitment of the CD45/CD26 complex by CD4 subsequently to gp120 binding (29) may explain the mechanism by wich a CD4 like transductional pathway (i.e. $p56^{lck}$ activation) is associated to HIV-induced apoptosis (25) although this apoptosis is independent of the CD4 cytoplasmic tail (26). In relation to this model, it has been shown that the species-specific HIV-1 fusion coreceptor CXCR4/fusin (31,32) is recruited by CD4 after gp120 binding (32,33). In this study, PMA treatment of cells expressing a CD4 devoid of its cytoplasmic tail leads to down regulation of fusin without any effect on CD4 expression. In contrast, if these cells are incubated with gp120 before PMA treatment they observe the down-regulation of a CD4 /fusin complex. Our recent result showing a CD4 independent inhibition of HIV-induced apoptosis by PMA (26) may be the consequence of fusin down regulation. Thus, we can suggest that fusin may be implicated in HIV-induced apoptosis by allowing optimal interactions between the gp120/41 complex and the CD4/cofactors complex,

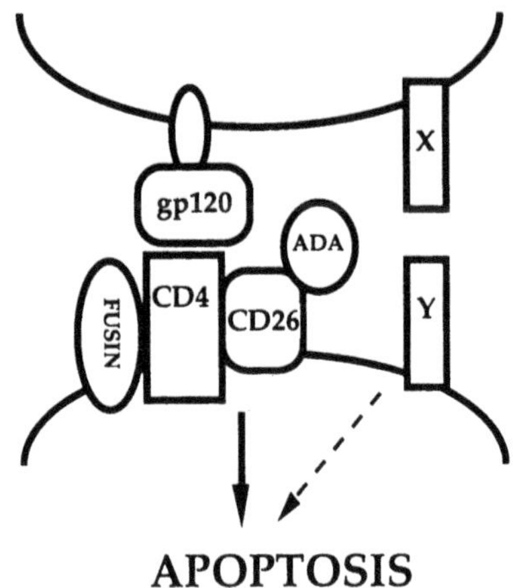

Figure 4. Model 1: CD26 is recruited by the gp120/CD4/fusin complex.

and/or by transducing specific protein G-dependent signals during these interactions (fig.4:Model 1).

Although there is data showing the possibility of a recruitement of CD26 in the presence of gp120 we cannot exclude the possibility in which CD26 could regulate apoptosis independently of a physical association with the gp120/CD4/fusin complex (Fig.5:Model 2).

At the moment, the precise role of CD26 in the mechanism of gp120/gp41 complex induced apoptosis remains still undefined, our results are consistent with the implication of CD26 directly or indirectly in this process. We suggest that the gp120/gp41 complex modifies CD4-dependent CD26-signaling pathway, and that this perturbation may be an early event which contributes to the induction of apoptosis.

In addition to the depletion of $CD4^+$ T helper cells during the evolution of AIDS, evidence has been provided for an intrinsic defect of such cells to recognize and respond to "recall antigens" (34), a property of $CD4^+$ $CD26^+$ T cells (35). In agreement with this latter, Schmitz et al. have demonstrated that this defective in vitro recall antigen response in HIV-1 infected individuals can be restored by the addition of soluble CD26 (36). Interestingly, several groups have reported a selective decrease in CD26 expressing $CD4^+$ T lymphocytes in HIV-1 infected individuals. Recent work from our laboratory has confirmed this and demonstrated that the decline of CD26 expressing cells is mainly observed in $CD4^+$ but not in $CD8^+$ cells (17). It is also of interest to note that adenosine deaminase (ADA) is presented on the surface of $CD26^+$ cells because CD26 is the ADA binding protein, and elevated levels of serum ADA have been reported in AIDS patients. In this regard, we have shown that soluble gp120 or HIV particles are able to block ADA binding on CD26 at the cell surface (see chapter from Valenzuela et al.). Thus, it is conceivable that this latter property of gp120 is connected to its capacity to generate apoptosis. Consistent with the in vivo observations in HIV-infected individuals, CD26 appears to be implicated in the cytopathic effect of HIV infection in cell cultures. Indeed, high CD26 expressing cells were reported to be eliminated preferentially during infection with monocytotropic HIV-1 isolates (19), and here we have shown that high CD26 expressing cells undergo increased rate of apoptosis initiated by a lymphocytotropic HIV-1 isolate. In conclusion, it is now clear that CD26 is implicated in many aspects of HIV infection such as: virions entry, direct gp120/gp41-induced apoptosis TAT immunosupressive effect (37)

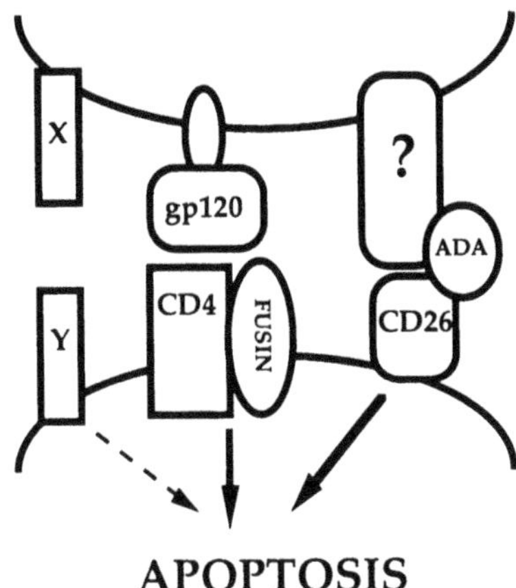

Figure 5. Model 2: CD26 acts separately from the gp120/CD4/fusin complex.

and ecto-ADA disregulation (see chapters from Callebaut et al. and from Valenzuela et al.). Although several mechanisms could be responsible for the depletion of $CD4^+$ $CD26^+$T lymphocytes in patients, the in vitro observations point out the contribution of HIV infection in this selective decrease.

ACKNOWLEDGMENTS

This work was supported by grants from Institut Pasteur, "Agence Nationale de la Recherche sur le SIDA" (ANRS). E.J. was supported by "Association des Artistes Contre le SIDA". J. B. was a recipient of a post-doctoral fellowship from *Spanish Ministerio de Educacion y Ciencia*. We thank Dr. C. Morimoto from Dana Farber Cancer Institute, Boston for the different Jurkat cell lines and mAb 1F7. The lysine-[Z(NO_2)]-pyrrolidide was a gift of Drs. K. Neubert and A. Barth, Halle, Germany. We thank Dr R. Franco for critical reading of the manuscript.

REFERENCES

1. Ameisen, J.C. (1992). Immunol. Today 13: 388–391.
2. Sheppard H.W. and Ascher M.S. J of AIDS. 1992: 5,143–147.
3. Wei X. et al. Nature . 1995; 373,117–22.
4. Ho D.D. et al..Nature . 1995; 373,123–6.
5. Hovanessian A.G., in Apoptosis II : the Molecular Basis of Apoptosis in Disease E. D. L. T. a. F. O. Cope., Ed. 1994) pp. pp. 21–42.
6. Sodroski J.G. et al.Nature . 1986; 322,470–474.
7. Terai, C. et al. J. Clin. Invest. 1991 87: 1710–1715.
8. Laurent-Crawford, A.G. et al.Virology 1991 185: 829–839.
9. Laurent-Crawford, A.G. et al. AIDS Res. Hum. Retroviruses 1993;9:761–773.
10. Laurent-Crawford, A.G., et al. Res. Virol. 1995;146: 5–17.
11. Cohen D.I. et al. Science . 1992; 256,542–545.
12. Martin S.J. et al. J Immunol . 1994; 152,330–42.
13. Rey-Cuillé, M.A. et al. Virology 1994; 202: 471–476.
14. Corbeil, J. and D.D. Richman. J. Gen. Virol. 1995; 76: 681–690.
15. Maldarelli, F. et al. J. Virol. 1995; 69: 6457–6465.

16. Nardelli, B. et al. Proc. Natl. Acad. Sci. USA. 1995; 92: 7312–7216.
17. Gougeon, M-L. Res. Immunol. 1996; 147: 5–8.
18. Gehri, R. et al. AIDS 1996; 10: 9–16.
19. Oravecz, T. et al. Nature Medicine 1995; 1: 919–926.
20. Fleischer, B. Immunol. Today 1994; 15: 180–184.
21. Tanaka, T. et al. Proc. Natl. Acad. Sci. USA. 1993; 90: 4586–4590.
22. Jacotot, E et al; Virology 1996; 223: 318–330.
23. Werner, A. et al. AIDS 1994; 8: 1348–1349.
24. Blanco, J. et al. Exp. Cell Res. 1996; 225: 102–111.
25. Tian, H. et al. International Immunol. 1996; 8: 65–74.
26. Jacotot, E. et al; submitted.
27. Jacotot, E. et al. Eur. J. Biochem. 1996; 239:248–258.
28. Nagata, S. and Suda, T. Immunol. Today 1995; 16: 39–42.
29. Dianzani, U. et al. Eur. J. Immunol. 1995; 25: 1306–11.
30. Di Somma M.M. et al. Febs Lett . 1995: 363,101–4.
31. Feng Y. et al. Science . 1996; 272,872–7.
32. Lapham C.K. et al. Science. 1996; 274,602–605.
33. Golding H. et al. J Virol. 1995; 69,6140–8.
34. Lane, H.C. et al. N. Eng. J. Med. 1985; 313: 79–84.
35. Morimoto, C. et al. J. Immunol. 1989; 143: 3430–3439.
36. Schmitz T. et al. J. Clin. Invest. 1996; 97:1545–49.
37. Subramanyam et al. J. Immunol. 1993; 150:2544–53.

28

COMPARATIVE STUDY OF CD26 AS A Th1-LIKE AND CD30 AS A POTENTIAL Th2-LIKE OPERATIONAL MARKER IN LEPROSY

Ulrike Seitzer, Dagmar Scheel-Toellner,* Margrit Hahn, Gesine Heinemann, Taila Mattern, Hans-Dieter Flad, and Johannes Gerdes

Department of Immunology and Cell Biology
Division of Molecular Immunology
Forschungszentrum Borstel, Parkallee 22, D-23845 Borstel, Germany

1. ABSTRACT

In the last years we have been able to establish CD26 as an operational marker for a human Th1-like reaction in various granulomatous diseases. Recently, CD30 was described as a marker for a Th2-type reaction, where CD30 is preferentially expressed and its soluble form released by human T cell clones producing Th2-type cytokines. To evaluate the possibility of CD30 as an eventual operational marker for a human Th2-like reaction *in vivo*, we performed immunohistological stainings on frozen sections of skin biopsies from patients with lepromatous and tuberculoid leprosy. A maximum of three to four CD30-positive cells was found per section, and there was no difference in the accumulation of CD30-positive cells between the tuberculoid and the lepromatous form of leprosy. With respect to CD26-positive cells, a high number was found in tuberculoid leprosy in contrast to a greatly reduced expression of CD26 in lepromatous leprosy. We conclude that, while CD26 was confirmed as an operational marker for a Th1-like reaction in leprosy, CD30 does not represent an operational Th2 marker in this disease.

2. INTRODUCTION

There is a general consensus that mice and humans express CD4+ T helper (Th) cells comprising functionally heterogeneous populations with specific profiles of cytokine

* Present address: Department of Rheumatology, University of Birmingham, UK.

production[1]. Th1 and Th2 cells are distinguished by their pattern of cytokines: Th1 cells produce interleukin 2, interferon γ and tumor necrosis factor-ß, whereas Th2 cells produce interleukin 4 and interleukin 5[2].

The spectral forms of leprosy represent a model system for the human Th1/Th2 reaction, since Th1 reactions predominate in the tuberculoid (TL) and Th2 reactions in the lepromatous (LL) form of this disease[3]. Since the same pathogen elicits a spectrum of different immune responses, ranging from a Th1-like to a Th2-like reaction, leprosy presents an interesting system for the analysis of T cell responses.

In a previous study we could show that staining for CD26 revealed high expression of CD26 in tuberculoid leprosy in contrast to no or very little expression in lepromatous leprosy[4,5]. In addition, double immunostaining demonstrated the coexpression of CD26 and IFNγ of T-cells in tuberculoid leprosy. Thus, high expression of CD26 indicated a Th1-like immune reaction.

Recently, Romagnani et al. described CD30 as a marker for a Th2-type reaction, based on observations that CD30 was preferentially expressed and its soluble form released by human T cell clones producing Th2-type cytokines. The *in vivo* relevance was discussed by showing that high numbers of CD30+ T cells were found in the lymph nodes of a patient suffering from Omenn's syndrome and the detection of circulating CD30+ T cells in atopic patients, both disorders being associated with a Th2-type immune response[6]. These observations led us to investigate whether an increased expression of CD30 is seen in the Th2-like lepromatous form of leprosy in contrast to the Th1-like tuberculoid form.

3. MATERIALS AND METHODS

3.1. Study Design

We investigated the application of CD30 as an operational Th2 marker *in vivo* by performing immunohistological staining on frozen sections of skin biopsies from untreated patients with lepromatous (n=15) and tuberculoid (n=4) leprosy. Hyperplastic human tonsils and five lymph node biopsies from patients suffering from Hodgkin's disease (nodular sclerosis) served as controls.

3.2. Patients

Skin biopsies of patients suffering from leprosy were obtained from patients of the leprosy eradication program performed by the Department of Leprosy of the Ministry of Health and Welfare, Paraguay. Patients were classified according to the clinicopathologic criteria of Ridley and Jopling[7]. All biopsies were taken from untreated patients and were snap-frozen in liquid nitrogen and stored until use at -80°C. Biopsies from Hodgkin patients were kindly provided by Prof. H. Stein, Institue of Pathology, University of Berlin.

3.3. Immunoenzymatic Staining

Cryostat frozen sections were fixed in acetone for 30 min, followed by fixation in chloroform for 30 min. After fixation the sections were preincubated with rabbit normal serum for 30 min to block nonspecific binding of the monoclonal antibodies to Fc receptors. Incubation with the primary monoclonal antibody was performed for 30 min, and immunostaining was undertaken according to the APAAP method[8] with New Fuchsin development. Finally, slides were counterstained with hematoxylin and mounted. The immunostainings were controlled by the use of only the secondary reagents to confirm their specificity.

3.4. Antibodies

The following monoclonal antibodies were used for the study: Leu2a (CD8, Becton Dickinson), Leu3a (CD4, Becton Dickinson), Okt 6 (CD1a, Ortho Diagnostic Systems) MIB-DS2/7 (CD26[9]) and Ber-H2 (CD30[10]). Rabbit anti-mouse immunoglobulin antiserum was a product of DAKO, and APAAP complexes were prepared according to Cordell et al.[8]

4. RESULTS

The results of the serial stainings are summarized in table 1. CD1a staining was included as an additional discriminating factor for the classification of lepromatous and tuberculoid leprosy besides the clinicopathological data. The subepidermal distribution of CD1a-positive Langerhans cells is characteristic for the different forms of leprosy. In tuberculoid forms, large amounts of subepidermal CD1a-positive cells are found in contrast to no subepidermal CD1a-positive cells in lepromatous leprosy[11]. Staining for CD4+ cells. CD8+ cells and BerMac3, a macrophage marker, were performed to confirm the presence of T-cells and macrophages in the granulomatous lesions.

As expected and described earlier[12], in frozen tonsil sections positive staining with the antibody Ber-H2 was restricted to some scattered large cells located in the T-zones and at the rim of germinal centers. In all Hodgkin cases investigated the tumor cells, i.e. Hodgkin and Reed-Sternberg cells, proved to be positive with the anti CD30 antibody. In contrast, in the leprosy cases a maximum of three to four CD30 positive T-cells was found per section, and there was no difference in the accumulation of CD30-positive cells between the tuberculoid and lepromatous form of leprosy. Staining with anti-CD26 confirmed the previous observation of high CD26 expression in tuberculoid compared to little or no expression in lepromatous leprosy.

Table 1. Summary of the results obtained from the serial stainings of tuberculoid and lepromatous leprosy skin biopsy sections with CD26 and CD30

	Immune response	Granuloma	CD1a	CD26	CD30
TLTh1-like (n=4)	Cell mediated, Th1-like	Organized, central core of epithelioid cells and few T-cells	Large amounts of subepidermal CD1a-positive Langerhans cells	High level of CD26 expression	Little or no CD30 expression
LLTh2-like (n=15)	Lack of cellular immunity, Th2-like	Poorly organized infiltrates, large number of macrophages, lower number of T-cells	No CD1a-positive subepidermal Langerhans cells	Dramatically reduced CD26 expression	Little or no CD30 expression

5. DISCUSSION

From the results of the serial stainings we conclude that, while CD26 was confirmed to be an operational marker for a Th1-like reaction in leprosy, CD30 does not represent an operational Th2 marker in this disease, since no preferential accumulation of CD30-positive cells was observed in lepromatous leprosy tissue. Other investigators were also unable to confirm the observations of Romagnani et al.[6], and reported that CD30 expression did not discriminate between human Th1- and Th2-type cells[13], and that not only Th2 cells but also Th1 and Th0 cells express CD30 after activation[14]. These results and our own observations imply that the eventual role of increased CD30 in a Th2-like response may not be detectable at the cellular level. Indeed, Del Prete et al. discuss whether the possible role of CD30 in a Th2-like response may be based on the CD30 triggering of activated Th cells by antigen-presenting cells expressing CD30 ligand and thus representing a costimulatory signal for the development of Th2-type responses[15]. Nonetheless, since high levels of sCD30 in serum are described to be associated with Th2-like diseases by Romagnani et al[6], it may be of interest to investigate this aspect in the different forms of leprosy as well.

6. ACKNOWLEDGMENTS

This research project was supported in part by a grant of the Deutsche Forschungsgemeinschaft (SFB367/C1).

7. REFERENCES

1. Romagnani S (1991) Human Th1 and Th2 subsets: doubt no more. Immunology Today 12:256–257.
2. Mosman TR, Cerwinski H, bond MW, Giedlin MA, Coffman RL (1986) Two types of murine helper T cell clone I. Definition according to profiles of lymphkine activities and secreted proteins. The Journal of Immunology 136:2348–2357.
3. Modlin RL (1994) Th1-Th2 paradigm: insights from leprosy. Journal of Investigative Dermatology 102: 828–832.
4. Scheel-Toellner D, Richter E, Toellner K-M, Reiling N, Wacker H-H, Flad H-D, Gerdes J (1995) CD26 expression in leprosy and other granulomatous diseases correlates with the production of interferon-γ. Laboratory Investigation 73: 685–690.
5. Mattern T, Ulmer AJ, Scheel-Toellner D, Flad H-D (1995) CD26 as a functional marker of TH1 lymphocytes. In "Dipeptidyl Peptidase IV (CD26) in Metabolism and the Immune Response" (ed. Fleischer B.) R.G. Landes Company, Austin, Texas, USA, 185–199.
6. Romagnani S, Del Prete G, Maggi E, Chilosi M, Caligaris-Cappio F, Pizzolo G (1995) CD30 and type 2 T helper (Th2) responses. Journal of Leukocyte Biology 57: 726–730.
7. Ridley DS, Jopling WH (1966) Classification of leprosy according to immunity. A five group system. International Journal of Leprosy 34:255–273.
8. Cordell JL, Fallini B, Erber WN, Ghosh AK, Abdulaziz Z, MacDonald S, Pulford KAF, Stein H, Mason DY (1984) Immunoenzymatic labeling of monoclonal antibodies using immune complexes of alkaline phosphatase and monoclonal anti-alkaline phosphatase (APAAP complexes). Journal of Histochemistry and Cytochemistry 32:219–229.
9. Scheel D, Richter E, Toellner K-M, Reiling N, Key G, Wacker H-H, Ulmer AJ, Flad H-D, Gerdes J (1995) Correlation of CD26 expression with Th1-like reactions in granulomatous diseases. In: Leucocyte Typing V "White Cell Differentiation Antigens" (SF Schlossmann, L Boumsell, W Gilks, JM Harlan, T Kishimoto, C Morimoto, J Ritz, S Shaw, R Silverstein, T Springer, TF Tedder, RF Todd, eds), Oxford University Press, Oxford, 1111–1114.
10. Schwarting R, Gerdes J, Dürkop H, Falini B, Pileri S, Stein H (1989) Ber-H2: A new anti-K1 (CD30) monoclonal antibody directed at a formol-resistant epitope. Blood 74: 1678–1689.

11. Collings LA, Poulter LW (1985) The involvement of dendritic cells in the cutaneous lesions associated with tuberculoid and lepromatous leprosy. Clinical and Experimental Immunology 62:458–467.
12. Schwab U, Stein H, Gerdes J, Lemke H, Kirchner H, Schaadt M, Diehl V (1982) Production of a monoclonal antibody specific for Hodgkin and Sternberg-Reed cells of Hodgkin's disease and a subset of normal lymphoid cells. Nature 299: 65–67.
13. Hamann D, Hilkens CMU, Grogan JL, Lens SMA, Kapsenberg ML, Yazdanbaksh M, van Lier RAW (1996) CD30 expression does not discriminate between human Th1- and Th2-type cells. The Journal of Immunology 156:1387–1391.
14. Bengtsson A, Johansson C, Linder MT, Hallden G, van der Ploeg I, Scheynius A (1995) Not only Th2 cells but also Th1 and Th0 cells express CD30 after activation. Journal of Leukocyte Biology 58:683–689.
15. Del Prete G, de Carli M, D'Elios MM, Daniel KC, Almerigogna F, Alderson M, Smith CA, Thomas E, Romagnani S (1995) CD30-mediated signaling promotes the development of human T helper type 2-like cells. Journal of Experimental Medicine 182:1655–1661.

REGULATION OF NEUTROPHIL ACTIVATION BY PROTEOLYTIC PROCESSING OF PLATELET-DERIVED α-CHEMOKINES

Hans-Dieter Flad, Luc Härter, Frank Petersen, Jan-Erik Ehlert, Andreas Ludwig, Lothar Bock, and Ernst Brandt

Department of Immunology and Cell Biology
Research Centre Borstel
D-23845 Borstel, Germany

1. INTRODUCTION*

In recent years evidence has been accumulated that platelets besides their function in coagulation play an important role in inflammation and wound repair. Upon activation platelets release a variety of mediators, among which members of the α-chemokine subfamily of proinflammatory cytokines have been identified. These platelet-derived polypeptides do not only comprise members of the so-called ß-thromboglobulin family, such as platelet basic protein (PBP), connective tissue-activating peptide III (CTAP-III), and neutrophil-activating peptide 2 (NAP-2)[1], but also platelet factor 4 (PF4)[2,3] and the ß-chemokine RANTES (Regulated upon activation normal T cell expressed and probably secreted)[4]. While ß-chemokines have been shown to activate monocytes, T lymphocytes and eosinophils, α-chemokines such as IL-8, NAP-2 and melanoma growth-stimulating activity (MGSA/gro-α) appear to represent rather selective activators of polymorphonuclear leukocytes (PMN)[5]. Importantly, their biological activity, such as chemotaxis and degranulation-inducing capacity, has been demonstrated to be closely connected with the presence of an N-terminal glutamic acid-leucine-arginine (ELR) motif. Only recently, attention has been paid to the regulatory properties of α-chemokines. In the present article we will focus on two aspects of regulation of PMN functions, namely 1. the proteolytic processing of platelet-derived α-chemokines as a regulatory event in the induction and modulation of PMN activation, and 2. the phenotypic and functional consequences for the PMN under the constraints of such regulatory events.

* This work was supported in part by Deutsche Forschungsgemeinschaft SFB 367, project C4.

2. NEUTROPHILS GENERATE THEIR ACTIVATOR NAP-2 BY PROTEOLYTIC PROCESSING OF PLATELET-DERIVED CTAP-III

In the years of 1989 and 1990 three different laboratories described independently from each other the presence in cell culture supernatants of so far unrecognized ß-thromboglobulin antigen isoforms, in particular CTAP-III isoforms[6-9]. It soon became apparent that CTAP-III and truncated isoforms of down to a length of 79 amino acids had no chemotactic activity and that its biological activity gradually increased with the degree of N-terminal truncation. Finally, our group succeeded in showing that the biological activity of this molecule was associated with a polypeptide of 70 amino acids and that only a few additional residues at the N-terminus drastically reduced the biological activity, determined as degranulation response (elastase release), from PMN. In subsequent studies the question was raised which cells were involved in the generation of NAP-2 from its inactive precursors. Although platelets activated by thrombin were found to generate some biological activity associated with NAP-2, the biological activity detectable in supernatants was much higher when PMN were also present. In fact, comparing different leukocyte preparations, it turned out that PMN were by far the most potent cells in generating NAP-2 and superior compared to monocytes and lymphocytes[10].

These findings were at variance with those of other investigators describing monocytes as the predominant cell type processing CTAP-III into NAP-2[11]. To identify the PMN-associated enzyme processing CTAP-III into NAP-2 various inhibitors specific for different classes of proteases were tested. Among these inhibitors only phenyl methylsulfonyl fluoride (PMSF) reduced the generation of NAP-2, indicating that a serine protease was involved in proteolytic processing. Among the proteases known to be present in PMN cathepsin G, a chymotrypsin-like protease, was found to generate PMN degranulation-inducing activity within 30 minutes. Isoelectric focussing and measurement of elastase release revealed that the appearance of degranulation-inducing biological activity paralleled the presence of immunoreactive NAP-2.

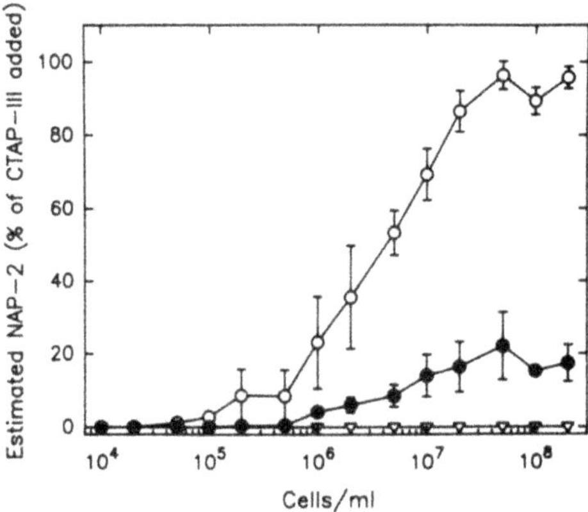

Figure 1. Cell concentration-dependent processing of CTAP-III to NAP-2 by purified PMN (O—O), monocytes (●—●), or lymphocytes (▽—▽). The cells in PBS/0.1% BSA were incubated with 10 μg/ml of CTAP-III for 1 h at 37°C. NAP-2 in the recovered supernatants was determined in the elastase release assay. Data represent ±SD from three different experiments with cells from different donors. (From ref. 10, Copyright 1994. The American Association of Immunologists.)

From these data we concluded that it is the PMN, the target cell itself, which by means of a cathepsin G-like serine protease generates NAP-2 from its inactive precursor CTAP-III[12].

3. C-TERMINALLY TRUNCATED ISOFORMS OF NAP-2 EXHIBIT ENHANCED BIOLOGICAL ACTIVITY

Studies on the structure-function relationship of α-chemokines have revealed that the ELR motif, localized N-terminally of the CXC configuration, is absolutely essential for receptor-mediated activation of PMN. This motif is lacking in PF4 and interferon-induced protein 10 (IP10) and, therefore, these α-chemokines do not stimulate PMN. Likewise, insertion of the ELR motif into the N-terminus of PF4 led to PMN-activating

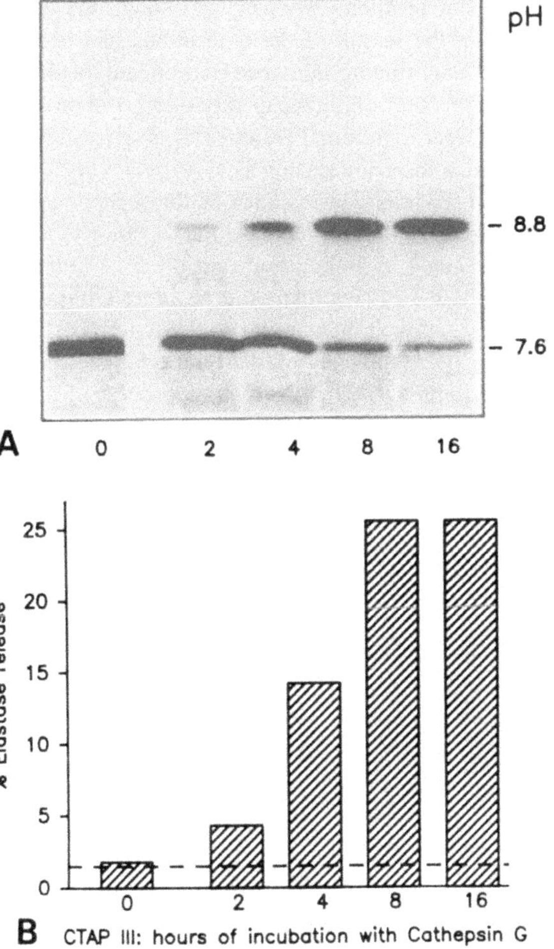

Figure 2. Enzymatic cleavage of CTAP-III into NAP-2 by cathepsin G. CTAP-III (300 μg/ml) was digested with cathepsin G (1 μg/ml) for the time periods indicated. (A) IEF and Western blot analyses of immunopurified polypeptide samples reveal newly generated NAP-2 (pI 8.8 band). (B) Biological activity of corresponding samples as measured in the elastase release assay. ——, assay background. (From ref. 12.)

properties[13, 14]. Surprisingly, insertion of the ELR motif into IP10 did not result in a biologically active peptide[14], which suggested to us that other regions of the molecule such as the C-terminus might also be involved in receptor binding. The discovery of C-terminally truncated isoforms of NAP-2 in supernatants of activated mononuclear cells enabled us to investigate structure-function relationships of these isoforms of NAP-2. The first C-terminally truncated isoform detected was determined to lack at least one and maximally four residues and, to our surprise, exhibited three to four times higher biological activity than native NAP-2[15, 16]. Since the degree of C-terminal truncation could not exactly be defined, a series of C-terminally deleted isoforms of NAP-2 were prepared by recombinant technology, which resulted in NAP-2 1–70 and variants 1–69 to 1–64. With these preparations it could be clearly established that the biologial activity, degranulation capacity, and receptor binding potency was increased four times in isoforms deleted by up to four residues.

Furthermore, subsequent sequence analyses of the previously discovered C-terminally truncated native variant revealed that its structure corresponded to the recombinant NAP-2 1–66 lacking the four terminal residues[16]. Although the dependency of the functional activity of IL-8 from the length of the C-terminus had been described previously with synthetic analogues[17], our finding demonstrated unequivocally for the first time that the biological activity and receptor binding capacity of an α-chemokine were significantly increased by deletion of single C-terminal residues. It is tempting to speculate that in the native molecule C-terminal acid residues such as aspartic acid (70) and glutamic acid (67) may be repulsed by negatively charged residues of the receptor and, thus, interfere with the binding. Their deletion may, in fact, increase the receptor binding. Since these C-terminal residues are not protected by integration into the C-terminal α-helix, they may be more susceptible to proteolytic cleavage than other residues within this sequence.

From these data we concluded that similar to the N-terminal processing limited proteolytic processing at the C-terminus may play an important role in ligand-receptor interaction and, thus, may contribute to the final activation process of the PMN target cell.

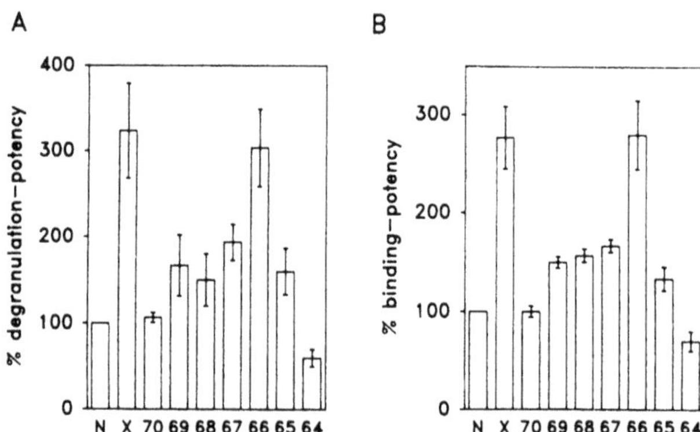

Figure 3. Relative potencies of native and recombinant NAP-2 variants for receptor binding and degranulation. N = native NAP-2; x = native C-terminally truncated NAP-2; 70–64: recombinant rNAP-2 (1–70) to rNAP-2 (1–64). A: relative potencies to induce degranulation in PMN (potency of native NAP-2 = 100%). B: relative potencies to bind to IL-8 receptors (potency of native NAP-2 = 100%). (From ref. 16.)

4. LOW CONCENTRATIONS OF NAP-2 DESENSITIZE PMN AND RENDER THEM UNRESPONSIVE TO A SUBSEQUENT STIMULUS OF α-CHEMOKINES

Our studies on the interaction of NAP-2 and IL-8 with the IL-8 receptor A and IL-8 receptor B and in particular cross-competition experiments with radioactively labeled NAP-2 or IL-8 demonstrated unequivocally that both ligands bind to the same receptors, although with different affinity[18]. Whereas IL-8 is able to bind to both receptors with high affinity (Kd ~ 3.0 nM), NAP-2 engages in the binding to IL-8 receptors with two affinities, a high affinity binding (Kd ~ 0.7 nM) to the IL-8 receptor B and a low affinity binding (Kd ~ 22 nM) probably to the IL-8 receptor A.

Since NAP-2 had been shown to be generated by proteolytic processing from its precursor CTAP-III, it could be envisaged that the increasing concentration of NAP-2 in a local environment would first lead to down-modulation of high affinity receptors. To imitate such a situation in vitro, PMN were preincubated for five minutes with a low concentration (2 nM) of NAP-2. This procedure led to a desensitization of the PMN with the consequence that the cell was practically refractory to a subsequent stimulus by NAP-2 and to a somewhat lower degree also to IL-8[18].

This phenomenon of desensitization was reflected as a complete absence of high affinity binding of NAP-2 and a partial reduction of low affinity binding of NAP-2 as well as a reduction of binding of IL-8[18].

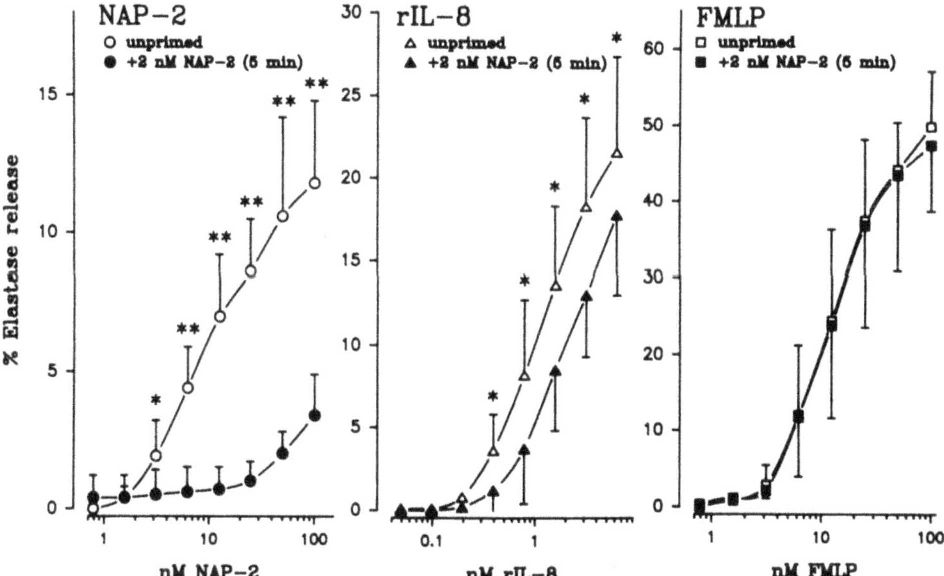

Figure 4. Short-term priming of PMN with low concentrations of NAP-2 desensitizes the cells to a subsequent stimulus of NAP-2 or IL-8, but not FMLP. Preincubation ("priming") for 5 min. with 2 nM NAP-2. Data given as mean ± SD of three independent experiments. Statistically significant differences between samples from "primed" versus "unprimed" PMN * $p < 0.05$, ** $p < 0.01$. (From ref. 18, Copyright 1994. The American Association of Immunologists.)

The fact that the cells responded to formyl-methyl-leucyl-phenylalanine (fMLP) in a similar fashion as the non-preincubated control cells suggested that the desensitization state was associated with a reduction of binding of α-chemokines to the common IL-8 receptors and, in particular, to the IL-8 receptor B. In subsequent studies the precursor molecule CTAP-III could be shown to desensitize PMN to an even stronger degree provided that the proteolytic processing to NAP-2 was maintained. More specifically, if the processing had been prevented by pretreatment of the cells with the serine protease inhibitor aprotinin the desensitizing effect of CTAP-III was completely abolished. This phenomenon could be observed at the level of the degranulation response of PMN. Furthermore, in parallel experiments CTAP-III was shown to down-modulate the binding of NAP-2 and IL-8 to common high affinity IL-8 receptors B, and receptor down-modulation was again abolished by treatment of the cells with aprotinin. From these findings it was concluded that the high affinity binding of newly generated NAP-2 to IL-8 receptor B was the active principle responsible for the desensitizing effect.

5. SUMMARY AND CONCLUSION: THE CTAP-III/NAP-2 SYSTEM REGULATES PMN FUNCTIONS

As mentioned above, low concentrations of NAP-2 would occupy a part of IL-8 receptors selectively. Furthermore, C-terminally truncated isoforms of NAP-2 exhibit higher binding capacity than full-size NAP-2. These findings enable to deduce the concept that NAP-2 due to its unique mode of generation by proteolysis might be a predominantly regulatory α-chemokine. Indeed, in human plasma the inactive precursors CTAP-III and PBP are present in micromolar concentrations. These inactive precursors would be con-

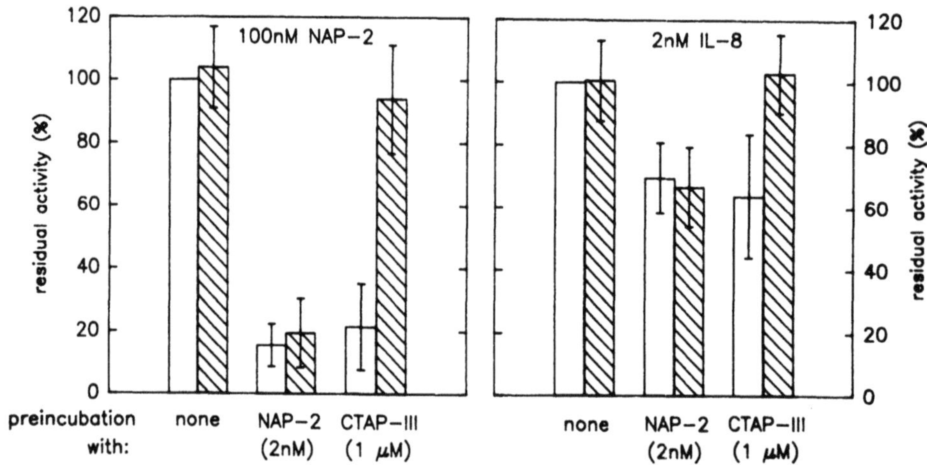

Figure 5. CTAP-III and NAP-2 down-regulate NAP-2- or IL-8-induced degranulation of PMN, and the protease inhibitor aprotinin abolishes the down-regulating effect of CTAP-III. PMN (1 x 10^7/ml) were treated with aprotinin (10 µg/ml) (hatched bars) or were left untreated (white bars). Before stimulation with 100 nM NAP-2 (left graph) or 2 nM IL-8 (right graph) cells were preincubated with 2 nM NAP-2 or 1 µg CTAP-III or were left unexposed. Data given as percentage of elastase release rate of stimulated but unexposed cells not receiving aprotinin. Mean ± SD from three different experiments. (From ref. 10, Copyright 1994. The American Association of Immunologists.)

stantly processed to biologically active NAP-2 unless this process would be controlled by the PMN target itself. It thus appears that the CTAP-III/NAP-2 system represents a physiologic regulatory control mechanism which prevents intravascular PMN from premature degranulation. Whether such a desensitized and regulated PMN would still be capable of responding to a strong chemotactic stimulus exerted by IL-8 or other chemotactic mediators to perform transendothelial diapedesis and phagocytosis at the inflammatory site is presently under intensive investigation by our group.

6. ACKNOWLEDGMENTS

We gratefully acknowledge the expert technical assistance of Mrs. G.Kornrumpf and Mrs. C.Pongratz. We also thank Mrs. R.Hinz for secretarial help with the preparation of the manuscript.

7. REFERENCES

1. Brandt,E., H.-D.Flad: Structure and function of platelet-derived cytokines of the ß-thromboglobulin/interleukin 8 family. Platelets 3, 295, 1992
2. Niewiarowski,S., B.Rucinski, A.Z.Budzynski: Low affinity platelet factor 4 and high affinity platelet factor 4: two antiheparin factors secreted by human platelets. Thromb.Haemost. 42, 1679, 1979
3. Files,J.C., T.W.Malpass, E.K.Yee, J.L.Ritchie, L.A.Harker: Studies of human platelet α-release in vivo. Blood 58, 607, 1981
4. Kameyoshi,Y., A.Dērschner, A.I.Mallet, E.Christophers, J.-M.Schrēder: Cytokine RANTES released by thrombin-stimulated platelets is a potent attractant for human eosinophils. J.Exp.Med. 176, 587, 1992
5. Oppenheim, J.J., C.O.Zachariae, N.Mukaida, K.Matsushima: Properties of the novel proinflammatory supergene "intercrine" cytokine family. Annu.Rev.Immunol. 9, 617, 1991
6. Van Damme,J., M.Rampart, R.Conings, B.Decock, N.Van Osselaer, J.Willems, A.Billiau: The neutrophil-activating proteins interleukin 8 and ß-thromboglobulin: in vitro and in vivo comparison of NH_2-terminally processed forms. Eur.J.Immunol. 20, 2113, 1990
7. Walz,A., B.Dewald, V.von Tscharner, M.Baggiolini: Effects of the neutrophil-activating peptide 2 (NAP-2), platelet basic protein, connective tissue-activating peptide III, and platelet factor 4 on human neutrophils. J.Exp.Med. 170, 1745, 1989
8. Brandt,E., M.Ernst, H.Loppnow, H.-D.Flad: Characterization of a platelet-derived factor modulating phagocyte functions and cooperating with interleukin 1. Lymphokine Res. 8, 281, 1989
9. Brandt,E., M.Ernst, H.-D.Flad: Enzymatic cleavage of CTAP-III from human platelets generates neutrophil-activating and anti-proliferative activities. In: Molecular and Cellular Biology of Cytokines (Eds.: J.J.Oppenheim, M.C.Powanda, M.J.Kluger, C.A.Dinarello), Wiley-Liss Inc., New York, pp. 357, 1990
10. Härter,L., F.Petersen, H.-D.Flad, E.Brandt: Connective tissue-activating peptide III desensitizes chemokine receptors on neutrophils. Requirement for proteolytic formation of the neutrophil-activating peptide 2. J.Immunol. 153, 5698, 1994
11. Walz,A., M.Baggiolini: Generation of the neutrophil-activating peptide NAP-2 from platelet basic protein or connective tissue-activating peptide III through monocyte proteases. J.Exp.Med. 171, 1797, 1990
12. Brandt,E., J.Van Damme, H.-D.Flad: Neutrophils can generate their activator neutrophil-activating peptide 2 by proteolytic cleavage of platelet-derived connective tissue-activating peptide III. Cytokine 3, 311, 1991
13. Yan,Z., J.Zhang, J.C.Holt, G.J.Stewart, S.Niewiarowski, M.Poncz: Structural requirements of platelet chemokines for neutrophil activation. Blood 84, 2329, 1994
14. Clark-Lewis,I., B.Dewald, T.Geiser, B.Moser, M.Baggiolini: Platelet factor 4 binds to interleukin 8 receptors and activates neutrophils when its N terminus is modified with Glu-Leu-Arg. Proc.Natl.Acad.Sci.USA 90, 3574, 1993
15. Brandt,E., F.Petersen, H.-D.Flad: A novel molecular variant of the neutrophil-activating peptide NAP-2 with enhanced biological activity is truncated at the C-terminus: identification by antibodies with defined epitope specificity. Mol.Immunol. 30, 979, 1993

16. Ehlert,J.E., F.Petersen, M.H.G.Kubbutat, J.Gerdes, H.-D.Flad, E.Brandt: Limited and defined truncation at the C terminus enhances receptor binding and degranulation activity of the neutrophil-activating peptide 2 (NAP-2). J.Biol. Chem. 270, 6338, 1995
17. Clark-Lewis,I., C.Schumacher, M.Baggiolini, B.Moser: Structure-activity relationships of interleukin-8 determined using chemically synthesized analogs. Critical role of NH_2-terminal residues and evidence for uncoupling of neutrophil chemotaxis, exocytosis, and receptor binding activities. J.Biol.Chem. 266, 23128, 1991
18. Petersen,F., H.-D.Flad, E.Brandt: Neutrophil-activating peptides NAP-2 and IL-8 bind to the same sites on neutrophils but interact in different ways. Discrepancies in binding affinities, receptor densities, and biologic effeccts. J.Immunol. 152, 2467, 1994

SELECTIVE PROTEOLYTICAL CLEAVAGE OF THE LIGAND-BINDING CHAINS OF THE IL-2-RECEPTOR AND IL-6-RECEPTOR BY NEUTROPHIL-DERIVED PROTEASES

U. Bank,[1] D. Reinhold,[1] D. Kunz,[2] and S. Ansorge[1]

[1]Institute of Experimental Internal Medicine
Center of Internal Medicine
[2]Institute of Clinical Chemistry
Otto-von-Guericke-University Magdeburg
Leipziger Str. 44, D-39120-Magdeburg, Germany

Soluble forms of usually membrane bound molecules were identified in blood or other body fluids. Among these soluble forms of membrane proteins, which have been found to lack the transmembrane and the intracellular domain, the soluble cytokine receptors are of interest, because they have been found to be capable of binding the ligand and of influencing the biological activity of the cytokines [1,2].

The origin of soluble forms of membrane proteins *in vivo* is unclear as yet. Two principally different mechanisms for the release of soluble receptors have been discussed: One mechanism is thought to be the release of the soluble receptor molecules as a product of an alternative spliced mRNA [3,4]. The other mechanism discussed is the solubilization of the membrane bound molecule by proteolytic cleavage [5–7]. However, for both mechanisms hints were provided by several groups [1].

The contribution of proteolytic enzymes on the release of soluble cytokine receptor molecules seems to be of special interest, because several clinical studies demonstrated that the concentrations of soluble cytokine receptors, especially those of immunostimulating cytokines as IL-2, TNF-α, and IL-6, increase in disease states which are characterized by an elevated proteolytic potential [8–10].

A specific way of proteolytic solubilization of membrane bound proteins seems to be the action of special cellular proteases, which are located on the surface of the same cell as the shedded receptor molecule and which cleave the membrane bound protein in response to a distinct signal. *In vitro* this process was found to be inducable by the PKC-affecting agent phorbolmyristat acetate [11–13]. Inhibitor studies provided evidence that this specific way of proteolytic receptor cleavage is catalyzed by cell surface bound metalloproteinases [1,14–16], but distinct enzymes could not be identified as yet.

Cellular Peptidases in Immune Functions and Diseases, edited by Ansorge and Langner
Plenum Press, New York, 1997

Inflammatory processes were found to be associated with the release of high amounts of proteolytic enzymes by several cell types as monocytes, macrophages, mast cells and granulocytes into the extracellular space [17–19]. Among the extracellularly released enzymes the serine proteases elastase, cathepsin G and proteinase 3 derived from the azurophilic granules of neutrophil granulocytes are of special interest because these enzymes are catalytically active at physiological pH-values and characterized by a broad substrate specificity [20,21]. At local sites of inflammation the neutrophil-derived serine proteases were found to be inhibited inefficently, because the natural inhibitors are often low concentrated and furthermore inactivated [22–24]. Several structural and biological active proteins were identified as substrates of inefficiently inhibited neutrophil serine proteases [25–28]. The proteolysis of these substrates *in vivo* is thought to cause secondary tissue damage and dysregulation of the inflammatory process [29–34]. The ability of neutrophil-derived serine proteases to cleave several cell surface bound proteins [35, 36] suggests a putative role of these enzymes in releasing soluble cytokine receptors at local sites of inflammation.

Here we demonstrate a close temporal correlation between the release of soluble IL-2 and IL-6-receptors and elevated concentrations and activities of neutrophil-derived serine proteases *in vivo* as well as the capability of elastase, cathepsin G and proteinase 3 in catalyzing the cleavage of the ligand-binding chains of functional IL-2- and IL-6-receptors from the cell surface of vital cells *in vitro*.

TEMPORAL CORRELATION BETWEEN THE RELEASE OF SOLUBLE IL-2R AND IL-6R LIGAND-BINDING CHAINS AND ELEVATED CONCENTRATIONS OF NEUTROPHIL-DERIVED SERINE PROTEASES IN INFLAMMATORY PROCESSES

A putative role of the neutrophil-derived serine proteases in the proteolytical cleavage of soluble cytokine receptors is thought to be reflected by a temporal correlation in the concentration changes of the soluble cytokine receptors and the extracellularly released proteases during the time course of inflammation. Clinical investigations focused on two models of inflammatory diseases: acute pancreatitis as an example for of a severe and exceeded inflammatory process with systemic consequences and isolated head injury as an example for trauma-induced, locally restricted inflammatory process.

The *in vivo* levels of soluble IL-2 and IL-6 receptors in the plasma of patients with severe acute pancreatitis were determined by commercially available immunoassays (BIOSOURCE, USA). The plasma concentrations of elastase as one of the three serine proteases released from the azurophilic granules of polymorphonuclear neutrophils were measured by using a modified ELISA basing on a test system purchased from MERCK, Germany.

Figure 1 demonstrates the close temporal correlation between the elevated proteolytic potential and increasing levels of the soluble forms of the IL-2 and IL-6 receptors in one of the patients with acute pancreatitis. This temporal correlation was exemplary for all investigated patients (N=7). However, while no significant changes in the relative amount of IL-2 and IL-6 receptor expressing mononuclear cells could be measured by flow cytometry, a remarkable decrease in IL-6 receptor expressing neutrophils was found in the circulation corresponding to the elevated protease concentrations.

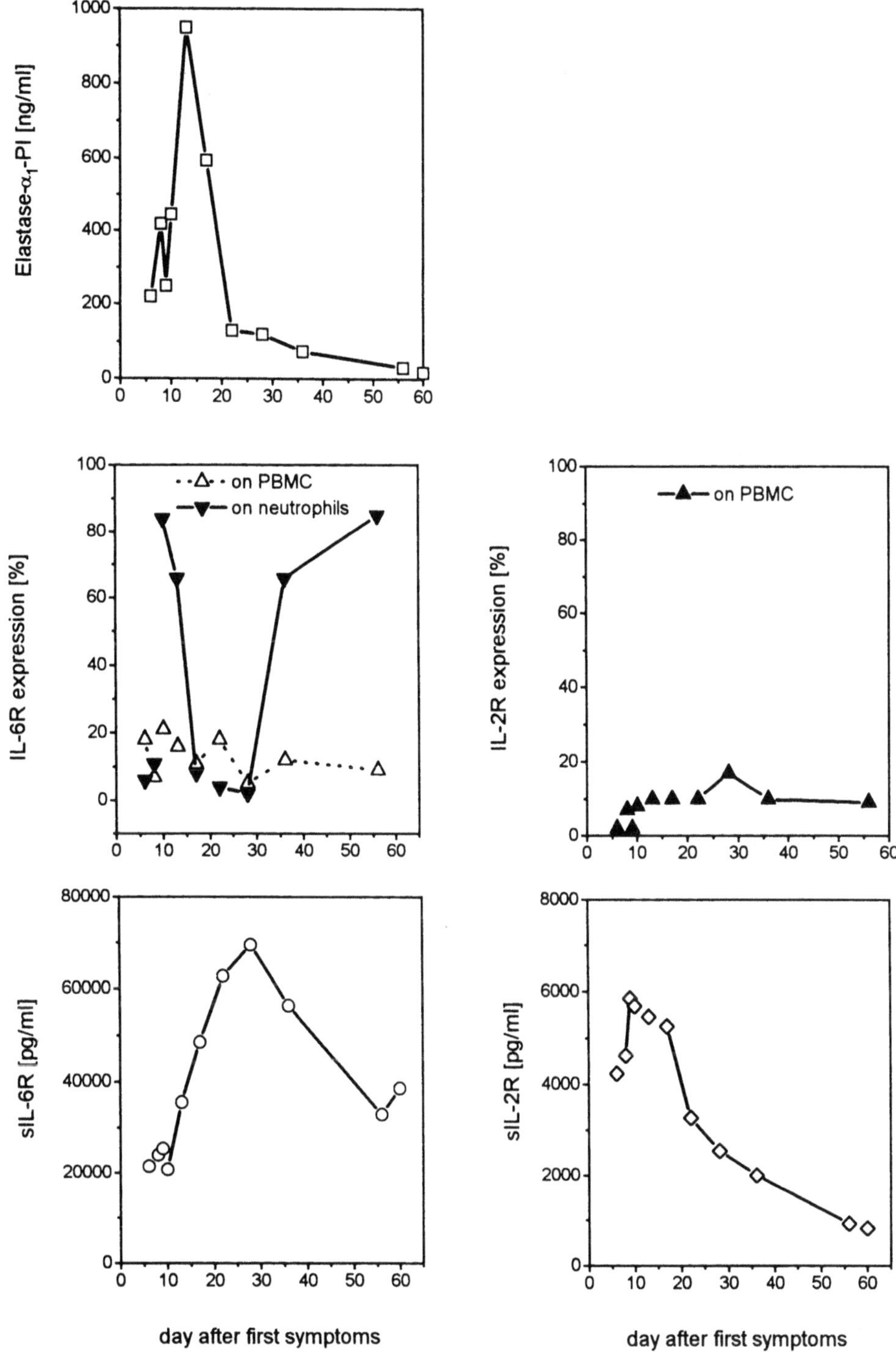

Figure 1. Correlation between the concentration of PMN-elastase and the release of soluble cytokine receptors in acute pancreatitis. This figure represents the data of a 31-year-old patient with severe pancreatitis (sepsis, artificial respiration needed). The plasma concentrations of the PMN-elastase-α_1-proteinase inhibitor-complex, sIL-2R and IL-6R were determined by enzymimmunoassays. The cellular expression of the IL-2R and IL-6R ligand-binding chains was analyzed by flow cytometry.

A similar temporal correlation between cytokine receptor release could be observed in locally restricted inflammatory processes, as shown in figure 2 exemplary for patients undergoing isolated head injury. The concentrations of the neutrophil-derived proteases and the soluble IL-2 and IL-6 receptors were determined in fluids obtained via subarachnoidal drains (routinely established for monitoring of intracerebral pressure) directly from the local site of tissue destruction. In the early posttraumatic phase, which was found to be associated with rapid increasing protease concentrations and very low concentrations of the natural serine protease inhibitor α_1-proteinase inhibitor, a simultaneous rapid increase in the local concentrations of the soluble IL-2R and IL-6R could be observed. The protease-antiprotease-imbalance in this phase of inflammation was confirmed by the measurement of proteolytic

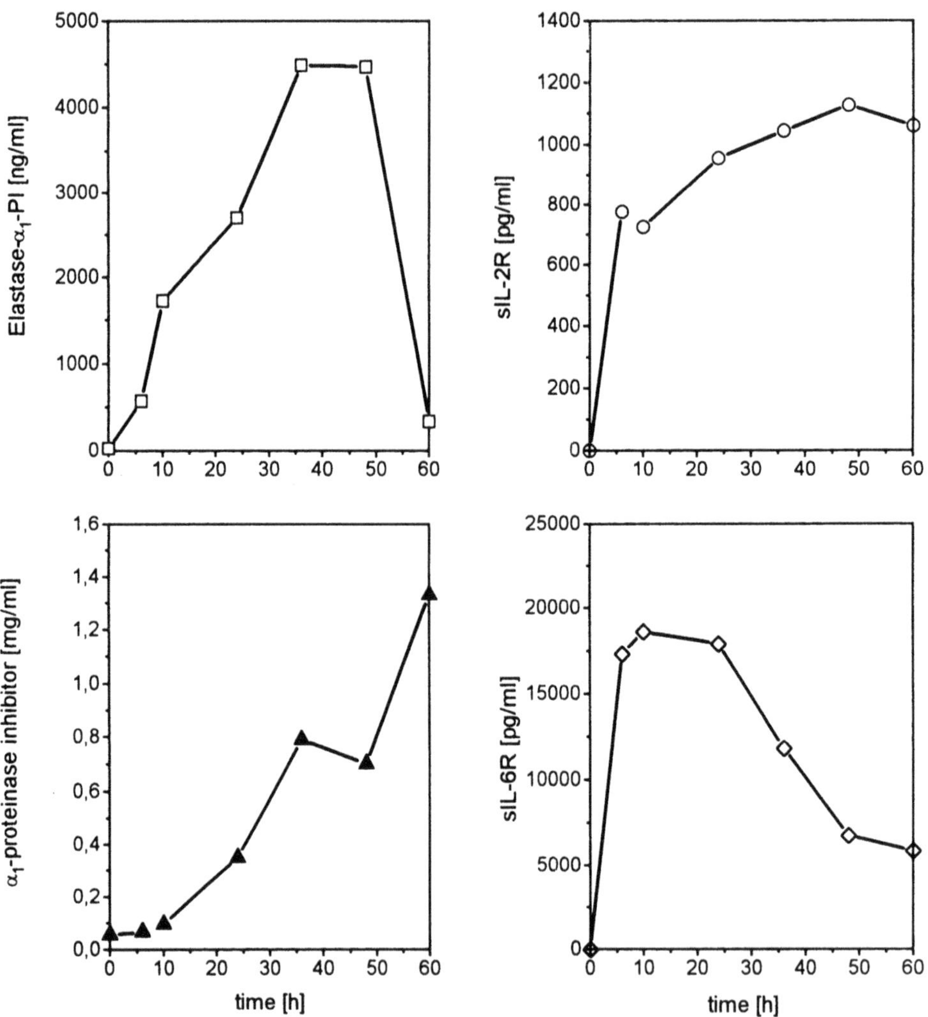

Figure 2. Correlation between the elevated proteolytical potential and the release of soluble cytokine receptors at local sites of inflammation, demonstrated by means of the posttraumatic intracerebral concentrations of elastase, the natural inhibitor and the soluble IL-2- and IL-6-receptor in patients with neurotrauma. The data of a 24-year-old patient with isolated head injury are shown exemplary for 31 patients investigated in this study. Iatrogene neurotrauma (surgery of benign tumor). Time point zero represents induction of neurotauma immediately dura opening!

activities of elastase and cathepsin G in the inflammatory exudates from the local site of the neurotrauma using specific synthetic substrates. The rather slower increase or decrease of the sIL-2R- and IL-6R-concentrations in the further time course, which was associated with higher inhibitor concentrations suggests direct interactions between unefficently inhibited neutrophil-derived serine proteases and the solubilization of the cytokine receptors.

In contrast to plasma samples of healthy donors non-inflammatory control liquor samples were found to contain undetectable or very low concentrations of the soluble IL-2-receptor as well as of the soluble IL-6-receptor. By calculating the relative amount of infiltrating plasma proteins into the intracerebral inflammatory area (via the quotient of liqour and plasma albumin concentrations as a degree of blood-brain-barrier-disorders), it becames evident that the increase of the soluble cytokine receptor concentrations detectable in the inflammatory exudates is in fact the result of a local process and is not caused by the infiltration of soluble receptor molecules present in the cirulation.

CLEAVAGE OF THE IL-2Rα-CHAIN FROM ACTIVATED HUMAN T-CELLS BY NEUTROPHIL-DERIVED PROTEASES

1. Effect of Proteases-Rich Culture Supernatants and Lysates of Neutrophils

At local sites of inflammation high amounts of catalytically active proteases were released from degranulating neutrophils, suggesting that these enzymes might contribute to the solubilization of functional cytokine receptors. Therefore the task of *in vitro*-studies was to prove possible effects of neutrophil-derived proteases on the cytokine receptor solubilization.

Freshly isolated polymorphonuclear neutrophils of healthy donors were stimulated with the chemotactic tripeptide formyl-Met-Leu-Phe (fMLP, 10^{-7}M) for 30 min under serumfree conditions in order to induce the release of high amounts of proteolytic enzymes by degranulation. The incubation of PHA-activated T-cell blasts for 60 min in the presence of various dilutions of these protease-rich neutrophil culture supernatants resulted in an increase of the concentrations of the soluble IL-2Rα in the culture supernatant of the T-cell blasts as shown in table 1. The solubilization was found to be dependent from the protease amount. This was confirmed by the remarkable release of the IL-2Rα-chain detectable following the incubation

Table 1. Effect of protease-rich neutrophil-derived culture supernatants and lysates on the solubilzation of the IL-2Rα-chain from PHA-activated T cells. Results from 6 experiments given as mean±SD

Treatment	sIL-2Rα [pg/ml]	Cellular IL-2Rα expression (mean of fluorescence intensity of the IL-2Rα-labelling with anti CD25-PE)
Control	499 ± 193	7.49 ± 1.02
+ Neutrophil-derived culture supernatant	826 ± 401	7.32 ± 0.79
1:10	1930 ± 1246	5.25 ± 2.48
1:2		
+ Neutrophil-derived culture supernatant	892 ± 195	6.99 ± 1.21
1:2 + 100μM DFP		
+ Neutrophil-lysate		
1:2	6332 ± 1569	1.07 ± 0.98

of the T cell-blasts in the presence of neutrophil lysates which contain the approximately five-fold protease amount in comparison to the culture supernatants. The increasing concentrations of the soluble IL-2Rα in the T cell culture supernatants after the short time incubation in the presence of the protease-rich neutrophil-derived culture supernatants and lysates were accompanied by a remarkable reduction of number of IL-2Rα-molecules on cell surface of the T cell blasts (see also table 1) as determined by flow cytometry using a phycoerythrin-labelled monoclonal anti-CD25 antibody (BECTON-DICKENSON).

In the presence of the serine protease inhibitor diisopropylfluorophosphate (DFP, 100μM) the solublization of the IL-2Rα-chain was found to be abolished almost completely. Metalloprotease inhibitors (EDTA, ortho-Phenanthrolin) had only slight inhibitory effects at higher concentrations.

2. Effect of Purified Neutrophil-Derived Serine Proteases

The almost complete inhibition of the release of the IL-2Rα-chain from the activated T cells in the presence of a serine protease inhibitor led to the conclusion, that catalytically active serine proteases derived from neutrophils are capable to cleave the cell surface bound IL-2Rα-chain. The task of further *in vitro*-investigations was to evaluate the distinct role of the serine proteases elastase, cathepsin G and proteinase III released from the azurophilic granules of neutrophils.

As demonstrated in figure 3, the serine proteases elastase and proteinase III, both characterized by a trypsin-like activity, were found to catalyze the cleavage of the cell surface bound IL-2Rα-chain in a concentration-dependent manner. The proteolytical solubilization of the IL-2Rα-chain was reflected by the increasing concentration of the soluble IL-2-receptor in the culture supernatant and by a remarkable decrease in the number of cell surface bound IL-2-receptor molecules per cell in dependence on the protease concentration.The loss of cell surface-bound IL-2Rα was detectable by means of the fluorescence intensity of the binding of the phycoerythrin-labelled anti-CD25-antibody to cell surface-bound receptor molecule.

Proteinase III elicits significantly lesser effects in comparison to the neutrophil-derived elastase, while cathepsin G—the neutrophil-derived protease with a chymotrypsin-like activity—had none.

The protease concentrations used in these experiments were within the range of concentrations detectable in various biological fluids derived from local sites of inflammation, which often reached 20μg/ml elastase and more. The short time treatment of the T cell blasts in the presence of the purified proteases was found to cause no decrease of the viability rate as well as no significant change in the cellular expression of the T cell antigens CD3 and CD4 (data not shown).

CLEAVAGE OF THE IL-6Rα-CHAIN FROM U937 CELLS BY NEUTROPHIL-DERIVED PROTEASES

1. Effect of Proteases-Rich Culture Supernatants and Lysates of Neutrophils

Comparable experiments were performed using a subclone of the U937 cell line, which expressed high levels of the gp80 ligand-binding IL-6 receptor chain, constitutively. After the

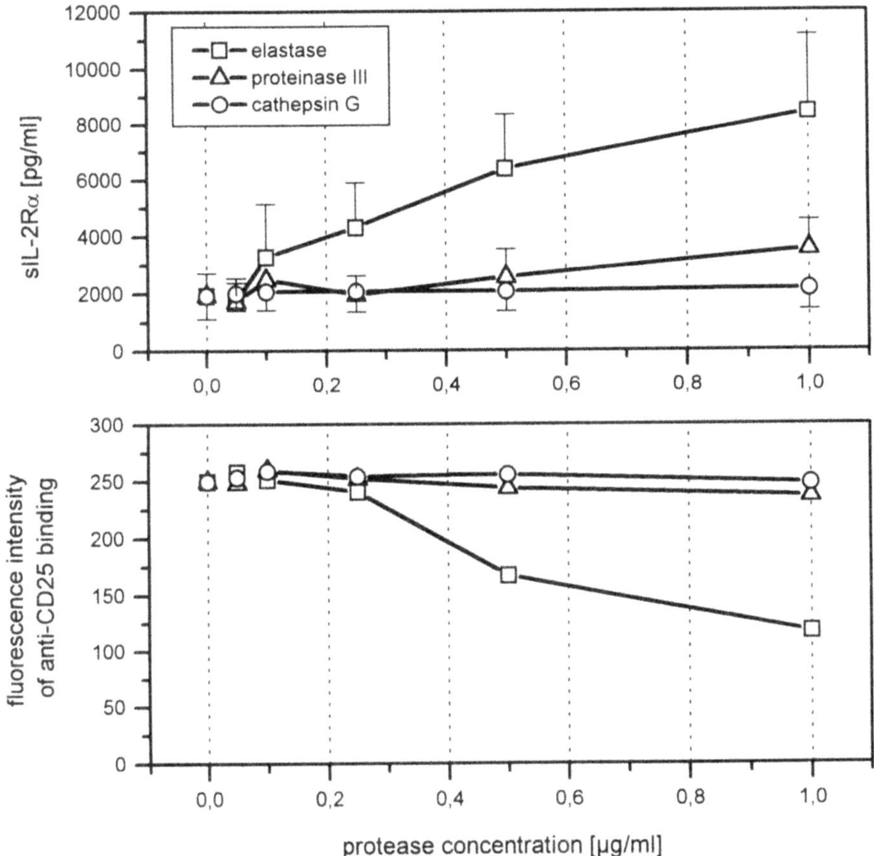

Figure 3. Cleavage of the IL-2Rα-chain from activated T cells catalyzed by neutrophil-derived serine proteases. T cells were isolated and PHA -activated under serum free conditions in order to induce a high IL-2 receptor α -expression. T cell blast ($2x10^6$ per ml) were incubated for 60 min in presence of increasing concentrations of the purified proteases elastase, cathepsin G and proteinase III incubation was stopped by adding a protease inhibitor mixture followed by a centrifugation step. The separated cells were prepared for flow cytometric analysis using a phycoerythrin-labelled monoclonal anti-CD25 antibody. The concentrations of sIL-2R in the culture supernatants was determined by enzymimmunoassay.

incubation of $2x10^6$/ml U937cells for 60 min in the presence of various dilutions of the protease-rich neutrophil-derived culture supernatants and lysates increased concentrations of the soluble IL-6R ligand-binding chain were detectable in the culture supernatant (table 2). While after the treatment only slight changes in the number of IL-6R expressing cells were detectable by flow cytometry using a monoclonal anti-gp-80 IL-6R antibody (BIOSOURCE, data not shown), a remarkable reduction of the fluorescence intensity which reflects the number of IL-6-receptor molecules per cell could be observed. In presence of neutrophil lysates which contained approximately the 5-fold protease amount as the neutrophil culture supernatants, these effects were strengthened (see also table 2).

In analogy to the results described above the IL-6R-solubilization was found to be prevented by adding of the serine protease inhibitor DFP, suggesting that indeed enzymatically active serine proteinases from neutrophils are capable in catalyzing the cleavage of both cell surface bound IL-2 and IL-6 receptors.

Table 2. Effect of protease-rich neutrophil-derived culture supernatants and lysates on the solubilzation of the IL-6R gp80-chain from U937 cells. Results from 6 experiments given as mean±SD

Treatment	sIL-6R [pg/ml]	Cellular IL-6R expression (mean of fluorescence intensity of the IL-6R-labelling with anti-gp80)
Control	335 ± 134	1.97 ± 0.94
+ neutrophil-derived culture supernatant		
1:10	1056 ± 616	1.75 ± 0.80
1:2	2772 ± 1292	1.40 ± 0.95
+ neutrophil-derived culture supernatant		
1:2 + 100μM DFP	798 ± 373	1.85 ± 0.73
+ neutrophil-lysate		
1:2	6950 ± 2214	1.15 ± 0.42

2. Effect of Purified Neutrophil-Derived Serine Proteases

Proving the capability of the purified serine proteases from the azurophilic granules of neutrophils to catalyze the cleavage the IL-6-receptor gp80-chain from the surface of U937 cells, it became evident that the neutrophil-derived serine proteases act selectively with regard to the solubilization of several cell surface bound receptor molecules. Whereas elastase and proteinase III were found to be not capable of catalyzing the IL-6-receptor solubilization, the cleavage of the IL-6- receptor gp80-chain was observable in the presence of cathepsin G. This was reflected by means of the increasing concentrations of soluble IL-6-receptor molecules in the culture supernatant of cathepsin G treated cells as well as by means of the decreasing number of membrane bound IL-6-receptor molecules per cell, as indicated in figure 4.

The highly purified proteases in the concentrations used in these experiments were found to elicit no effect on the viability of the cells. Furthermore, no changes in the expression of myeloid cell surface antigens CD15 and CD66b on the cell surface of U937 cells were detectable by flow cytometry after the protease treatment.

DISCUSSION

Cytokine receptors have been found to be one of the targets of the complexe control mechanisms, which determine the biological activities of cytokines: The response of a distinct cell to a cytokine is considerably influenced by the number of receptor molecules per cell, the affinities of several types of receptors, the presence of associated proteins as well as receptor antagonists. The modulation of the cytokine receptor expression in response to changes of enviromental conditions was described to be associated with the solubilization of cell surface bound receptor molecules. Proteolytic enzymes were implicated to be capable in catalyzing the cleavage of the cell surface bound receptors.

Evidences were provided that specific metalloproteinases catalyze the shedding of membrane bound receptor molecules in response to distinct signals.[1,14-16] Besides these metalloproteinases which are located on the surface of the same cell as the shedded receptor, proteolytic enzymes released from several cell types into the extracellular milieu were

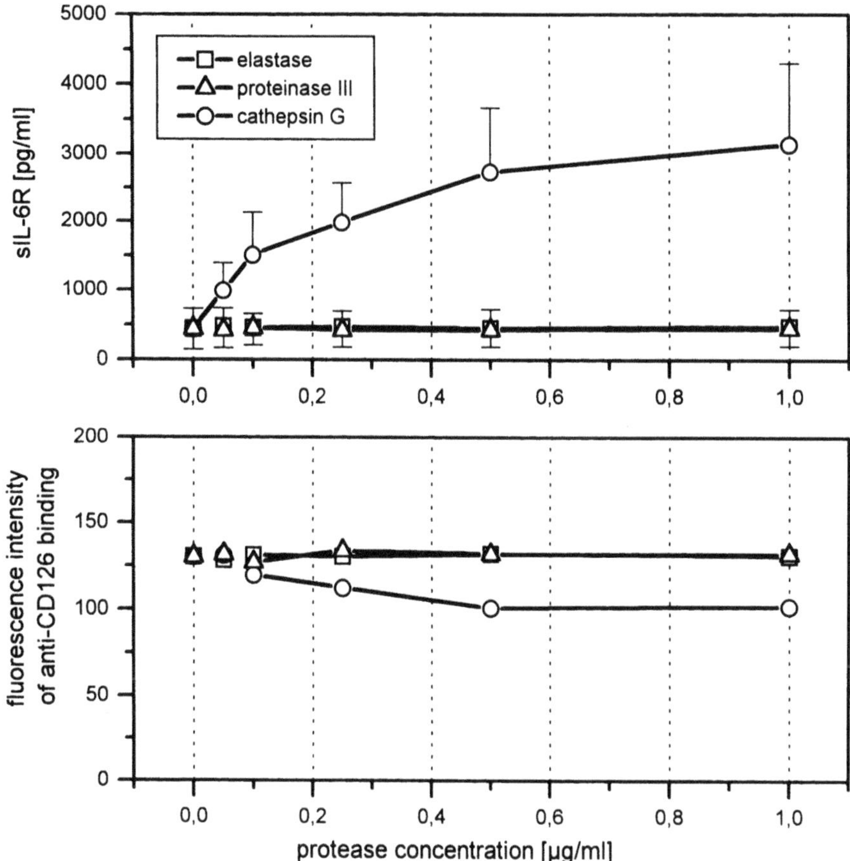

Figure 4. Cleavage of the IL-6R gp80-chain from U937 cells catalyzed by neutrophil-derived serine proteases. U937 cells ($2x10^6$/ml) were incubated for 60 min under serumfree conditions in the presence of the purified proteases. The sIL-6R concentrations in the culture supernatants of the U937 cells were measured by enzymimmunoassay. the cellular expression of the gp80-chain was detected by flow cytometry using a monoclonal anti-IL-6R gp80-antibody.

discussed to contribute to the cleavage of cell surface bound proteins. In this context especially the action of proteases released from activated neutrophils might have pathophysiological relevance.

The presented data show that serine proteases derived from the azurophilic granules of polymorphonuclear neutrophils are also capable to catalyze the clevage of the ligand-binding chains of the IL-2- and IL-6-receptor. Although the proteases namely elastase, proteinase and cathepsin G were known to be characterized by a broad substrate specificity, we demonstrate here, that they act selectively with regard to both cytokine receptors. Considering the substrate specificity of these enzymes - whereas elastase and proteinase III were shown to catalyze preferentially the hydrolysis of peptide bonds with small, uncharged non-aromatic amino acids in P_1-position [37], cathepsin G the cleavage of peptide bonds with aromatic amino acids in P_1-position [21]- the amino acid sequence of both receptor molecules [38–41] contains several potential cleavage sites susceptible for these proteases. However, the task of further investigations has to be the proteinbiochemical analysis of the receptor fragments released by the action of the serine proteases with regard to the cleaved peptide bonds and the ligand-binging characteristics.

The selectivity of the action of the rather unspecific neutrophil-derived proteases on cell surface bound molecules was also confirmed by the finding that the short time treatment of the T cells or U937 cells in the presence of the purified proteases had no influence either on the cellular expression of the T cell specific antigens CD3 and CD4 nor on the myeloid antigens CD15 and CD66b on the surface of the U937 cells in the concentration range used here. In contrast to that, Döring and collaborators demonstrated effects of elastase and cathepsin G on the T cell antigens CD4 and CD8 [42]. Several other cell surface bound molecules - among them membrane-bound precursors of cytokines - were identified to be substrates of neutrophil-derived serine proteases [35, 36, 43].

The cellular IL-2R- and IL-6R expression on protease treated cells were found to reach the level of untreated control cells within 24 hours after the short time protease treatment, indicating that the cells were not damaged irreversibly by the proteolytic activities. The temporary decreased expression of IL-2- and IL-6- ligand-binding chains induced by the proteolytic activities might result rather in a temporary functional impairment of the cells to response to the immunostimulation cytokines at local sites of inflammation. Especially the immunostimulatory activities of IL-2 should be strongly disturbed, when the cell surface bound IL-2Rα molecules are shedded by proteases. With high probability the cells would be capable only with a lesser degree to form the high affinity-α -receptor trimer. Furthermore, the ligand-binding soluble IL-2R-molecules could prevent the binding of the ligand IL-2 to cell surface bound receptors. On the other hand the complex of the soluble IL-6R gp80-chain and the ligand IL-6 was shown to be capable to bind on the gp130 signal transducing protein and to trigger signalling on target cells lacking the IL-6R gp 80-chain, but expressing the gp130 molecule [44]. Undoubtedly, the state of responsiveness of cells to the cytokines IL-2 and IL-6 is closely connected with the cleavage of cell surface bound ligand-binding chains of the cytokine receptors. Nevertheless, the knowledge concerning the regulatory mechanisms underlying the contol of the cytokine activities and cytokine-receptor interactions is full of gaps, so it seems to be impossible to overview all the effects associated with the receptor solubilization.

The temporal correlation between the elevated proteolytic potential and increasing concentrations of soluble cytokine receptors *in vivo* and the demonstrated capability of neutrophil proteinases to catalyze the receptor cleavage support the hypothesis of a putative involvement of neutrophil proteases in the proteolytic cleavage of cell surface bound cytokine receptor molecules. Taken into consideration the high concentrations of these enzymes and the low concentrations of natural inhibitors at local sites of inflammation a *in vivo* relevance of this way of receptor shedding is absolutely conceivable. This protease-induced cytokine receptor cleavage and the associated impairmant of the cells to response to the immunostimulating cytokines IL-2- and IL-6 could be considered as an additional example for intervention of neutrophil-derived proteases into regulatory processes of the inflammatory response.

ACKNOWLEDGMENTS

This work was supported by the Deutsche Forschungsgemeinschaft grant A3 of SFB 387.

The authors wish to thank Mrs. Ines Meinert and Mrs Michaela Daum for the exellent technical assistance and Dr. H.-U. Schulz and Dr. Chr. Schneemilch for supporting the clinical investigations.

REFERENCES

1. Rose-John, St. and Heirich, P.C. (1994) Soluble receptors for cytokines and growth factors: generation and biological function. Biochem. J., 300, 281–290
2. Fernandez-Botran, R. (1991) Soluble cytokine receptors: their role in immunoregulation. FASEB J. 5, 2567–2574
3. Fernadez-Botran, R and Viletta, E.S. (1990) Proc. Natl. Sci. U.S.A. 87, 4202–4206
4. Goodwin, R.G., Friend, D., Ziegler, S.F., Jerzy, R., Falk, B.A., Gimple, S., Cosman, D., Dower, S.K., March, C.J., Namen, A.E. and Park, L.S. (1990) Cell 60,, 941–951
5. Müllberg, J., Schooltink, H., Stojan, T., Günter, M., Graeve, L., Buse, G., Mackiewicz, A., Heinrich P.C. and Rose-John, St. (1993) The soluble IL-6- receptor is generated by shedding. Eur. J. Immunol. 23, 473–480
6. Bazil, V. and Strominger, J.L. (1991) Shedding is a mechanism of down modulation of CD14 on stimulated human monocytes. J. Immunol. 147/5, 1567–1574
7. Arribas, J., Coodly, L., Vollmer, P., Kishimoto, T.K., Rose-John, St. and Massague, J. (1996) Diverse cell surface protein ectodomains are shed by a system sensitive to metalloproteinase inhibitors. J. Biol. Chem. 271/19, 11376–11382
8. Ching-Hon Pui (1989)Serum interleukin-2-receptor: clinical and biological implications. Leukemia, 3/5, 323–327
9. Van Zee, K.J., Kohno, T., Fischer, E., Rock, C.S., Moldawer, L.L. and Lowry, S.F. (1992) Tumor necrosis factor soluble receptors circulate during experimental and clinical imflammation and can protect against excessive tumor necrosis factor in vitro and in vivo. Proc. Natl. Acad. Sci. U.S.A. 89, 4845–4849
10. Frieling, J.T.M., Sauerwein, R.W., Wijdenes, J., Hendriks, T., and van der Linden, G.J. (1994) Soluble Interleukin 6 receptor in biological fluids from human origin. Cytokine 6,376–381
11. Fritz, S., Striggow, F., Reinhold D., Schluter, T., Schonfeld, P., Ansorge, S. Bohnensack, R.(1996) Phorbol ester-induced shedding of intercellular adhesion molecule-1 (ICAM-1) on erythroleukemic K 562 cells. Biochim-Biophys-Acta. 1312/3, 255–261
12. Galve-de-Rochemonteix, B., Nicod, L.P., Dayer J.M.(1996) Tumor necrosis factor soluble receptor 75: the principal receptor form released by human alveolar macrophages and monocytes in the presence of interferon(gamma). Am. J. Respir. Cell. Mol. Biol. 14/3, 279–87
13. Porteu, F., Brockhaus, M., Wallach, D., Engelmann, H. and Nathan, C.F. (1991) Human neutrophil elastase releases a ligand-binding fragment from the 75-kDa Tumor Necrosis Factor (TNF) receptor. J. Biol. Chem. 266/28, 18846–18853
14. Kayagaki, N., Kawasaki, A., Ebate, T., Ohmoto, H., Ikeda, S., Inoue, S., Yoshino, K., Okomura, K. and Yagita, H. (1995) Metalloproteinase-mediated release of human Fas-ligand J. Exp. Med. 182 1777–1783
15. Crowe, P., Walter, B.N., Mohler, K.M., Otten-Evans, C., Black, R.A., Ware, C.F. (1995) A metalloproteinase inhibitor blocks shedding of 80kD TNF receptor and TNF processing in t-lymphocytes. J. Exp. Med. 181, 1205–1210
16. Müllberg, J., Durie, F.H., Otten-Evans, C., Alderson, M.R., Rose,-John, St., Cosman, D., Black, R.A. and Mohler, K., M. (1995) A metalloproteinase inhibitor blocks the shedding of then IL-6 receptor and the TNF receptor. J. Immunol. 155, 5198–5205
17. Campbell, E.J., Cury, J.D., Shapiro, S.D., Goldberg, G.I. and Welgus, H.G. (1991) Neutral proteinases of human mononuclear phagocytes, J. Immunol. 146, 1286–1293
18. Trabandt, A., Gay, R.E., Fassbender, H.-G. and Gay, S. (1991) Cathepsin B in synovial cells at the site of joint destruction in rheumatoid arthritis. Arthritis and Rheumatism **34**, 1444–1451
19. Nadel, J.A. (1991) Role of mast cell and neutrophil proteases in airway secretion. Am. Rev. Respir. Dis. **144**, 48–51
20. Olssen, I. and Venge, S. 1974) Cationic proteins of human granulocytes. II. Separation of the cationic proteins of granules of leukemic myeloid cells. Blood **44**, 235–246 (1974)
21. Travis, J (1978) Neutral proteinases of human polymorhonuclear leukocytes, Biochemistry, physiology and clinical significance. Eds. Havemann, K., Janoff, A. Urban & Schwartzenberg Baltimore 118–1128

22. Johnson, D and Travis, J. (1978) The oxidative inactivation of human α_1-proteinase inhibitor. Further evidence for methionine at the reactive center. J. Biol. Chem. 254, 4022–4026
23. Chidwick, K., Winyard, P.G., Zhang, Z., Farrell, A.D., Blake, D.R. (1991) Inactivation of the elastase inhibitory activity of α_1-antitrypsin in fresh samples of synovial fluid from patients with rheumatoid arthritis. Ann. Rheumat. Dis. 50, 915–916
24. Michaelis, J., Vissers, M.C.M. and Winterbourn, C.C. (1990) Human neutrophil collagenase cleaves α_1-antitrypsin. Biochem. J. 270, 809–814
25. Janusz, M.J., Doherty, N.S.: Degradation of cartilage matrix proteoglycan by human neutrophils involves both elastase and cathepsin G. J. Immunol. **146**, 3922–3928 (1991)
26. Eckle, I., Seitz, R., Egbring, R., Kolb, G. and Havemann, K. (1991) Protein C degradation in vitro by neutrophil elastase. Biol. Chem. Hoppe-Seyler **372**, 1007–1013
27. Turkington, S.T.(1991) Degradation of human factor X by human polymorphonuclear leucocyte cathepsin G and elastase. Hämostasis **21**, 111–116
28. Padrines, M., Wolf, M., Walz, A. and Baggiolini, M. (1994) Interleukin-8 processing by neutrophil elastase, cathepsin G and proteinase-3. FEBS Letters **352**, 231–235
29. Hannah, S. and Ryle, A.S. (1991) Proteolysis of lung elastin by human neutrophil elastase. Biochemical Society Transactions **19**, 294
30. Eckle, I., Kolb, G. and Havemann, K.(1991) Inhibition of neutrophil chemotaxis by elastase-generated IgG fragments. Scand. J. Immunol. **34**, 359–364
31. Kanayama, N., Terao, T.: Inactivation of human tumor cell pro-urokinase by granulocyte elastase. Jpn. J. Cancer Res. **81**, 994–1002 (1990)
32. Janoff, A (1985) Elastase in tissue injury. Ann. Rev. Med. 36, 207–216
33. Brown, D.M., Brown, G.M., Macnee, W. and Donaldson, K. (1992) Activated human peripheral blood neutrophils produce epithelial injury and fibronectin breakdown in vitro. Inflammation 16, 21–30
34. Jochum, M., Gippner-Steppert, C., Machleidt, W. and Fritz, H. (1994) The role of phagocyte proteinases and proteinase inhibitors in multiple organ failure. Am. J. Crit. Care Med. 150, S123–130
35. Johnstone, R.M. (1996) Cleavage of the transferrin receptor by human granulocytes: differential proteolysis of the exosome-bound TFR. J. Cell. Physiol. 168/2, 333–345
36. Bjornberg, F., Lantz, M., Gullberg, U. (1995) Metalloproteases and serineproteases are involved in the cleavage of the two tumour necrosis factor (TNF) receptors to soluble forms in the myeloid cell lines U-937 and THP-1. Scand-J-Immunol. 42/4, 418–24
37. Nakajima, K., Power, J.C., Ashe, B.M. and Zimmermann, M. (1979) Mapping the extended substrate binding site of cathepsin G and human leukocyte elastase. J. Biol.Chem. 254, 4027–4032
38. Leonhard, W.J., Deppner, J.M., Crabtree, G.R., Rudikoff, S., Pumphrey, J., Robb, R.J., Kronke, M., Svetlik, P.B., Pfeffer, N.J., Waldmann, T.A. (1994) Molecular cloning and expression of cDNA for the human interleukin-2 receptor. Nature 311, 626–631
39. Cosman, D. Cerreti, D.P., Larsen, A., Park, L., March, C., Dower, S., Gillis, S. and Urdal, D. (1984) Cloning, sequence and expression of the human interleukin-2 receptor. Nature 312, 768–771
40. Nikaido, T., Shimizu, A., Ishida, N., Sabe, H., Teshigwara, K., Maeda, M., Uchiyama, , Yodoi, J., Honjo, (1984) Molecular cloning of cDNA encoding human interelukin-2 receptor. Nature 311, 631–635
41. Yamasaki, K., Taga, T., Hirata, Y, Yamata, H., Kawanishi, Y., Seed, B., Taniguchi, T., Hirano, T., Kishimoto, T. (1988) Cloning and expression of the human interleukin-6 (BSF-2/IFN-ß2) receptor Science 241 825–828
42. Döring, G., Frank, F., Boudier, C., Herbert, Silvia, Fleischer, B. and Bellon, G. (1995) Cleavage of the lymphocyte surface antigens CD2, CD4, and CD8 by polymorphonuclear leukocyte elastase and cathepsin G in Patients with cystic fibrosis. J. Immunol. 154, 4842–4850
43. Ugnotz, R.A., Kelly, B., Davis, R.J. and Massague, J. (1986) Biologically active precursorfor transforming growzh factor type alpha released by retroviral transformed cells. Proc. Natl. Acad. Sci. 83/17 6307–6311
44. Hibi, M., Murakami, M., Saito, M., Hirano, T., Taga, T. and Kishimoto, T. (1990) Molecular cloning and expression of an IL-6 signal transducer, gp 130. Cell 63, 1149–1157

31

IN VITRO EFFECTS OF γ-GLUTAMYLTRANSPEPTIDASE INHIBITOR ACIVICIN ON HUMAN MYELOID AND B LINEAGE CELLS

Brigitte Bauvois*

UNITÉ 365 INSERM
Institut Curie, Section De Recherche
26 rue d'Ulm, 75231 Paris Cedex 05, France

1. EXPRESSION AND REGULATION OF γ-GT

1.1. Expression

Gamma-glutamyltranspeptidase (γ-GT, EC 2.3.2.2) (Figure 1) is a widely distributed enzyme that acts as a transferase in the transfer of the γ-glutamyl group from a wide variety of peptide donors to other amino acids and peptide acceptors, and as a hydrolase in removing the γ-glutamyl residue from such peptides [1]. γ-GT activity was initially detected at the surface of human monocytes [2]. Using a spectrophotometric method, we showed it to be a sensitive assay of γ-GT activity detected on the surface of intact cells, as well as in highly purified cell membrane fractions. Cell-surface γ-GT (assayed with γ–Glu-para-nitroanilide as substrate and Gly-Gly as acceptor and abolished by the specific inhibitor of γ-GT i.e. acivicin) was confirmed in the myeloblastic (HL-60) and monoblastic (U937) cell lines [3]. Peripheral blood monocytes, granulocytes as well as macrophages developed *in vitro* from monocytes, were also found to exhibit γ-GT activity [3]. Another line of evidence that human myeloid cells express γ-GT was obtained from Northern blot analysis. In humans, there are at least four potential genes for γ-GT located on chromosome 22 [4]. Using a liver γ-GT probe, we have detected an mRNA species of 2.4 kb, corresponding to that previously described for the HepG2 cell line [5] in both cell lines and in granulocytes and macrophages [3]. Resting human B cells purified from blood as well as plasma cell lines (U266, RPMI 8226, Eskol) expressed γ-GT activity. However, no γ-GT transcripts were detected in any cells of the B lineage.

* Fax : 00.01.44.07.07.85; phone : 00.01.42.34.67.20; E-mail : bbauvois@curie.fr.

Cellular Peptidases in Immune Functions and Diseases, edited by Ansorge and Langner
Plenum Press, New York, 1997

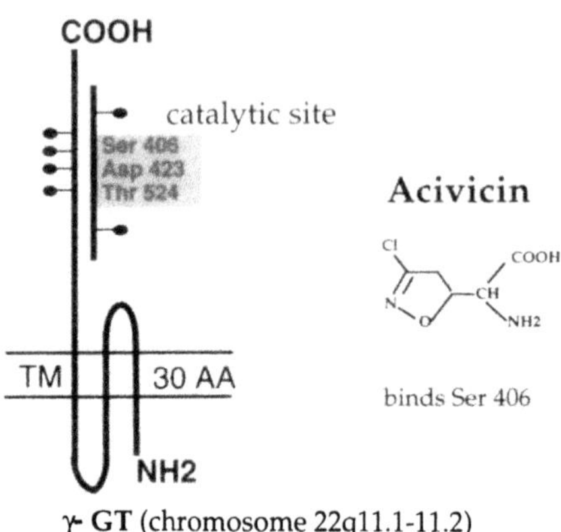

Figure 1. Structures of human γ-GT and its inhibitor acivicin. TM, transmembrane; —●, glycosylation sites.

1.2. Regulation

HL-60 and U937 cells differentiate into macrophages by phorbol esters, granulocyte-macrophage colony stimulating factor or interferon-γ whereas HL-60 cells are induced to differentiate into a mature neutrophil-like phenotype by retinoic acid or dimethylsulfoxide. We observed that the levels of 2.4 kb γ-GT mRNA species increased in HL-60 or U937 cells undergoing granulocyte or macrophage differentiation upon addition of any of the inducers of granulocyte-macrophage differentiation tested [3] and this was accompanied by a significant enhancement of γ-GT activity in the cytosol of stimulated cells [3]. However, no up-modulation of γ-GT expression was observed at the surface of stimulated cells nor were detectable levels of soluble γ-GT released into the culture supernatants of differentiating cells [3].

2. γ-GT AND ITS SIGNALLING FUNCTIONS IN HUMAN MYELOID AND B LINEAGE CELLS

Several recent *in vitro* studies, employing monoclonal antibodies and/or synthetic inhibitors of ectopeptidases have suggested that these molecules are involved in regulating cell activation, proliferation and cytokine secretion [5–8]. We tempted here to define the role of γ-GT in the process of activation of human myeloid and B cells using the specific inhibitor of γ-GT acivicin (Figure 1).

2.1. Effect of Acivicin on Myeloid Cells

When incubated with HL-60 cells, acivicin produced a dose- and time-dependent inhibitory growth [3] associated with the withdrawal of most cells from the cell cycle, which accumulated in the G0-G1 phase. Acivicin respectively up- and down-modulated the mRNA levels encoded by retinoblastoma (Rb) tumour suppressor gene and c-myc gene (see Figure 2.A) whose nuclear proteins are involved in the control of the cell cycle and transcription. Acivicin-stimulated HL-60 cells exhibited a marked shift towards a macrophage morphology, associated with an enhancement of the maturation markers Mac-1/CD11b, CD14 and CD71 (transferrin receptor) [3]. When cultured in the presence of acivicin, U937 cells and freshly isolated monocytes also underwent characteristic changes

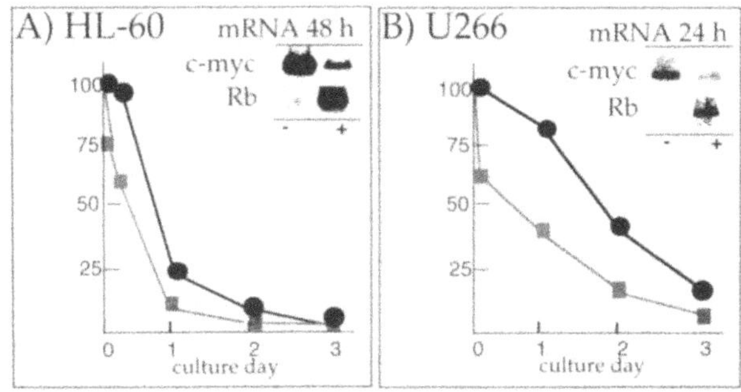

Figure 2. Effects of acivicin on growth, cell surface γ-GT activity and mRNA expression in HL-60 and U266 cells. HL-60 (A) or U266 (B) cells were cultured in the absence (control) or in the presence of 10 μg/ml acivicin for up to 3 days. After each day of incubation, cells were assayed for cell number (●-●) and γ-GT activity (■-■) which was assayed by incubating cells (10^6/ml) with 1 mg/ml γ-Glu-paranitroanilide and 20 mM Gly-Gly for 30 min at 37°C. Results are expressed relative to the controls (=100). Total RNAs extracted from unstimulated (-) or stimulated cells (+) for 1 or 2 days with 10 μg/ml acivicin were prepared by acid guanidinium thiocyanate-phenol-chloroform extraction. Northern blots were hybridized with [^{32}P]-labelled c-myc, Rb or GAPDH cDNAs. GAPDH mRNA (not shown) was taken as a control because it did not change upon treatment with acivicin.

in morphology and antigenic phenotype consistent with their differentiation into macrophages [3]. As expected, macrophage maturation correlated with the up-regulation of γ-GT and FcγRI transcripts known to be enhanced along the macrophage pathway.

We have investigated the possibility of an indirect effect of acivicin on myeloid cells that could involve growth factors or cytokines operating by an autocrine mechanism. Unstimulated HL-60 cells, U937 cells and monocytes, as well as cells stimulated by acivicin, did not secrete detectable levels of IL-1β, IL-10, TNF-α, GM-CSF and low levels of IL-6 (<10 pg/ml). In contrast, significant levels of latent TGF-β (neutralized by an antibody against TGF-β1/TGF-β2) were released by HL-60 and U937 cells (up to 1200 pg/ml) and by monocytes (up to 280 pg/ml [3]. Steady state levels of TGF-β1 mRNA species were apparently not affected by acivicin treatment. Moreover, the neutralizing antibody, anti-TGF-β1/β2 was unable to prevent cell differentiation induced by acivicin [3].

2.2. Effect of Acivicin on B Lineage Cells

Like HL-60 and U937 cells, treatment of U266 cells with acivicin resulted in a block of cell growth associated with the concomitant modulation of Rb and c-myc genes (see Figure 2.B). U266 cells released detectable levels of interleukin-6 (IL-6) (around 100pg/ml). Although IL-6 mRNA levels were not affected by acivicin, we found that acivicin exerted a remarkable efficacy in U266 cells by inhibiting IL-6 production (>80% inhibition at day 2). As U266 cells do not express the IL-6 receptor (CD126), it is unlikely that the IL-6 synthesized by U266 cells themselves serve as a spontaneous growth factor, and absence of IL-6 release by acivicin-treated cells could not therefore account fot the block in cell growth. In the presence of acivicin, U266 cells underwent morphologic changes (inversion of the nucleus/cytoplasm ratio, appearance of an arcoplasmic area and enlargement in cell size) characteristic of mature plasma cells. U266 cells expressed at their surface low levels of the CD40 antigen (known to be lost during terminal differentiation to plasma cells). However, as measured by FACS analysis, no changes in CD40 levels were observed in acivicin-treated cells for 3 days.

The potential effect of acivicin in U266 cells was recently investigated by immunoblotting, to detect tyrosine phosphorylation of intracellular proteins. Acivicin was shown to induce in U266 cells a transient increase (from 10 to 30 min) in tyrosine phosphorylation of a 52 kDa protein.

3. CONCLUSIONS

Apart from its role in the regulation of cytokine levels, γ-GT, present at the surface of human myeloid and B lineage cells, appears to be functionally involved in cell growth arrest, induction of maturation and modulation of cytokine secretion. A positive correlation has already been observed between over-expression of an ectopeptidase and hyperproliferative properties of leukaemia cells, as in acute myeloblastic and myelomonocytic leukaemias, which express higher γ-GT activity than do normal monocytes [9]. Thus, the inhibition of enzymes such as γ-GT, could be a new approach to suppressing the growth or metastasis of cancer. γ-GT has a very short intracytoplasmic tail and it is unclear whether this enzyme is connected to functional molecules through which might be channelled activation signals. Our preliminary data obtained with tyrosine phosphorylation of a 52 kDa protein in U266 cells stimulated by a specific inhibitor of γ-GT i.e. acivicin, led to the speculation that (a) specific tyrosine kinase(s) could be involved in signal transduction pathways. Our future challenge will be to elucidate the involvement of tyrosine kinases in cell activation by acivicin.

4. ACKNOWLEDGMENTS

This work was supported by grants from Institut National de la Santé et de la Recherche Médicale (INSERM) and Association pour la Recherche sur la Cancer (ARC No 6689). The author gratefully acknowledges E. Barthélémy, Dr. A. Laouar and D. Rouillard for their participation in this work. The author thanks Dr. G. Guellaën for the gift of human liver γ-GT cDNA.

5. REFERENCES

1. Tate SS, Meister A. (1981) γ-glutamyl transpeptidase : catalytic, structural and functional aspects. *Mol Cell. Biol.* **39**:357–368.
2. Grisk O, Küster U, Ansörge S. (1993) The activity of γ-glutamyl transpeptidase (γ-GT) in populations of mononuclear cells from human peripheral blood. *Biol. Chem. Hoppe-Seyler* **374**:287–290.
3. Bauvois B, Laouar A, Rouillard D, Wietzerbin J. (1995) Inhibition of γ-glutamyl transpeptidase activity at the surface of human myeloid cells is correlated with macrophage maturation and transforming growth factor-β production. *Cell Growth & Differentiation* **6**:1163–1170.
4. Guellaën C. (1989) Rôle physiologique de la γ-glutamyl transpeptidase. *Med. Sci.* **5**:637–644.
5. Bulle F, Matte MG, Siegrist S, Pawlak A, Passage E, Chobert MN, Laperche Y, Guellaën G. (1987) Assignment of the human γ-glutamyl transferase gene to the long arm of chromosome 22. *Hum. Genet.* **76**:283–286.
6. Fleischer B. (1992) CD26 : a surface protease involved in T-cell activation. *Immunol. Today* **15**:180–184.
7. Mari B, Checler F, Ponzio G, Peyron JF, Manie S, Farahifar D, Rossi B, Auberger P. (1992) Jurkat T cells express a functional neutral endopeptidase activity (CALLA) involved in T cell activation. *EMBO J.* **11**:3875–3885.
8. Shipp M, Look AT. (1993) Hematopoietic differentiation antigens that are membrane-associated enzymes: cutting is the key. *Blood* **68**: 1052–1070.
9. Morell A, Losa G, Carrel S, Heumann D, Von Fliedner VE. (1986) Determination of ectoenzyme activities in leukemic cells and in established hematopoietic cell lines. *Am. J. Hematol.* **21**:289–298.

EXPRESSION OF SEVERAL MATRIX METALLOPROTEINASE GENES IN HUMAN MONOCYTIC CELLS

U. Machein and W. Conca

Department of Rheumatology and Clinical Immunology
University of Freiburg
D-79106 Freiburg, Germany

1. ABSTRACT

Matrix metalloproteinases (MMP) are a family of structurally related endopeptidases that resorb macromolecules of the extracellular matrix (ECM)[1]. They are involved in normal tissue remodeling and wound repair as well as in pathological processes such as the irreversible destruction of joints observed in rheumatoid arthritis (RA)[2]. In addition, MMP catalyze the cleavage of the transmembrane form of tumor necrosis factor (TNF)[3,4]. Since cells of the monocyte lineage are major producers of TNF in the rheumatoid synovium we analysed the expression of MMP genes in these cells. To examine the transcriptional activity of MMP genes in undifferentiated monocytic cell lines (MonoMac6, U937) and in nature human monocytes isolated from peripheral blood, we developed an assay that is based on reverse transcription (RT) followed by a polymerase chain reaction (PCR). This screening procedure demonstrates that several MMP genes are transcriptionally active in the cells tested after exposure to a variety of stimuli such as phorbol ester, lipopolysaccharide (LPS) and staphylococcal enterotoxin B (SEB). The data were confirmed by quantitative Northern blot analysis. In conclusion, cells of the monocyte lineage produce high mRNA levels of at least six members of the MMP gene family that could participate in joint destruction by resorption of the ECM and secretion of TNF.

2. INTRODUCTION

The gene family of matrix metalloproteinases (MMP) comprises of 14 structurally related members that play an important role in remodeling and repair of connective tissues. Abnormal expression of MMP may contribute to a variety of destructive diseases including tumor invasion[5] and rheumatoid arthritis (RA)[2]. In this disease the MMP degrade irreparably the cartilage of the arthritic joint leading to invalidity. A novel function of

Cellular Peptidases in Immune Functions and Diseases, edited by Ansorge and Langner
Plenum Press, New York, 1997

MMP has recently been described which consists in the shedding of transmembrane molecules such as the tumor necrosis factor. This is of particular importance in RA because TNF probably represents the major proinflammatory mediator in this disease. Since monocytes are major producers of TNF in the rheumatoid synovium we conducted experiments to determine which members of the MMP gene family are expressed in cells of the monocyte lineage under basal conditions as well as after exposure to different stimuli using a two-step molecular strategy. The rapid screening procedure revealed the presence of transcripts of several MMP genes whose levels were measured by Northern blot analysis.

3. MATERIALS AND METHODS

3.1 Cell Culture

The monocyte cell lines U937 and MonoMac6 are of human origin and were purchased from Deutsche Sammlung von Mikroorganismen und Zellkulturen (DSM). Cultures were grown in RPMI 1640 supplemented with 10% fetal calf serum, L-glutamine, antibiotics, vitamins, amino acids, mercaptoethanol and pyruvate (Seromed). Mature monocytes were derived from peripheral blood mononuclear cells (PBL) by adherence and cultured as described. 10 ng/ml phorbol myristate acetate (PMA, Sigma), 1 μg/ml lipopolysaccharide (LPS, Sigma) or 2 ng/ml staphylococcal enterotoxin B (SEB, Serva) were used for stimulation.

3.2 Rapid Gene Family Screening

The screening procedure is shown schematically in Fig.1. It consists of RT-PCR and Southern blot hybridisation and was performed as follows: Total cellular RNA was extracted using the RNeasy kit (Qiagen) according to the manufacturer's instruction. A reverse transcription system (Promega) was used to generate the first cDNA strand that was subsequently amplified by PCR. PCR was performed with super Taq polymerase (P.H. Stehelin & Cie AG) and a pair of degenerate primers derived from two conserved motifs in MMP genes which represent the "cysteine switch domain" and the "zinc-binding site"[6].

The amplification products were separated in a 1.5% agarose gel. The amplificated species was visualized with Ethidium Bromide and UV light, retrieved from the gel with the Jet sorb gel extraction kit (Genomed) and labeled by random priming (Stratagene) and dCTP (Amersham). Partial or full-length cDNAs of collagenase 1 and 3, gelatinase A and B, stromelysin 1, 2 and 3, human metalloelastase (HME), matrilysin and membrane-type1-MMP (mt1-MMP) were separated in a 1% agarose gel. cDNAs were transferred onto Duralon UV-membranes (Stratagene) and cross-linked with UV light (Stratagene). Hybridizations were carried out in QuickHyb hybridization solution (Stratagene) using the labeled PCR product as a probe.

3.3 Northern Blot Analysis

Total cellular RNA (10 μg) was separated in a 1.4% denaturating agarose gel containing 2.2 M formaldehyde, transferred onto Duralon UV membranes and cross-linked with UV light (Stratagene). cDNA sequences of collagenase 1, gelatinase A and B, HME, matrilysin and mt1-MMP were labeled by random priming and used as probes for Northern blot analysis. Hybridization was carried out and followed by autoradiography.

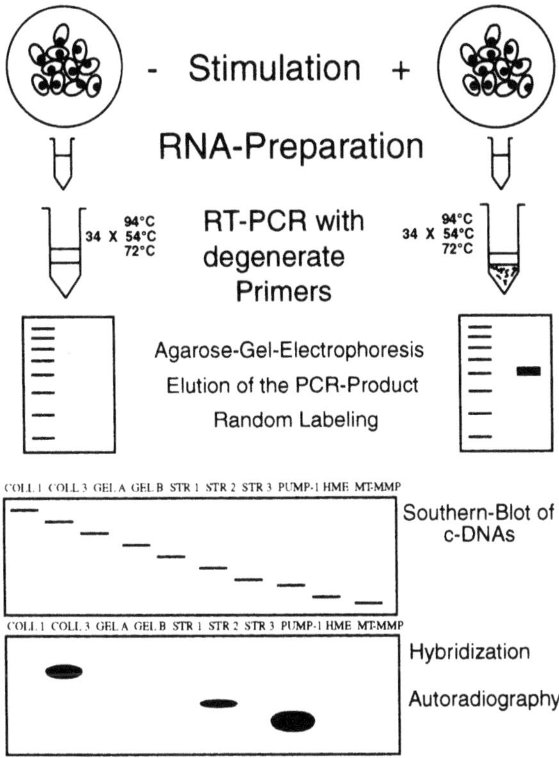

Figure 1. Schematic model of the rapid gene family screening.

4. RESULTS AND DISCUSSION

Cells of the monocyte lineage have been shown to synthesize various MMP. The production of these enzymes is dependent on the stage of maturation as well as on the experimental conditions used[7]. Recently, cell membranes of stimulated T-lymphocytes have been demonstrated to induce the release of MMP in monocytes and THP1 cells which is an undifferentiated monocyte cell line[8]. These observations suggest that direct contacts between cells may modulate the expression of MMP genes in cells of the monocyte lineage. To complete our knowledge on the expression of MMP genes in monocytes we developed a molecular strategy to define further transcripts of MMP genes. The two-step procedure is based on gene family specific RT-PCR, which takes the advantage of the presence of two conserved motifs in all 14 MMP genes, and is followed by Southern blot analysis where the cDNA products obtained by RT-PCR are used as probes (Fig.1). The results of such a screening procedure are shown in Fig.2. In panel A, RNA of U937 was processed, in panel B, we analysed the further differentiated cell line MonoMac6. PMA was used as stimulus for 48 hours. The intensity of the signals obtained varied and was strongest for matrilysin in both cell types. A strong signal was also observed for mt1-MMP in MonoMac6 cells.

Since the assay does not measure mRNA levels of MMP genes and did not result in the detection of transcripts of gelatinase A and B, we performed Northern blot analysis whose results are summarized in Table 1.

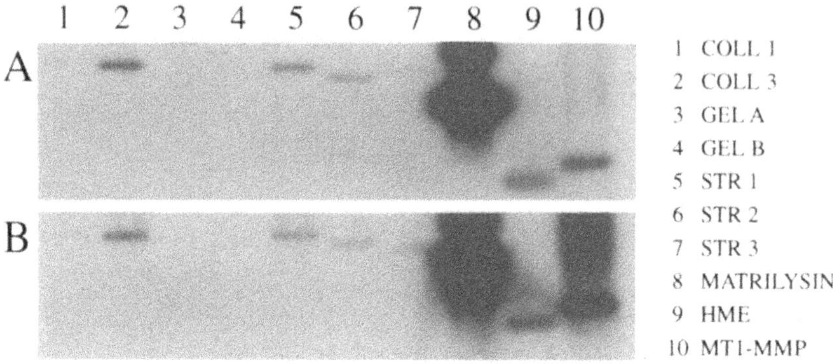

Figure 2. Autoradiography of a rapid gene family screening: A: U 937 cells, B: MonoMac6. The rapid screening was performed after 48 h of stimulation with 10 ng/ml PMA (for details see text).

Stimulation with PMA leads to an increase in the mRNA level of mt1-MMP in both cell lines. The expression of mt1-MMP is constitutive in MonoMac6 cells and after stimulation the mRNA levels are higher than in U937 cells. In contrast, inducibility of matrilysin with PMA was more prominent in U937. Stimulation with LPS had no effect on the expression of both genes. It should be noted that the results obtained by the Northern blot analysis confirmed those of the rapid gene family screening. Gelatinase A is constitutively expressed in both cell lines and does not respond to stimulation. In contrast, the gelatinase B gene, whose expression is similar to that of gelatinase A in U937 cells, is strongly activated in MonoMac6 cells. The effect of LPS on these genes are negligible. This may be due to the low density of CD14 molecules on the surface of these cells. (not shown).

We tested human monocytes that were isolated from PBL by adherence. These cells were also exposed to the superantigen SEB. mRNA levels of MMP genes are shown in Table 2. Peripheral blood monocytes were responsive to LPS, which was a potent inducer of mt1-MMP and gelatinase B. SEB, which binds to MHC class II molecules, leads to the expression of the HME and the collagenase 1 gene.

Table 1. Levels of MMP mRNA in MonoMac6 (A) and in U937 (B). - not detectable, + weak expression, ++ intermediate expression, +++ strong expression

	Control	PMA	LPS
A. MonoMac6			
mt1-MMP	++	+++	
matrilysin	–	++	–
gelatinase A	++	++	++
gelatinase B	–	+++	+
B. U937			
mt-1MMP	–	++	–
matrilysin	+	+++	+
gelatinase A	++	++	++
gelatinase B	++	++	++

Table 2. Levels of MMP mRNA in human monocytes. HME = metalloelastase. - not detectable, + weak expression, ++ intermediate expression, +++ strong expression

	Control	PMA	LPS	SEB
mt1-MMP	++	++	+++	++
matrilysin	+	++	–	+
gelatinase A	+	–	++	–
gelatinase B	++	++	+++	+
collagenase 1	–	++	–	++
HME	–	–	–	+++

5. SUMMARY

A molecular approach was used to define the pattern of expression of MMP genes in cells of the monocyte lineage. These cells possess the capacity to synthesize at least 6 known members of the MMP gene family. The pattern of transcriptional activity of these genes is dependent on the stage of differentiation and on the conditions of stimulation chosen. We conclude that the products of several MMP genes are involved in the resorption of the ECM or the processing of transmembrane molecules in human monocytic cells.

6. REFERENCES

1. Matrisian, L. M. *Bioessays* **14**, 455–463 (1992).
2. Krane, S. M., Conca, W., Stephenson, M. L., Amento, E. P. & Goldring, M. B. *Ann. N. Y. Acad. Sci.* **580**, 340–354 (1990).
3. McGeehan, G. M., Becherer, J. D., Bast, R. C.,Jr., et al. *Nature* **370**, 558–561 (1994).
4. Gearing, A. J., Beckett, P., Christodoulou, M., et al. *Nature* **370**, 555–557 (1994).
5. Stetler-Stevenson, W. G., Hewitt, R. & Corcoran, M. *Semin. Cancer Biol.* **7**, 147–154 (1996).
6. Conca, W. & Willmroth, F. *Arthritis Rheum.* **37**, 951–956 (1994).
7. Lacraz, S., Dayer, J. M., Nicod, L. & Welgus, H. G. *J. Biol. Chem.* **269**, 6485–6490 (1994).
8. Lacraz, S., Isler, P., Vey, E., Welgus, H. G. & Dayer, J. M. *J. Biol. Chem* **269**, 22027–22033 (1994).

LYSOSOMAL CYSTEINE PEPTIDASES AND MALIGNANT TUMOURS

Heidrun Kirschke

Institute of Physiological Chemistry
Faculty of Medicine
Martin-Luther University of Halle-Wittenberg
D-06097 Halle (Saale), Germany

Putative biological functions of lysosomal cysteine peptidases especially of cathepsins B and L in tumour progression have recently been considered in several reviews and the respective literature has been compiled[1–5]. Here, only some aspects of the extracellular and intracellular functions of these enzymes and the correlation of expression of cathepsins L and H with the malignancy of tumours should be discussed.

1. PUTATIVE EXTRACELLULAR FUNCTIONS

All steps of tumour growth such as local invasion, intravasation, extravasation, angiogenesis and metastasis require the activity of proteolytic enzymes. The involvement in these processes of peptidases from the classes of aspartic, cysteine, serine and metallo endopeptidases has been shown by several reports, but individual contributions of each enzyme are not yet clear.

Dissolution of the basement membrane by degradation of matrix proteins may require the action of several enzymes which are mainly secreted as inactive precursors. Several proteases of different classes are connected with one another by activating the precursors and thus initiating a proteinase cascade[6]. Cysteine proteinases are involved in these processes and in addition an interference with normal antigen processing by uptake of procathepsin L in antigen presenting cells has been described[7]. Cysteine peptidases are supposed to modify contact inhibin, receptors and adhesion proteins such as the integrins which may be of importance to cell motility. Some authors reported on another special role of procathepsin L to act like a growth factor or progression factor on cell proliferation and to be involved in differentiation processes, as well[8–10].

These functions of lysosomal cysteine peptidases seem to be restricted to their action outside the lysosomal-endosomal compartment after secretion from the cell. Indeed, several malignantly transformed cells and tumour cell lines were characterized by an enhanced secre-

Cellular Peptidases in Immune Functions and Diseases, edited by Ansorge and Langner
Plenum Press, New York, 1997

tion of mainly procathepsins B and L. Tumour cells adherent to basement membranes are possibly capable of forming an acidic microenvironment similar to that formed around macrophages or between adherent osteoclasts and bone. This may provide conditions for activation of the precursors and for proteolytic action of the mature cysteine peptidases.

From *in vitro* experiments cathepsin L and H and to some extent also cathepsin B are known to be irreversibly inactivated at pH 7 and above. New observations[11] showed that the stability of cathepsin L at pH 7 is dependent on the ionic strength of the solution. A half-life of about 3 min has been measured for cathepsin L activity at pH 7.2 in physiological salt solutions.

The enzymes also react with endogenous inhibitors - in the extracellular space mainly with cystatin C - but the enzyme-inhibitor complexes dissociate partially at acidic pH values[12].

Binding of mature cysteine peptidases to proteins at outer cell surfaces may render their sensitivity to neutral pH values as has been shown for cathepsins B and L bound to thyroid epithelial cells[13] but not yet to tumour cells. The clearance of the secreted precursors and enzyme complexes withα_2-macroglobulin and cystatins from the extracellular space must be very rapid, because the enzymes can hardly be detected outside cells by immunohistochemical methods.

2. INTRACELLULAR FUNCTIONS

The intracellular function of the enzymes is obviously to take part in the overall degradation of proteins in the lysosomal-endosomal compartment of tumour cells, where catabolic processes have increased. Activity measurements of cathepsins B, H and L in tumour tissue extracts mainly determines this part of the mature enzymes together with the part of fully processed cathepsins B and L in small vesicles associated with the plasma membrane of several tumour cells. This translocation was obviously dependent on acidic pH values outside the cells and was highest in invasive edges of the tumours[14].

The activities of cathepsins B, H and L have been shown to be increased in nearly all human tumours in comparison to normal tissues. Renal cell carcinomas were the only exception: the specific activities of the cathepsins B, H, L and dipeptidyl peptidase I (cathepsin C) were lower in the tumour tissue than in normal human kidney (Kirschke *et al.*, unpublished results).

3. CORRELATION OF THE OVEREXPRESSION WITH MALIGNANCY

The elevated concentration of lysosomal cysteine peptidases detected in most tumours and transformed cells was mainly estimated by activity measurements and in some cases also by mRNA (northern blot or in situ hybridization) or protein (ELISA, immunohistochemical analysis) determination. Only in some of these reports, the degree of overexpression of the enzymes was compared with malignancy of the tumours. High malignant tumours show among other things an early high rate of metastasis and rapid growth.

The concentration of cathepsin B mRNA, protein and activity was correlated with malignancy in several cases[4]. Some studies on the correlation of cathepsin L expression with malignancy are summarized in Table 1. The highest level of cathepsin L was seen in most of those tumour samples and transformed cell lines possessing the highest malignancy. In col-

Table 1. Correlation of cathepsin L expression with malignancy of different tumours. (a) higher activity in Dukes' stages A and B than in the more metastatic tumours (Dukes' stages C and D); (b) immunohistochemical analysis; (c) concentration of cathepsin L is of prognostic value

Tumor type	Level	Correlation with malignancy	Ref.
Breast, human			
cell lines	Activity	+	16
Colorectal carcinomas, human			
tumour samples	Activity	- (a)	17
tumour samples	Activity	-	18
Melanomas, human & murine			
cell lines	Activity	+	19
Melanomas, murine			
cell lines	mRNA	-	20
Melanomas, human			
tumour samples	Protein	+ (b)	21
Gliomas, human			
tumour samples	Activity	+	22
	Protein	+	22
Lung, human			
tumour samples	Activity	(c)	15
Fibroblasts, murine			
cell lines (ras-transfected)	Protein	+	23
	mRNA	+	23
NIH 3T3			
cell lines (ras-transfected)	Protein	+	24
	mRNA	+	24
Lymphomas, human			
tumour samples	Activity	+	Kirschke *et al.*, unpubl.

orectal carcinomas, however, the activity of cathepsin L did not increase in accordance with pathological prognostic variables such as stage (later stages represent metastatic tumours). Regardless of the pathological stage, cathepsin L seems to be overexpressed in high amounts by the most malignant phenotype of lung carcinomas, because high concentration of ca-thepsin L activity in lung was associated with short survival times of the patients[15].

Cathepsin H has only been included in a few studies on malignant tumours. This en-zyme seems also to be linked to tumour progression, because elevated activity has been de-tected in most of the tumour samples. But comparison of cathepsin H concentration with the degree of malignancy showed only a correlation in gliomas by activity assay and ELISA[25], breast carcinomas by ELISA[26] and in melanomas by immunohistochemical analysis[21] (Table 2). Cathepsin H activity was highest in colorectal carcinomas of Dukes' stage C, but again low in those of Dukes' stage D[27] although both, stages C and D, represent metastatic tumours. Cathepsin H may act mainly as an aminopeptidase. Special functions of this enzyme in the endosomal-lysosomal compartment or in the extracellular space are not yet known. In con-trast to cathepsins B and L the overexpression of cathepsin H seems not to be strongly corre-lated with the malignancy of tumours or transformed cells as it is shown in Table 2.

Table 2. Correlation of cathepsin H expression with malignancy of different tumours (a) immunohistochemical analysis; (b) highest activity in metastatic tumours of Dukes' stage C; (c) increased activity - but no correlation with malignancy

Tumor type	Level	Correlation with malignancy	Ref.
Gliomas, human			
Tumour samples	Activity		25
	Protein		25
Breast carcinomas, human			
Tumour samples	Protein		26
Breast, human			
Cell lines	Activity	-	16
Melanomas, human			
Tumour samples	Protein	+ (a)	21
Melanomas, murine			
Cell lines	mRNA	-	20
Colorectal carcinomas, human			
Tumour samples	Activity	+ (b)	27
Lung carcinomas, human			
Tumour samples	Activity	- (c)	Plehn & Kirschke, unpubl.
Lymphomas, human			
Tumour samples	Activity	-	Kirschke *et al.*, unpubl.

How the expression of lysosomal cysteine peptidases is regulated in normal and transformed cells is not yet known. More experimental data are needed on the altered concentration of these enzymes at more than one level in tumour tissues. Some experiments with synthetic inhibitors[28], antibodies[9] and antisense mRNA (Kirschke *et al.*, unpublished results) showed that inhibition of the increased concentration especially of cathepsin L rendered the high malignant phenotype of some tumour cells. These results give an idea that downregulation of the lysosomal cysteine peptidase expression may possibly be of value in the future cancer therapy.

REFERENCES

1. Kane, S.E., and Gottesman, M.M., 1990, The role of cathepsin L in malig nant transformation, *Semin. Cancer Biol.* 1: 127–136.
2. Sloane, B.F., Moin, K., Krepela, E., and Rozhin, J., 1990, Cathepsin B and its endogenous inhibitors: the role in tumor malignancy, *Cancer Metast. Rev.* 9: 333–352.
3. Gottesman, M.M., 1993, Cathepsin L and cancer, in: *Proteolysis and Protein Turnover*, Bond, J.S., and Barrett, A.J., eds., Portland Press, London and Chapel Hill, 247–251.
4. Berquin, I.M., and Sloane, B.F., 1996, Cathepsin B expression in human tumors, in: *Intracellular Protein Catabolism*, Suzuki, K., and Bond, J.S., eds., Plenum Press, New York and London, 281–294.
5. Kirschke, H., Barrett, A.J., and Rawlings, N.D., 1995, Proteinases 1: Lyso somal cysteine proteinases, *Protein Profile* 2: 1587–1643.

6. Schmitt, M., Goretzki, L., Jänicke, F., Calvete, J., Eulitz, M., Kobayashi, H., Chucholowski, N., and Graeff, H., 1991, Biological and clinical relevance of the urokinase-type plasminogen activator (uPA) in breast cancer, *Biomed. Biochim. Acta* 50: 731–741.
7. McCoy, K., Gal, S., Schwartz, R.H., and Gottesman, M.M., 1988, An acid protease secreted by transformed cells interferes with antigen processing, *J. Cell Biol.* 106: 1879–1884.
8. Kasai, M., Shirasawa, T., Kitamura, M., Ishido, K., Kominami, E., and Hi- rokawa, K., 1993, Proenzyme form of cathepsin L produced by thymic epi thelial cells promotes proliferation of immature thymocytes in the presence of IL-1, IL-7, and anti-CD3 antibody, *Cell. Immunol.* 150: 124–136.
9. Weber, E., Günther, D., Laube, F., Wiederanders, B., and Kirschke, H., 1994, Hybridoma cells producing antibodies to cathepsin L have greatly reduced potential for tumour growth, *J. Cancer Res. Clin. Oncol.* 120: 564–567.
10. Homma, K., Kurata, S., and Natori, S., 1994, Purification, characterization, and cDNA cloning of procathepsin L from the culture medium of NIH-Sape- 4, an embryonic cell line of *Sarcophaga peregrina* (flesh fly), and its invol vement in the differentiation of imaginal discs, *J. Biol. Chem.* 269: 15258- 15264.
11. Dehrmann, F.M., Coetzer, T.H.T., Pike, R.N., and Dennison, C., 1995, Ma ture cathepsin L is substantially active in the ionic milieu of the extracellular medium, *Arch. Biochem. Biophys.* 324: 93–98.
12. Turk, B., Dolenc, I., Turk, V., and Bieth, J.G., 1993, Kinetics of the pH-indu ced inactivation of human cathepsin L, *Biochemistry* 32: 375–380.
13. Brix, K., Lemansky, P., and Herzog, V., 1996, Evidence for extracellularly acting cathepsins mediating thyroid hormone liberation in thyroid epithelial cells, *Endocrinology* 137: 1963–1974.
14. Rozhin, J., Sameni, M., Ziegler, G., and Sloane, B.F., 1994, Pericellular pH affects distribution and secretion of cathepsin B in malignant cells, *Cancer Res.* 54: 6517–6525.
15. Ebert, W., Werle, B., Ebert, E., Kos, J., Lah, T., and Abrahamson, M., 1996, Cathepsins and cystatins: prognostic factors in human lung cancer, *11th International Conference on Proteolysis and Protein Turnover, Turku* (Poster Abstract №132).
16. Scaddan, P.B., and Dufresne, M.J., 1993, Characterization of cysteine pro teases and their endogenous inhibitors in MCF-7 and adriamycin-resistant MCF-7 human breast cancer cells, *Invas. Metast.* 13: 301–313.
17. Sheahan, K., Shuja, S., and Murnane, M.J., 1989, Cysteine protease ac tivities and tumor development in human colorectal carcinoma, *Cancer Res.* 49: 3809–3814.
18. Adenis, A., Huet, G., Zerimech, F., Hecquet, B., Balduyck, M., and Peyrat, J.P., 1995, Cathepsin B, L, and D activities in colorectal carcinomas: rela tionship with clinico-pathological parameters, *Cancer Lett.* 96: 267–275.
19. Rozhin, J., Wade, R.L., Honn, K.V., and Sloane, B.F., 1989, Membrane-as- sociated cathepsin L: a role in metastasis of melanomas, *Biochem. Biophys. Res. Commun.* 164: 556–561.
20. Qian, F., Bajkowski, A.S., Steiner, D.F., Chan, S.J., and Frankfater, A., 1989, Expression of five cathepsins in murine melanomas of varying metas tatic potential and normal tissues, *Cancer Res.* 49: 4870–4875.
21. Kageshita, T., Yoshii, A., Kimura, T., Maruo, K., Ono, T., Himeno, M., and Nishimura, Y., 1995, Biochemical and immunohistochemical analysis of cathepsins B, H, L and D in human melanocytic tumours, *Arch. Dermatol. Res.* 287: 266–272.
22. Sivaparvathi, M., Yamamoto, M., Nicolson, G.L., Gokaslan, Z.L., Fuller, G.N., Liotta, L.A., Sawaya, R., and Rao, J.S., 1996, Expression and immu nohistochemical localization of cathepsin L during the progression of human gliomas, *Clin. Exp. Metast.* 14: 27–34.
23. Denhardt, D.T., Greenberg, A.H., Egan, S.E., Hamilton, R.T., and Wright, J.A., 1987, Cysteine proteinase cathepsin L expression correlates closely with the metastatic potential of H-*ras*-transformed murine fibroblasts, *On cogene* 2: 55–59.
24. Chambers, A.F., Colella, R., Denhardt, D.T., and Wilson, S.M., 1992, In-creased expression of cathepsins L and B and decreased activity of their in hibitors in metastatic, *ras*-transformed NIH 3T3 cells, *Mol. Carcinog.* 5: 238- 245.
25. Sivaparvathi, M., Sawaya, R., Gokaslan, Z.L., Chintala, K.S., and Rao, J.S., 1996, Expression and the role of cathepsin H in human glioma progression and invasion, *Cancer Lett.* 104: 121–126.
26. Gabrijelčič, D., Svetic, B., Spaić, D., Škrk, J., Budihna, M., Dolenc, I., Po povič, T., Cotič, V., and Turk, V., 1992, Cathepsins B, H and L in human breast carcinoma, *Eur. J. Clin. Chem. Clin. Biochem.* 30: 69–74.
27. Murnane, M.J., Cai, J., Shuja, S., Coté, L., Del Re, E., Iacobuzio-Donahue, C., Kim, K., and Sheahan, K., 1995, Changing patterns of proteolytic ex pression with colorectal tumor progression, in: *Proteases Involved in Cancer*, Suzuki, M., and Hiwasa, T., eds., Monduzzi Editore, Bologna, 11–26.
28. Yagel, S., Warner, A.H., Nellans, H.N., Lala, P.K., Waghorne, C., and Den hardt, D.T., 1989, Suppression by cathepsin L inhibitors of the invasion of amnion membranes by murine cancer cells, *Cancer Res.* 49: 3553–3557.

34

EXPRESSION OF CYSTEINE PROTEASE INHIBITORS STEFIN A, STEFIN B, AND CYSTATIN C IN HUMAN LUNG TUMOR TISSUE

Eileen Ebert,[1] Bernd Werle,[1] Britta Jülke,[1] Natasa Kopitar-Jerala,[2] Janko Kos,[2] Tamara Lah,[3] Magnus Abrahamson,[4] Eberhard Spiess,[5] and Werner Ebert[1*]

[1]Thoraxklinik Heidelberg-Rohrbach
Amalienstr. 5, D-69126 Heidelberg, Germany
[2]Jozef Stefan Institute
Department of Biochemistry and Molecular Biology
Jamova 39 and
Krka, p.o., R&D Division
Cesta na Brdo 49, SLO-61111 Ljubljana, Slovenia
[3]National Institute of Biology
Vecna pot 111, SLO-61111 Ljubljana, Slovenia
[4]University Hospital
Department of Clinical Chemistry
University of Lund, S-22185
Lund, Sweden
[5]Deutsches Krebsforschungszentrum
Biomedizinische Strukturforschung INF 280,
D-69120 Heidelberg, Germany

ABSTRACT

In human lung tumor tissue specimen (n=73) concentrations of stefins A and B were found to be increased 2.0-fold ($p<0.01$) and 1.3-fold ($p<0.01$), respectively, as compared to matched normal tissue. Stefin A and B concentrations were higher in primary tumors than in secondary tumors, i.e. metastases from other organs to the lung ($p<0.01$; $p<0.05$, respectively). Cystatin C concentrations were rather low and did not differ between tumor and normal tissue. Both concentrations of stefins did not correlate with TNM stages. Ste-

* Address for correspondence: Dr. W. Ebert, Thoraxklinik Heidelberg-Rohrbach, Amalienstr. 5, D-69126 Heidelberg-Rohrbach, Germany. Fax: ++ 49-6221-396415.

Cellular Peptidases in Immune Functions and Diseases, edited by Ansorge and Langner
Plenum Press, New York, 1997

fin A was higher in squamous cell carcinoma than in adenocarcinoma ($p<0.01$), while stefin B did not show such a difference. At investigation of a relationship between survival probability of patients with primary tumors it was found that increased stefin B concentrations and total cysteine-protease-inhibitory activities but not stefin A concentrations were positively correlated with survival probability.

It is concluded that stefins A and B are major contributors to the cysteine protease inhibitory activity in primary lung tumors. Stefin B proved to be a prognostic factor, especially in squamous cell carcinoma.

1. INTRODUCTION

There are several lines of evidence that the cysteine proteases cathepsin B and cathepsin L play a crucial role in the process, when cancer cells invade host tissue and metastasize to secondary sites. Cathepsins B and L have been found to be overexpressed in a variety of animal and human tumors (reviewed in 1, 2, 3). The activitiy of these cathepsins is regulated by physiological cysteine protease inhibitors (CPIs), i.e. stefins A and B, cystatin C, and kininogens (reviewed in 4). It can, therefore, hypothesized that the increased proteolytic activity of the cathepsins may be caused by an underexpression of CPIs or by their reduced inhibitory activity (5).

Several studies, on both the protein and the mRNA level, of the CPIs in various malignant tissues or cell lines up to now agreed with this hypothesis (reviewed in 6).

However, in a recent study with lung tumor tissue, we found an increased expression of CPI activity in the majority of samples. Nevertheless, we could demonstrate an imbalance between cathepsin B activity and total CPI activity in favour of the protease (7). In addition, the increased ratio of cathepsin B to CPI activity between tumor and normal tissue was related to poorer prognosis (7).

The increase in CPI activity is consistent with earlier observations by Kyllönen et al. (8) who reported an increased stefin A immunostaining in lung tumors. Very recently Pyykkönen et al. (9) confirmed these findings.

To extend our knowledge on the expression of individual CPIs we assessed the amounts of stefins A and B, and cystatin C in matched pairs of lung tumor tissue and normal lung parenchyma. The concentration of the CPIs were correlated with clinical factors of prognostic significance and survival probability of patients suffering from this malignant disease.

2. MATERIALS AND METHODS

2.1 Patients

Lung tumor tissue and adjacent normal lung parenchyma was obtained from 73 patients with newly recognized lung tumors who were subjected to lung surgery. The age of the patients ranged from 15 - 81 years (median: 61 years). The cell type of lung cancer was classified according to the WHO recommendations and based on the predominant cell type (10). The tumor stage (pTNM) was classified according to the international staging system (11).

2.2 Tissue Prepraration and Determination of CPI Values

Tissue homogenization was carried out as described earlier by Werle et al. (12).

CPI activity was determined by using papain as target enzyme as described by Knoch et al. (7). The amounts of immunoreactive stefins A and B were assessed by using commercially available ELISAs. Stefin A assay was from BioAss, Dießen (Germany). Stefin B assay was from Dr. J. Kos (13). Cystatin C asssay was done with ELISA established by Dr. M. Abrahamson (14).

2.3 Protein Determination

Protein concentration was determined according to Bradford (15). Bovine serum albumin was used as a standard.

2.4 Statistical Anaylsis

The results of the CPI activity assays and the concentration of the inhibitors in the group under study are given as 5%, 50% and 95% percentiles. For comparing data in matched pairs, Wilcoxon's rank test was used. The correlation between cysteine proteases and the inhibitors was calculated by linear regression analysis. The significance of Spearman's correlation coefficient was evaluated by co-variance test. The calculation of survival probability was by the method developed by Kaplan and Meier (16) and the significance of a relationship between survival of patients and the levels of the biochemical parameters was tested by the log-rank test.

3. RESULTS

The results of CPI activity assays and the ELISA measurement of CPIs are summarized in Table 1.

With the exception of cystatin C, the median levels of CPI activity, stefin A, and stefin B in the tumor tissue were found to be increased 2.1-fold, 2.0-fold and 1.3-fold, respectively, as compared to the corresponding levels in normal lung tissue. The concentrations of stefin A and B were significantly higher in primary lung tumors than in secondary lung tumors ($p<0.01$; $p<0.05$, respectively). With respect to stefin levels the latter were within the range found in the corresponding normal lung parenchyma.

In tumor tissue, the concentration of stefin B was higher than that of stefin A ($p<0.01$) and that of cystatin C ($p<0.01$). There was a poor but significant positive correlation between CPI activity and stefin A ($r= 0,48$; $p<0.01$) and between CPI and stefin B ($r= 0.31$; $p<0.01$), suggesting that the stefins are major contributors to the observed increased CPI activity.

There were no significant differences in CPI activities and concentrations of the inhibitors with respect to pTNM stages.

Regarding histologic cell types, CPI activity and stefin A concentration were significantly higher in squamous cell carcinoma than in adenocarcinoma ($p<0.05$; $p<0.01$, respectively).

Regarding cell differentiation, only stefin A concentration was higher in poorly differentiated cells as compared to well and moderately differentiated cells ($p<0.05$).

In the investigation of a relationship between survival probability of patients with primary tumors and CPI activity levels it was found that patients with CPI activity levels

Table 1. Median 5% and 95% percentiles of specific inhibitory activities of CPI (μIU/mg) and concentrations of stefins A and B and cystatin C in pairs of lung tumor tissue and normal lung parenchyma

	n	CPI activity [μIU/mg] Median (5%, 95%)	Stefin A [ng/mg] Median (5%, 95%)	Stefin B [ng/mg] Median (5%, 95%)	Cystatin C [ng/mg] Median (5%, 95%)
Normal lung tissue (primary tumors)	55	427 (190, 1775)	27 (1, 229)	94 (13, 180)	4.81 (1.65, 18.00)
Normal lung tissue (secondary tumors)	18	452 (209, 2242)	16 (1, 122)	60 (23, 481)	5.03 (3.25, 10.74)
Total primary tumors	55	891 (341, 4125)	52 (6, 602)	107 (6, 383)	3.39 (0.68, 19.55)
Squamous cell carcinoma	28	1156 (455, 4545)	98 (24, 650)	134 (0.0, 437)	2.45 (0.68, 19.55)
Adeno carcinoma	24	764 (249, 2135)	32 (4, 177)	101 (34, 199)	5.35 (1.26, 16.12)
Large cell carcinoma	3	874 (560, 1054)	69 (66, 91)	206 (127, 209)	1.7 (0.36, 3.35)
Secondary tumors	18	860 (282, 3318)	17 (2, 96)	68 (19, 286)	3.94 (0.72, 50.70)
Well or moderately differentiated (G1/G2)	22	714 (341, 2401)	29 (4, 290)	96 (24, 221)	3.77 (1.24, 16.54)
Poorly differentiated (G3)	51	945 (368, 4125)	45 (10, 602)	111 (19, 383)	3.27 (0.68, 26.70)
TNM 1	18	1017 (341, 4125)	44 (1, 658)	106 (6, 294)	2.83 (0.68, 26.7)
TMN 2	8	1085 (368, 4987)	38 (13, 650)	148 (83, 495)	2.96 (0.68, 9.42)
TNM 3	19	855 (198, 2401)	41 (6, 448)	90 (0, 383)	3.56 (0.61, 48.60)
TNM 4	10	900 (560, 3783)	65 (10, 143)	195 (63, 256)	2.45 (0.36, 6.61)

The values are further subdivided according to histology, cell differentiation and TNM staging.

below 1108 [IU/mg protein] had a significantly shorter survival rate than those with levels above 1108 [IU/mg protein] ($p<0.05$) (Fig. 1). 17 out of 29 (58.6%) patients died versus 7 out of 21 (33.3%) patients within the observation period of 2 years. There was also a significant correlation of stefin B with survival probability ($p<0.01$) (Fig. 2). A cut-off level of 132 [ng/mg protein] stefin B was calculated by means of a computer program (17). In the group with stefin B concentrations below this level 18 out of 27 patients (66.7%) died within the 2-years-observation-time; but only 6 out of 24 patients (25.0%) died with stefin B concentrations above the cut-off level.

In contrast, stefin A levels in tumor tissue were not related to survival probability.

4. DISCUSSION

The results of the present study demonstrate a substantial difference in the expression of stefins A and B between primary and secondary lung tumors. In the primary tumors we found significantly higher levels of stefins A and B, while in the secondary

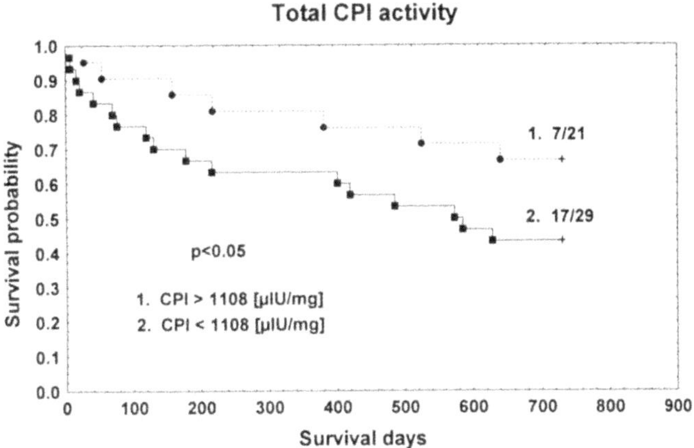

Figure 1. Probability of survival of patients with primary tumors in relation to specific inhibitory activity of CPIs (µIU/mg protein) in tumor tissue.

tumors, i.e. metastases from other organ to the lung, the levels of both stefins were in the same range as in the normal counterparts.

Cystatin C was present in lung tumors in much lower concentrations compared to the stefins and was not significantly elevated in comparison to the normal lung tissue counterparts. Our findings are in contrast to the most previous studies, which agreed that malignant tumors contain the same (13, 18, 19) or lower (3, 20) levels of stefins than normal tissue. The discrepancy can be explained by the differences in the types of tumors investigated as well as by the different methods and antidodies used in the quantitative measurements of the inhibitors.

Our data suggests that the levels of the inhibitors are increased to compensate for highly induced concentrations of cathepsin B and cathepsin L (7, 12). As these increases were much higher than was the increase in the levels of stefins, the imbalance between both antagonistic molecules results in a net increase in cysteine proteinase activity. There-

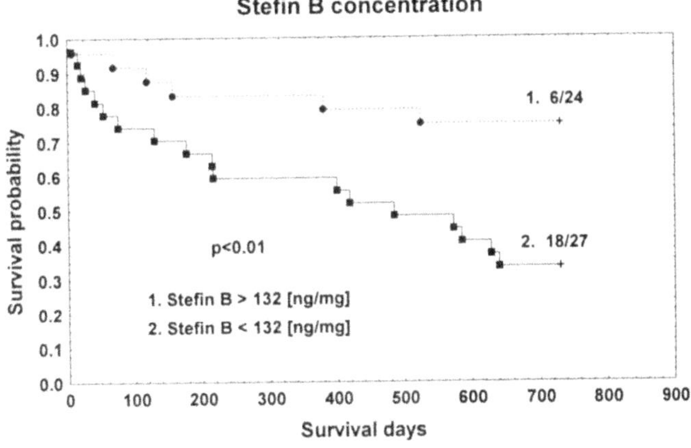

Figure 2. Probability of survival of patients with primary tumors in relation to stefin B concentration in tumor tissue. The cut-off value of 132 [ng/mg protein] was calculated by means of a computer program (17).

fore, all these studies are consistent with the hypothesis that a relative lower concentration of the endogenous cysteine proteinase inhibitor *vs* the cysteine proteinases in tumor compared to the normal tissues contributes to the imbalance between proteolytic and inhibitory activities, which might play an important role in tumor cell invasion and metastasis progression.

Our findings that stefin A was about three fould higher in squamous cell carcinoma than in adenocarcinoma are in line with Kyllönen et al. (8) who also reported high expression of stefin A in squamous cell carcinoma but not in adenocarcinoma of the lung. This is further supported by recent data of Pyykkönen et al. (9), who found an intense immunochemical staining of stefin A in the cytoplasma of squamous cell carcinoma, while the majority of adenocarcinomas did not stain.

In contrast, stefin B staining was much weaker than was stefin A and was only observed in the squamous cell carcinoma (8). This does not quite support our measurement of relatively higher concentrations of stefin B in the cytosols of lung tumors in comparison to stefin A. Also, we could not demonstrate a difference between stefin B concentrations in squamous cell carcinoma and the adenocarcinoma. This discrepancies may be explained by the very different methodology (antibodies) of stefin B evaluation, e.g. immunohistochemical staining and the immunochemical determination in tumor homogenates.

It is known that squamous cell carcinoma have the best prognosis, stage by stage, amongst the various types of bronchial carcinomas (21). Therefore, the high content of the stefins in squamous cell carcinoma would suggest that the inhibitors play a protective role in the progression of metastasis and that their high levels are related to a better prognosis of the postoperative survival.

However, stefin A was found of no prognostic significance neither in the patients with squamous cell carcinoma nor in the total lung tumor population. Contrary to stefin A, stefin B was clearly related to the survival probability. Patients with increased stefin B levels had a significant better prognosis than those with lower stefin B levels irrespective of the tumor histology. Since Kyllönen et al. (8) demonstrated also a strong staining of stefin B in the tumor surrounding macrophages, it is likely that the increase in stefin B levels parallels the capability of the host cells to resists the invading tumor cells proteases. This explains the positive correlation of increased stefin B levels with the survival probability in bronchial carcinomas.

Therefore, stefin B seems to be the more important cysteine protease inhibitor than stefin A, at least in lung cancer. Furthermore, as the increase in cysteine proteases, cathepsin B and cathepsin L was also found prognostic for lung cancer, one may conclude that both the tumor cell proteinases and one of their inhibitors, stefin B are of prognostic significance and that their elevated levels are associated with tumor progression.

5. REFERENCES

1. Sloane, B. F., K. Moin, E. Krepela, and J. Rozhin. 1990. Cathepsin B and its endogenous inhibitors: the role in tumor malignancy . *Cancer Met. Rev.* **9**, 333–352.
2. Kane E., and M. M., Gottesman. 1990. The role of cathepsin L in malignant transformation. *Semin. Cancer Biol.* **1**, 127–136.
3. Sloane B. F., K. Moin, and T. T. Lah. 1994. Regulation of lysosomal endopeptidases in malignant neoplasia. *In* G. Thomas, II. Pretlow, and Th. P. Pretlow, editors, Biochemical and molecular aspects of selected cancers, Academic Press, Toronto, Vol. 2, 411–466.
4. Turk V., and W. Bode. 1991. The cystatins: protein inhibitors of cysteine proteinases. *FEBS Lett.* **285**, 213–219.

5. Lah T.T., J. L. Clifford, K. Helmer, N. Day, K. Moin, K. V. Honn, J. D. Crissman, and B. F. Sloane.1989. Inhibitory properties of low molecular weight cysteine proteinase inhibitors from human sarcoma. *Biochim. Biophys. Acta* **993**, 63–73.
6. Calkins C. C., and B. F. Sloane. 1995. Mammalian cysteine protease inhibitors: biochemical properties and possible roles in tumor progression. *Biol. Chem. Hoppe-Seyler* **376**, 71–80.
7. Knoch H., B. Werle, W. Ebert, and E. Spiess. 1994. Imbalance between cathepsin B and cysteine proteinase inhibitors is of prognostic significance in human lung cancer. *Int. J. Oncol.* **5**, 77–85.
8. Kyllönen, A. P., M. Järvinen, V. K. Hopsu-Havu, A. Dorn, O. Räsänen, T. Larmi, and A. Rinne. 1984. The behaviour of small molecular cysteine proteinase inhibitors in lung cancers and in surrounding tissue. *Acta. Histochem.* **74**, 109–113.
9. Pyykkönen K., T. Räsänen, M. Järvinen, R. Rinne, and O. Räsänen. 1996. Immunolocalization of cystatin A in different types of the lung carcinomas. *In* INTERNATIONAL COMMITTEE ON PROTEOLYSIS, editors, 11th International Conference on Proteolysis and Protein Turnover, p. 135, Turku.
10. World Health Organization: Histological classification of lung tumors, WHO, Geneva (1981).
11. Hermanek P., and L. Sobin. TNM Classification of malignant tumors. *1987. In* INTERNATIONAL UNION AGAINST CANCER (UICC), edtors, Vol. 4, Springer Verlag, Berlin.
12. Werle B., W. Ebert, W. Klein, and E. Spiess. 1995. Assessment of Cathepsin L activity by use of the inhibitor CA-074 compared to cathepsin B activity in human lung tumor tissue. *Biol. Chem. Hoppe-Seyler* **376**, 157–164.
13. Kos J., A. Smid, M. Krasovec, B. Svetic, B. Lenarcic, I. Vrhovec. J. Skrk, and Turk V. 1995. Lysosomal proteases cathepsin D, B, H, L and their inhibitors stefins A and B in head and neck cancer. *Biol. Chem. Hoppe-Seyler* **376**, 401–405.
14. Olafsson I., H. Löfberg, M. Abrahamson, and A. Grubb. 1988. Production, characterization and use of monoclonal antibodies against the major extracellular human cysteine proteinase inhibitors cystatin C and kininogen. *Scand. J. Lab. Invest.* **48**, 573–582.
15. Bradford M. M. 1976. A rapid and sensitive method for the quantitation of microgramm quantities of protein utilizing the principle of protein-dye binding . *Anal. Biochem.* **72**, 248–254. 16. Kaplan E.L., and Meier P. 1958. Nonparametric estimation from incomplete observations. *J. Am. Stat. Assoc.* **53**, 457–481.
17. Abel U., J. Berger, and H. Wiebelt. 1984. Critlevel. An exploratory procedure for the evaluation of quantitative prognostic factors. *Methods Inf. Med.* **23,** 154–156.
18. Budihna M., P. Strojan, L. Smid, J. Skrk, I. Vrhovec, A. Zupevc, Z. Rudolf, M. Zargi, M. Krasovec, B. Svatic, N. Kopitar-Jerala, and J. Kos. 1996. Prognostic value of cathepsins B, H, L, D and their endogenous inhibitors stefin A and B in head and neck carcinoma. *Biol Chem Hoppe-Seyler* **377**, 385–390.
19. Sheahan K., S. Shuja, and M. J. Murnane. 1989. Cysteine protease activities and tumor development in human colorectal carcinomas. *Cancer Res.* **49**, 3809–3814.
20. Lah T.T., M. Kokalj-Kunovar, B. Strkelj, J. Pungercar, D. Barlic-Maganja, M. Drobnic-Kosorok, L. Kastelic, J. Babnik, R. Golouh, and V. Turk. 1992. Stefins and lysosomal cathepsins B, L and D in human breast carcinoma. *Int. J.Cancer* **50**, 36–44.
21. Mountain C.F. 1991. Surgical treatment of lung cancer. *Critic. Rev. Oncol./Hematol.* **11**, 179–207.

35

CONTRIBUTION OF THE PROTEASOME TO THE α-SECRETASE PATHWAY IN ALZHEIMER'S DISEASE

Philippe Marambaud,[1] François Rieunier,[2] Sherwin Wilk,[3] Jean Martinez,[2] and Frédéric Checler[1*]

[1]Institut de Pharmacologie Moléculaire et Cellulaire du CNRS
UPR 411, 660 route des lucioles, Sophia Antipolis, 06560 Valbonne, France
[2]URA CNRS 1845
Av. C. Flahault, 34060 Montpellier, France
[3]Department of Pharmacology
Mount Sinai School of Medicine
1 place Gustave Levy, New York 10029, New York

1. INTRODUCTION

It has been extensively documented that the β-amyloid precursor protein (βAPP) can undergo several proteolytic cleavages by various secretases, the activity of which leads to the production of either physiological or potentially pathological catabolites[1,2]. Thus, the concomittant and likely sequential action of β- and γ-secretases ultimately triggers the release of the 39–43 amino-acids long Aβ peptide that corresponds to the main component of the senile plaques invading the cortex in late stages of Alzheimer's disease neuropathology[3,4]. An alternative cleavage ascribed to an α-secretase occurs inside the Aβ sequence, thereby generating a secreted C-terminally truncated fragment, APPα[2]. A dense network of evidences indicates that this catabolite can regulate the activity of serine proteinases involved in blood coagulation and wound repair[5,6] but could also fulfill both cytoprotective and neurotrophic cell functions[7–9]. Interestingly, this α-secretase-derived product is generally accompanied by a down regulation of the production of Aβ[2]. Thus, it has been well established that effectors targeting the protein kinase C (PKC) enhance the APPα secretion and concomittantly decrease the Aβ production[10–13]. It is therefore of interest to identify α-secretase(s) candidates, the activators of which could ultimately lower the formation of Aβ. Here, we present evidences of a phosphorylation-dependent contribution of the proteasome to the α-secretase pathway in human cells.

* To whom correspondence should be addressed.

Cellular Peptidases in Immune Functions and Diseases, edited by Ansorge and Langner
Plenum Press, New York, 1997

2. RESULTS AND DISCUSSION

Several works have examined the proteasome as a putative secretase. These studies, by means of small peptides as well as chromogenic or fluorimetric substrates mimicking the sequence targeted by the various secretases, have led to the surprising and somewhat incoherent conclusions that the proteasome could behave as both α-, β- and γ- secretases[14–16]. HPLC analysis of the degradation of a 11-amino-acids peptide (peptide S) encompassing the sequence likely recognized by α-secretase indicates a rapid cleavage of this substrate by purified proteasome (Fig 1A). This hydrolysis is prevented by a 25μM concentration of the proteasome inhibitor, Z-IE(Ot-Bu)A-Leucinal[17] (Fig. 1A).

Several studies established that missense mutations introduced on the βAPP sequence could alter the rate of the α-secretase-derived production of APPα by various cell lines[18–20]. Interestingly, the replacement of several amino-acids of peptide S by the corresponding mutated residues influenced the catalysic efficiency of the proteasome in a parallel manner. Thus, the H94→D and K94→E modifications diminish or fully prevent hydrolysis by the proteasome (Fig. 1B), in agreement with a previous study[19]. In addition, the K94→V substitution slowed down the catalysis of the mutant peptide (Fig. 1B), in keeping with two independent studies[18,20].

The *in vitro* studies carried out with synthetic peptides[14–16] did not examine whether the proteasome could degrade the natural substrate of α-secretase, i.e βAPP. In this context, we have expressed and purified recombinant baculoviral βAPP751[21] and we establish that this protein undergo an efficient Z-IE(Ot-Bu)A-Leucinal-sensitive cleavage by purified proteasome (Fig. 1C).

The above data could suggest a role of the proteasome as α-secretase, *in vitro* . This prompted us to examine whether this multicatalytic complex could participate to the α-secretase pathway in human cells. Human kidney cells (HK293) secrete an about 120kDa fragment, the immunological characterization of which clearly identified it as APPα[22,23]. As expected from previous reports, APPα secretion appeared drastically increased upon PKC stimulation by PDBu but not by its inactive analog 4α-PDD (Fig.2B). Furthermore, concomittantly to another study[24] we established that APPα secretion can be augmented upon stimulation of the PKA pathway[22] by forskolin (Fig.2A) and 8-br-cAMP (not shown). Interestingly, the PDBu-and forskolin-stimulated APPα secretion appeared prevented by pretreatment of the cells with Z-IE(Ot-Bu)A-Leucinal (Fig. 2A,B). This was corroborated by the drastic inhibition of the PDBu-stimulated APPα secretion by lactacystin, a recently described specific inhibitor of the proteasome[25]. As stated in the introduction, any blocker of a genuine intermediate of the α-secretase pathway should be expected to display opposite effect on Aβ secretion and thus, to enhance the recovery of this catabolite. In this context, it is interesting that we reported on the drastic potentiation of the secretion of Aβ by HK293 cells upon lactacystin treatment[23].

An important issue was to examine whether the various kinases directly targeted the proteasome, thereby modulating its proteolytic activities. We were particularly interested in monitoring its chymotrypsin-like activity (ChTi-Li). First, it was demonstrated that lactacystin displayed a more potent inhibitory potency on the proteasome ChTi-Li than on the other catalytic activities of the enzyme[25]. Second, the various cerebral lesions observed in Alzheimer's disease display intense α1-antichymotrypsin-like immunoreactivity[26,27], suggesting a putative role of this protein in the regulation of endogenous ChTi-Li activities. Interestingly, we demonstrate (Fig.3) that, *in vitro*, the ChTi-Li of the proteasome was inhibited by α1-antichymotrypsin (IC50= 62μg/ml).

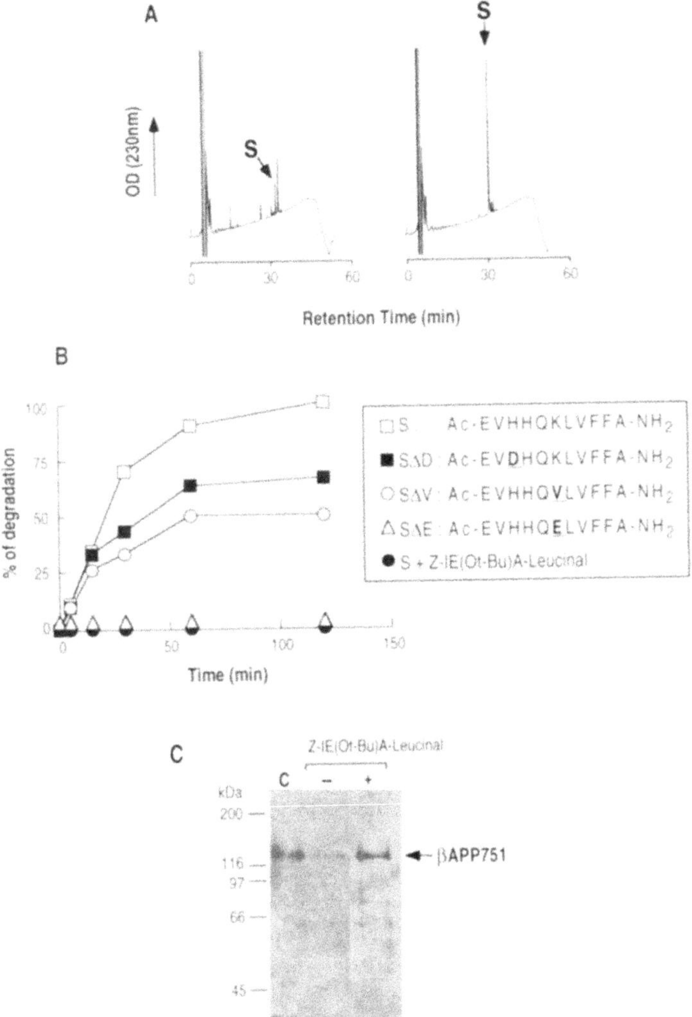

Figure 1. Cleavage of peptide S, its mutated analogs and recombinant βAPP751 by purified proteasome. Peptide S (A) and its indicated mutants (B) were incubated at 37°C for 1 hour (A) or the indicated times (B) in absence (A, left panel) or in the presence (A, right panel) of Z-IE(Ot-Bu)A-Leucinal (25 μM) in a final volume of 200 μl of 20 mM Tris-HCl buffer, pH 7.5 containing 15 μg of purified 20S proteasome. Incubations were stopped and HPLC analysed as described[31]. Degradation was estimated from remaining absorbing material after comparison with known amounts of the corresponding peptide run in the same HPLC conditions. In C, recombinant βAPP751 was incubated without (lane c) or with 20S purified proteasome in the above conditions in absence (-) or in the presence (+) of enzyme inhibitor. Incubations were stopped, submitted to 8% SDS-PAGE and then western blotted onto nitrocellulose using BR188 antibody (1/500), directed toward the C-terminus of βAPP751.

In HK293 cells, all the fluorimetric substrate-hydrolysing ChTi-Li was immunoprecipitated by specific antibodies directed towards the human proteasome[22] and thus could be ascribed to this proteolytic activity. Therefore, it was possible to assess the possible influence of PKA and PKC effectors on the ChTi-Li of the proteasome in HK293 cells. Forskolin drastically increased the ChTi-Li of the proteasome (Fig.4A) while PDBu appeared uneffective (Fig.4B). In agreement with such an observation, forskolin (Fig.4E) but

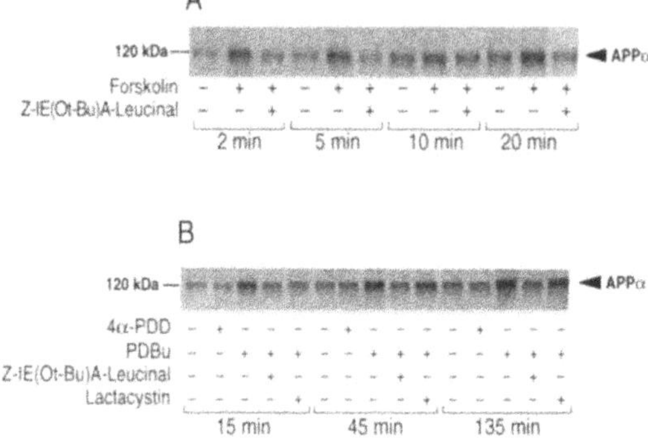

Figure 2. PKA and PKC stimulation of APPα secretion in HK293 cells and effect of proteasome inhibitors. HK293 cells were incubated for 15 hours in absence or in the presence of the indicated inhibitors then cells were metabolically labeled for 2 hours in the same conditons. Incubations were further continued for the mentioned times without or with the various PKA (A) and PKC (B) effectors. Immunoprecipitations of secreted APPα were carried out with the 207 antibody (1/800) and analysed as described[22,23].

not PDBu (Fig.4F) increased the phosphorylation state of a 30kDa subunit of the endogenous cellular enzyme. These observations were corroborated by the fact that *in vitro* phosphorylation experiments of the proteasome by purified PKA increased its ChTi-Li activity (Fig.4C) and trigger a PKI-sensitive phosphorylation of its 30 kDa subunit (Fig.4G) while PKC did not (Fig.4D,H). Altogether, these data seem to discriminate between a PKA-stimulated-APPα secretion involving a direct action of this kinase on the ChTi-Li of the proteasome while it should exist an indirect activation of this enzyme triggered by PKC.

Although our data clearly document the contribution of the proteasome in the α-secretase pathway in human cells, we can not yet conclude that the proteasome is, *per se*, α-secretase and one could also consider that the proteasome could target an intermediate effector taking place downstream in the cascade of events ultimately leading to APPα formation. However, it is possible to envision a potential therapeutic approach aimed at stimulating the proteasome activity that should ultimately lead to decreased formation of the Aβ peptide. It should be noted that such type of activators exist as endogenous entities and have been purified from several sources[28–30].

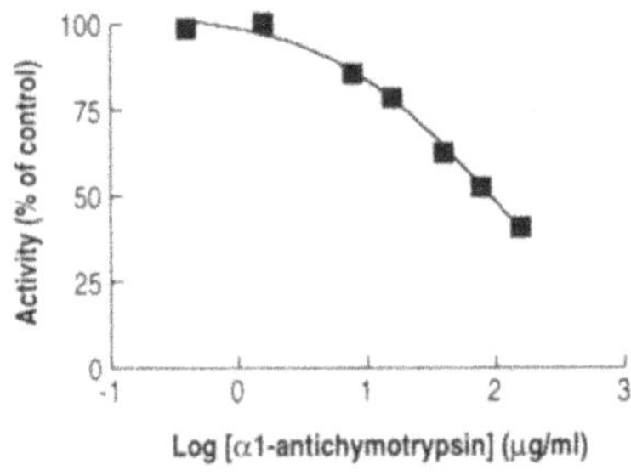

Figure 3. Effect of α1-antichymotrypsin on the chymotrypsin-like activity of the proteasome. 20S purified proteasome (3 μg) was incubated at 37°C for 1 hour in 100 μl of 20 mM Tris-HCl, pH 7.5 containing N-succinyl-LLVY-7-amido-4-methylcoumarin (0.2 mM) in the presence of the indicated concentrations of purified α1-antichymotrypsin. Chymotrypsin-like activity was then fluorimetrically recorded as described[23].

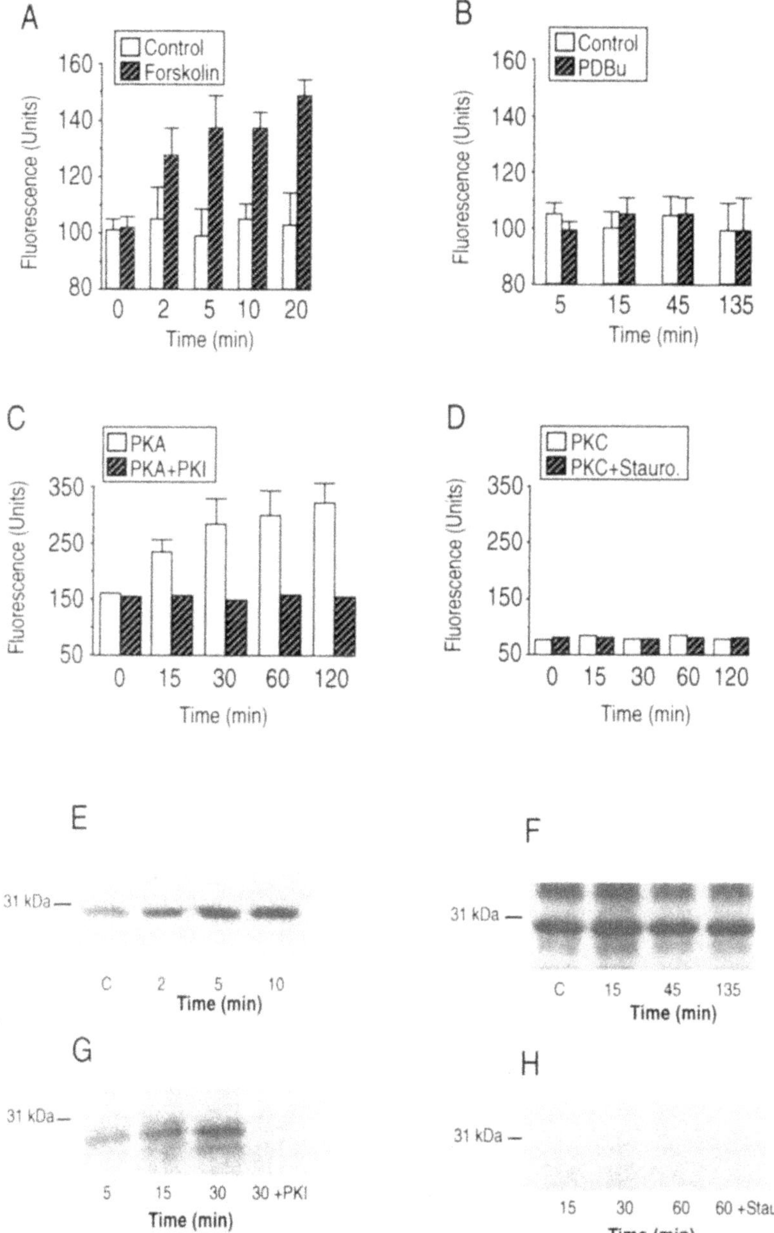

Figure 4. Influence of PKA- and PKC-mediated events on the chymotrypsin-like activity and phosphorylation state of the proteasome. HK293 cells were grown for 5 hours in phosphate-depleted DMEM containing 1 mCi of labelled inorganic phosphate (A,B,E,F), then cells were incubated for the mentioned time periods without or with 50μM forskolin (A and E) or 1 μM PDBu (B,F). HK293 cells were homogenized and assayed for chymotrypsin-like activity as described in the fig.3 (A,B) or submitted to immunoprecipitation by a specific antibody directed toward the human proteasome, H847 (800 fold dilution (E, F) and radioautographed as described[22]. *In vitro* phosphorylation of 20S purified proteasome (C,D,G,H) was performed for various times at 30°C with 0,1 unit of PKA (C,G) or PKC (D, H), in Tris-HCl buffer, pH 7,5, containing 1 mM ATP (C,D) or 1 μCi (20 μM) [γ-^{32}P]-ATP (G,H) in the absence or in the presence of PKI or staurosporine (Stauro.) At the end of incubations, samples were assayed for chymotrypsin-like activity (C,D) or electrophoresed on a 12% SDS-PAGE and autoradiographed (G,H).

ACKNOWLEDGMENTS

We wish to gratefully acknowledge Drs M. Goedert and C. P. Baur (MRC, Cambridge) for providing us with βAPP751cDNA in the baculoviral vector. Dr. Hendil (Taloa, Denmark) is thanked for his generous gift of proteasome antibodies and Dr. B. Greenberg (Cephalon, West Chester) for kindly providing us with the 207 antibody. This work was supported by the Centre National de la Recherche Scientifique et l'Institut National de la Santé et de la recherche Médicale.

REFERENCES

1. Selkoe D.J. (1994) *Annu. Rev. Neurosci.* 17, 489–517.
2. Checler F. (1995) *J. Neurochem.* 65, 1431–1444.
3. Glenner G.G. and Wong C.W. (1984) *Biochem. Biophys. Res. Commun* 120 (3), 885–890.
4. Masters C.L., Simons G., Weinman N.A., Multhaup G., Mc Donald B.L. and Beyreuther K. (1985) *Proc. Natl. Acad. Sci. USA* 82, 4245–4249.
5. Oltersdorf T., Fritz L.C., Schen D.B., Lieberburg I., Johnson-Wood K.L., Beattie E.C., Ward P.J., Blachen R.W., Dovey H.F. and Sinha S. (1989) *Nature* 341, 144–147.
6. Smith R.P., Higuchi D.A. and Broze G.J. (1990) *Science* 248, 1126–1128.
7. Saitoh T., Sundsmo M., Roch J.M., Kimura N., Cole G., Schubert D., Olsterdorf T. and Schenk D.B. (1989) *Cell* 58, 615–622.
8. Mattson M.P., Cheng B., Culwell A.R., Esch F.S., Lieberburg I. and Rydel R.E. (1993) *Neuron* 10, 243–254.
9. Qiu W.Q., Ferreira A., Miller C., Koo E.H. and Selkoe D.J. (1995) *J. Neurosc.* 15, 2157–2167.
10. Caporaso L., Gandy S.E., Buxbaum J.D., Ramabhadran T.V. and Greengard P. (1992) *Proc. Natl. Acad. Sci USA* 89, 3055–3059.
11. Gillespie S., Golde T.E. and Younkin S.G. (1992) *Biochem. Biophys. Res. Commun* 187 (3), 1285–1290.
12. Hung A.Y., Haass C., Nitsch R.M., Qiu W.Q., Citron M., Wurtman R.J., Growdon J.H. and Selkoe D.J (1993) *J. Biol. Chem.* 268 (31), 22959–22962.
13. Buxbaum J.D., Koo E.H. and Greengard P. (1993) *Proc. Natl. Acad. Sci. USA* 90, 9195–9198.
14. Ishiura S., Tsukahara T., Tabira T. and Sugita H. (1989) *FEBS Lett.* 257, 388–392.
15. Kojima I. and Omori M. (1992) *FEBS Lett.* 304, 57–60.
16. Mundy D.I. (1994) *Biochem. Biophys. Res. Commun* 204, 333–341.
17. Peirera M.E., Yu B. and Wilk S. (1992) *Arch. Biochem. Biophys.* 294, 1–8.
18. Sisodia S. (1992) *Proc. Natl. Acad. Sci. USA* 89, 6075–6079.
19. De Strooper B., Umans L., Van Leuven F. and Van der Berghe H. (1993) *J. Cell. Biol.* 121, 295–304.
20. Felsenstein K.M., Hunihan L.W. and Roberts S.B. (1994) *Nature Genetics* 6, 251–256.
21. Chevallier N., Marambaud P., Vizzavona J., Baur C.P., Spillantini M., Fulcrand P., Martinez J., Goedert M., Vincent J.P. and Checler F. (1997) *Brain Res.* in press.
22. Marambaud P., Wilk S. and Checler F. (1996) *J. Neurochem.* 67, 2616–2619.
23. Marambaud P., Chevallier N., Barelli H., Wilk S. and Checler F. (1997) *J. Neurochem.* 68, in press.
24. Xu H., Sweeney D., Greengard P. and Gandy S. (1996) *Proc. Natl. Acad. Sci. USA* 93, 4081–4084.
25. Fenteany G., Standaert R., Lane W.S., Choi S., Corey E.J. and Schreiber S.L. (1995) *Science* 268, 726–731.
26. Abraham C., Selkoe D.J. and Potter H. (1988) *Cell* 52, 487–501.
27. Gollin P.A., Kalaria R.N., Eikelenboom P., Rozemuller A. and Perry G. (1992) *Neuroreport* 3, 201–203.
28. Rivett A.J. (1989) *J. Biol. Chem.* 264, 12215–12219.
29. Rechsteiner M., Hoffman L. and Dubiel W. (1993) *J. Biol. Chem.* 268, 6065–6068.
30. Peters J.-M. (1994) *T.I.N.S.* 19, 377–382.
31. Ichai, C., Chevallier N., Delaere P., Dournaud P., Epelbaum J., Hauw J.J., and Checler, F. (1994) *J. Neurochem.* 62, 645–655.

36

DIPEPTIDYL PEPTIDASE IV (CD26) AND ALZHEIMER AMYLOID PROTEIN PRECURSOR (APP) IN POLYMYOSITIS

Walter Schubert, Karin Agha-Amiri, Oleg Mayboroda, and Christoph Rethfeldt

Otto-von-Guericke-University of Magdeburg
Institute of Medical Neurobiology
Neuroimmunology and Molecular Pattern Recognition Research Group, Medical Faculty
Hansa Park 1, Abtlg. 8, 39116 Magdeburg, Germany

1. INTRODUCTION

There are two major types of immune cell infiltration in human neuromuscular disorders: the polymyositis/inclusion body (PM/IBM) and the dermatomyositis (DM) type. In the PM/IBM type of immune cell invasion, T lymphocytes accumulating within the endomysium (space between muscle fibers) penetrate the basal lamina of intact muscle fibers (endomysial tube) and continuously displace and compress the muscle fiber plasma membrane. In the DM type of infiltration T and B lymphocytes accumulate in the perimysium and around blood vessels.

In the present investigation we have analyzed the PM/IBM type of infiltration. Patients belonging to these diagnostic categories may complain of myalgias and usually show chronic progressive symmetric weakness involving the muscles of the shoulder and the pelvic girdle. In muscle biopsies of PM patients the characteristic pathological feature implicated in the pathogenesis of the disease is represented by T lymphocytes surrounding and invading normal muscle fibers (1,2). The cause of the disease and the molecular mechanism(s) of the T cell invasion are not known. A large number of investigations have shown that these muscle-invasive T lymphocytes express CD8 or CD4 antigens. As yet, there is no evidence for the immunologic function of these T cells, although it has been suggested that antigen-specific cytotoxic events may take place (1). However, an (auto)antigen has not been identified, nor is there a clear morphological evidence for a cytotoxic lysis of muscle fibers by T cells (2,3).

Given that the muscle-invasive T cells (i) accumulate in the connective tissue between muscle fibers, (ii) then actively penetrate the endomysial tube of morphologically intact fibers, and (iii) progessively displace these fibers, these T cells must have a high mi-

Cellular Peptidases in Immune Functions and Diseases, edited by Ansorge and Langner
Plenum Press, New York, 1997

gratory potential (2). The latter must involve differential adhesive mechanisms at the T cell surface. These adhesive functions are likely to be different outside and inside the basal lamina cylinder surrounding muscle fibers, because the molecular components of these different microenvironments are different: Outside the endomysial tube, collagen and other extracellular matrix molecules characteristic for connective tissue, are present, whilst inside the basal lamina cylinder (within the endomysial tube) the cell surface molecules of the muslce fiber sarcolemma are directly applied to the internal surface of the basal lamina.

In an attempt to identify candidate molecules for those adhesive functions, we have studied a battery of proteins expressed at the T cell surface. Here we report on a differential expression of T cell-associated dipeptidyl peptidase IV (CD26) and Alzheimer amyloid protein precursor (APP) in PM/IBM type of muscle infiltration. The patterns, which were found, may provide clues for further functional studies on the organization of the T cell infiltrate and adhesive mechanisms mediating the migratory response.

2. SCREENING CANDIDATE T CELL SURFACE MOLECULES BY MULTI-EPITOPE IMAGING

It was shown earlier by our lab that it is possible to simultaneously label and selectively image a large number of different CD antigens in one and the same muscle tissue section by using a sequential multiparameter immunofluorescence approach (4). The method has led to the construction of a new microscope type, that automatically labels and recognizes 20 different CD antigens in a single tissue section or cellular probe (to be published). The underlying labelling method has allowed to examine different stages during the process of T lymphocyte invasion of endomysial tubes at the level of combinatorial CD antigen patterns expressed at the cell surface of muscle - invasive T cells (4). Interestingly these findings have shown, that the initially-invasive T cell (the front T lymphocyte), which penetrates the basal lamina of morphologically intact muscle fibers appears to downregulate a large number of CD antigens, which are frequently expressed by T cells accumulating behind the invasive front, outside the endomysial tube (4). These invasive front T lymphocytes may be CD8-positive (5,2) or CD4 -positive T cells (2). They displace muscle fibers expressing the neural cell adhesion molecule (NCAM) (2,4,6,7).

3. APP IS A MARKER FOR INVASIVE FRONT T LYMPHOCYTES

When muscle tissue sections showing abundant T cell infiltration are examined with antibodies against the N terminal domain of APP (i.e.monoclonal antibody 22C11) it is seen that approximately one third of all T cells are APP-positive (2). These APP^+ T cells are located at the invading front, consistently show the highest level of APP expression and are either $CD8^-CD4^+$, $CD8^+CD4^-$ T cell phenotypes or $CD8^+CD4^+$ T cell chimeras. In contrast, a paucity of all endomysial T lymphocytes show APP expression, both in PM and miscellaneous neuromuscular disorders (2). A comparison of all invading and endomysial T cells revealed that the most significant increase of invading APP^+ T cells is found within the $CD8^-CD4^+$ subset. None of the other APP^+ T cell subsets is increased at statistical significance (2). The APP^+ non - T cells are preferentially located in the endomysium (outside the endomysial tube, or basal lamina cylinder), part of which are endothelial cells. We also found a significant accumulation of the $CD8^+CD4^-APP^-$ T cell subset among the cells invading the endomysial tube. However, these cells are mainly accumulated behind the invasive front. It is interesting to note that the majority of the $CD8^{+}CD4^{+}$ T cells (T cell chimeras), representing only a minority of the mononuclear cell population in muscle tissue (between 1 and 5%), express APP.

4. CD 26 IS ASSOCIATED WITH T LYMPHOCYTES BEHIND THE INVASIVE FRONT

We have used 2 different monoclonal antibodies to localize CD26 in the infiltrated muscle tissue. We find that CD26 is localized at the cell surface of T lymphocytes. However, in striking contrast to APP, abundant CD26 immunofluorescence signal is predominantly found at the surface of T cells accumulating outside the endomysial tube, behind the invasive front T lymphocytes. The latter appear to be CD26-negative or may show only faint staining for CD26. Together we find 3 different types of CD26 expression patterns in T cells: single T cells near the endomysial tube show high (type 1) or low CD26 expression-levels (type 2). Type 1 is characterized by intense CD26 signal at the cell surface and around the T cell suggesting high accumulation of CD26 protein in the extracellular space near the T cell surface. This may suggest release of CD26 by single T lymphocytes expressing the CD4 or the CD8- cell surface antigen. Many other single T cells (T cells, which are not accumulating), are negative for CD26. Type 3 T cells are CD4-positive/CD26-positive T cells accumulating as dense networks behind the invasive front. We have found that single CD8-positive T cells may be present within the CD4 T cell network. These latter CD8 cells may be negative for CD26.

5. RT-PCR ANALYSIS FROM MUSCLE TISSUE SECTIONS

We have analyzed expression of APP and CD26 from muscle tissue sections using RT-PCR.

APP is a highly conserved molecule expressed by many different cell types. It exists as several isoforms that are produced by differential splicing of message from a single gene on chromosome 21 (8–10). The isoforms are named according to their length in amino acids. Briefly, there are three alternatively spliced APP exons: exon 7 (KPI: Kunitz protease inhibitor domain), exon 8 (MRC OX2-homology domain), and exon 15 (domain inserted into the glycosaminoglycan (GAG) attachment site of APP). Here we have used primers flanking APP exons 8 and 7 (11) to examine APP expression in normal muscle and muscle tissue infiltrated by T cells invading the endomysial sites in muscle. Both in normal muscle and muscle tiusue infiltrated by T cells we have identified here 3 major PCR products in the length of 443 bp, 386 bp, and 219 bp, corresponding to the 3 major APP splice isoforms 770, 751, and and 695. A fourth weak PCR product of 279 bp corresponding to the APP isoform 714 was also observed. There was no clear difference between normal and infiltrated muscle tissue. This may be in keep with earlier findings showing that APP is constitutively expressed in the motor end plate of normal muscle fibers (12) as well as by fibroblasts, endothelial cells and also T cells infiltrating the muscle tissue (2).

The CD26 gene comprizes 26 exons. The gene is transcribed without alternative splicing events (13). Here we have used primers flanking CD26 exons 17–20 to examine expression in muscle tissue. We find RT-PCR product in the length of 400 bp in infiltrated muscle tissue, but not in normal muscle tissue suggesting that CD26 is specifically associated with the process of immune cell invasion in muscle.

6. DISCUSSION AND CONCLUSIONS

Together, both CD26 and APP are expressed by muscle-invasive T lymphocytes. There is, however, a major difference between the expression patterns of CD26 and APP: whilst APP is associated with the invasive front T lymphocytes, CD26 is predominantly expressed by T cells behind the invasive front outside the endomysial tube. Whether the latter pattern represents a stable condition in that disease or may be a transient phenomenon or subject to variation will have to be examined in a larger number of cases.

It has been shown that all three major splice products of APP (APP_{695}; APP_{751}; APP_{770}) are synthesized and actively secreted by human peripheral blood leukocytes (PMBL's) following stimulation with several mitogens (14). The time course of induction was similar to that of interleukin 2 and interleukin 2 receptor, and the APP isoform predominantly secreted by T cells was APP_{751} containing the Kunitz-type proteinase inhibitor domain. In contrast to stimulated PMBL's, T cells invading muscle fibers do not show co - expression of APP and Il-2 receptors (2). Expression of transferrin receptors (4), which are early markers of proliferating T lymphocytes, as well as T cell secretion of Il-2 (15) is not observed in PM. Mitoses of T cells are also absent (2). Thus APP^+ T cells invading muscle fibers are non-proliferating lymphocytes with cell surface properties that are different from antigen - or mitogen - stimulated PMBL's in vitro. In addition, invasive T cells simply displace but not destroy sarcolemma membranes in PM (3). Together these findings are not well compatible with a cytotoxic action of the invasive T cells. CD26 is well defined as a molecule that can be involved in the activation of T cells, on the other hand we were unable to identify the battary of early response gene products in the muscle-invasive T lymphocytes, which are usually co-expressed during T cell activation (Il-2, Il-2R, transferrin receptor) (2,4). In addition CD26 is a multifunctional protein, that also serves as a cell-adhesion receptor which may mediate T cell- to extracellular matrix interaction (16). In particular CD26 binds to collagen. Functional analyzes on the role of APP and CD26 in the muscle-invasive T cells are in progress. The present data, however, together may suggest, that both APP and CD26 have a dominant role in the T cell invasion of muscle. Since there is a strict compartmentalization of these two molecules within the T cell infiltrate we suggest that the role of APP and CD26 may be a differential adhesive one, which would be important for the organization of the invasive process: whilst T cell-surface associated APP may mediate the interaction of the front T lymphocyte with the muscle fiber surface, CD26, which is expressed in dense T cell networks behind the invasive front (at the collagen-tissue sites), may be involved in T cell-to-extracellular matrix interaction at these sites outside the endomysial tube. The latter may be important for the formation of a dense lymphoid tissue behind the APP-expressing T cells, which penetrate the basal lamina cylinder.

ACKNOWLEDGMENT

Supported by grants from the Deutsche Forschungsgemeinschaft through SFB 387 and INK15/A1, as well as DFG (Schu 627/2–2, and 8–1) and the Land Sachsen-Anhalt.

7. REFERENCES

1. Engel, A.G., Arahata, K.: Monoclonal antibody analysis of mononuclear cells in myopathies. II. Phenotypes of autoinvasive cells in polymyositis and inclusion body myositis. Ann. Neurol. 16, 209–215 (1984).
2. Schubert, W., Masters, C.L., Beyreuther, K.: APP$^+$ T lymphocytes selectively sorted to endomysial tubes in polymyositis displace NCAM-expressing muscle fibers. EJCB 62, 333–342 (1993).
3. Arahata, K., Engel, A.G.: Monoclonal antibody analysis of mononuclear cells in myopathies. III. Immunoelectron microscopy aspects of cell-mediated muscle fiber injury. Ann. Neurol. 19, 112 - 125 (1984).
4. Schubert, W.: Multiple antigen-mapping microscopy of human tissue. In: G. Burger, M, Oberholzer, G.P. Vooijs (eds.) Advances in analytical cellular pathology. Elsevier/Excerpta medica ICS, 99. Amsterdam. 1990.
5. Arahata, K., Engel, A.G.: Monoclonal antibody analysis of mononuclear cells in myopathies. I. Quantitation of subsets according to diagnosis and sites of accumulation and demonstration and counts of muscle fibers invaded by T cells. Ann. Neurol. 16, 193–208 (1984).
6. Schubert, W., Zimmermann, K., Cramer, M., Starzinski-Powitz, A.: Lymphocyte antigen Leu19 as a molecular marker of regeneration in human skeletal muscle. Proc. Natl. Acad Sci. USA 86, 307–311 (1989).
7. Mundegar, R.R., J. von Oertzen, S. Zierz: Increased laminin A expression in regenerating myofibers in neuromuscular disorders. Muscle Nerve 18, 992–999 (1995).
8. Kang, J., Lemaire, H.-G., Unterbeck, A., Salbaum, J.M., Masters, C.L., Grzeschik, K.-H., Multhaup, G., Beyreuther, K., Müller-Hill, B.: The precursor of Alzheimer's disease amyloid A4 protein resembles a cell surface receptor. Nature 325, 733–736 (1987).
9. Tanzi, R.E., Gusella, J.F., Watkins, P.C., Bruns, G.A.P., St. George-Hyslop, P., Van Keuren, M.L., Patterson, D., Pagan, S., Kurnit, D.M., Neve, R.L.: Amyloid β protein gene: cDNA, mRNA distribution, and genetic linkage near the Alzheimer locus. Science 235, 880–883 (1987).
10. Kitaguchi, N., Takahashi, Y., Tokushima, Y., Shiojiri, S., Itoh, H.: Novel precursor of Alzheimer's disease amyloid precursor shows protease inhibitor activity. Nature 331, 530–532 (1988).
11. Golde, T.E., Estus, St., Usiak, M., Younkin, L.H., Younkin, G.St.: Expression of β amyloid protein precursor mRNA: Recognition of a novel alternatively spliced form and quantitation in Alzheimer's disease using PCR. Neuron 4, 253–267 (1990).
12. Schubert, W., Prior, R., Weidemann, A., Dircksen, H., Multhaup, G., Masters, C.L., Beyreuther, K.: Localization of Alzheimer βA4 precursor protein at central and peripheral synaptic sites. Brain Res. 563, 184–194 (1991).
13. Marguet, D., Bernard, A.-M., David, F., Lazaro-Trueba, I., Pierres, M.: Structural organization of the DPIV gene and its relationship with DPX and FAPα transcripts. In: Fleischer, B. (ed.) Dipeptidyl peptidase IV (CD26) in metabolism and the immune response. pp 37–53. R.G. Landes Company. Austin, Texas, USA. 1995.
14. Mönning, U., König, G., Prior, R., Mechler, H., Schreiter-Gasser, U., Masters, C.L., Beyreuther, K.: Synthesis and secretion of Alzheimer β precursor protein by stimulated human peripheral blood leukocytes. FEBS Lett. 277, 261–266 (1990).
15. Isenberg, D. A., Rowe, D., Shearer, M., Novic, D., Beverley, P.C.L.: Localization of interferons and interleukin 2 in polymyositis and muscular dystrophy. Clin. Exp. Immunol 63, 450–458 (1986).
16. Dang, N.H., Torimoto, Y., Schlossmann, S.F., Morimoto, C.: Human helper T cell activation: functional involvement of two distinct collagen receptors 1F7 and integrin family. J. Exp. Med. 172, 649–652 (1990).

37

THE HIV PROTEASE AND THERAPIES FOR AIDS

Bruce D. Korant and Christopher J. Rizzo

Virus Laboratory, Molecular Biology Department
DuPont Merck Pharmaceutical Co.
Experimental Station E336
Wilmington, DE 19880-0336

ABSTRACT

New, potent therapies for HIV disease are available, based on synthetic inhibitors of the viral protease, an essential viral enzyme. The results in clinical trials have been impressive with most treated individuals benefiting in terms of reduced quantity of detectable virus, enhanced numbers of CD4 lymphocytes and improvements in quality and duration of life. However, there are some remaining negatives associated with the new drugs, including high cost, side effects and appearance of drug-resistant strains of HIV. Problems and future prospects for use of protease inhibitors and alternate approaches in AIDS are discussed.

1. INTRODUCTION

The advance of the AIDS virus, HIV, continues, with the World Health Organization currently estimating that 30 million people are infected world-wide. There are reported large increases in case numbers in Africa and South Asia, especially India and Thailand, but with spread predicted of emerging virulent strains of the virus to virtually all parts of the world.[1] Progress toward a vaccine has been lagging, and it is hoped that new drugs targeting the virus will control the disease until adequate immunization procedures or some form of gene therapy can protect susceptible populations.

A concerted effort has been made to develop inhibitors of key viral enzymes, in particular the enzymatic functions within the pol gene products; namely the reverse transcriptase, the integrase and the protease. The protease is an essential enzyme for the virus, and is well-known to be a homo-dimeric structure with a pair of aspartic acids at the active site. The hydrolytic action of the enzyme gives rise to the mature core proteins within the virus particles and several enzymes needed for nucleic acid biosynthesis. Within the past year, three new and potent inhibitors of the HIV protease have been approved for use in HIV-infected individuals,

Cellular Peptidases in Immune Functions and Diseases, edited by Ansorge and Langner
Plenum Press, New York, 1997

and most early clinical studies have shown highly promising results, although several problems remain. I will review the current status and look toward the future use of protease inhibitors to treat AIDS, as well as an alternate non-traditional drug-based approach.

2. CURRENT STATUS

Figure 1 shows the structures of the three inhibitors approved in the U.S.A. They are competitive inhibitors of the HIV protease, with Ki values for the enzyme in the picomolar range. Designed based on known substrate preferences of the enzyme, they are quite specific for the viral enzyme versus distantly-related cellular enzymes, such as pepsin, renin, and cathepsin D, and able to discriminate among them by several orders of magnitude. Although they are very potent virus inhibitors in cultured cells, they are only able to reduce virus levels about 90–99% in infected people, partly because of metabolic instability of the drugs in the patients, but also because of dose-limiting toxicities and minimal bioavailability of the molecules, meaning that large oral doses give only moderate levels of useful free drug in the circulation. There may also be privileged compartments in the

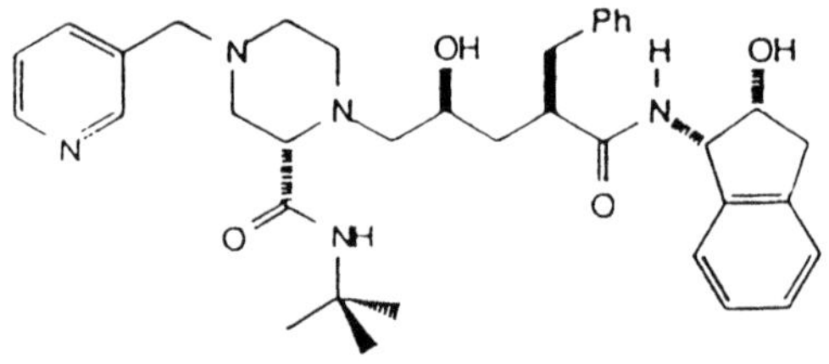

Indanivir
MK639

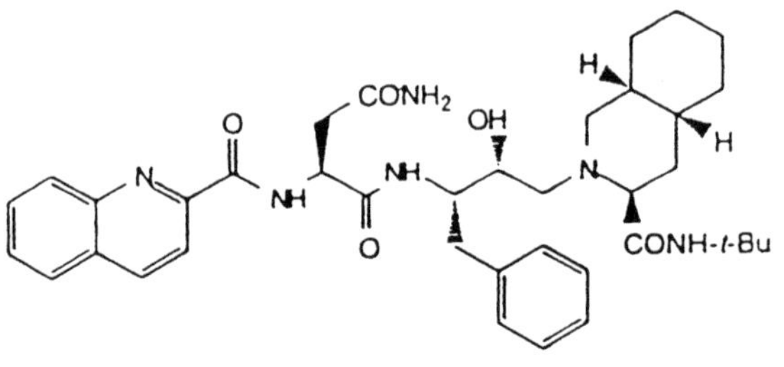

Saquinavir
Ro 31-8959

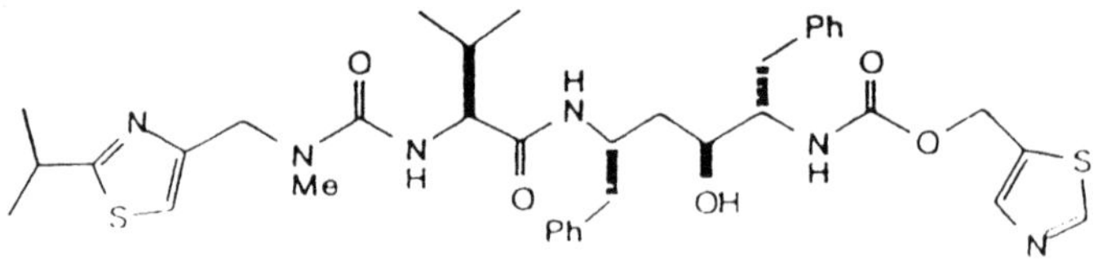

Ritonavir
ABT538

Figure 1. Chemical structures of HIV protease inhibitors recently approved for use in the United States.

infected individuals, eg the cells of the nervous system which harbor virus, but this is presently only speculative.

A major concern is the emergence, after mid-long duration treatment, of drug-resistant versions of the HIV protease. This phenomenon is common with anti-infectives, affecting antibiotics as well as antivirals. The ability to select for changes in the HIV protease as well as other more subtle variations elsewhere in the virus were first observed in tissue culture infections in the laboratory [2] but there is substantial documentation of this result now in drug-treated individuals. An approach to deal with this which is under current intense investigation is the use in combination of several drugs, eg two protease inhibitors or a protease inhibitor and a reverse transcriptase inhibitor, to try to fully suppress virus replication using distinct targets simultaneously. Anecdotal evidence suggests this may be a useful approach. However, the problem may then arise of unfavorable drug interactions with several of these molecules. For example, Ritonavir, a protease inhibitor, increases the amount of cytochrome P450 synthesized by the liver, and also partially inhibits it. This makes for a complex clinical picture, when other drugs used in combination may be susceptible to metabolism by the cytochrome.

Another general feature of protease inhibitors and of new reverse transcriptase inhibitors is their complex structure and multi-step synthesis, and the high doses required for extended periods (years?), leading to substantial drug treatment costs (USD 10,000–15,000 per patient per year), which may be unsustainable, even in the wealthiest societies.

There is some hope however from results of several, independent investigators that brief, high dose therapy using drug combinations, particularly early in infection, may reduce viral levels so much that the immune system may be able to clear the remaining HIV; thereby effecting a cure. If this result can be achieved, with improved treatment protocols based on a fuller understanding of the human pathobiology of HIV infection and therapy, the costs should be more limited and a larger number of infected people will have access to treatment.

3. RESISTANCE TO HIV PROTEASE INHIBITORS

Table 1 shows the changes reported in various amino acids of the HIV protease which are associated with reduced susceptibility to the approved drugs shown in Figure 1, as well as to our cyclic-urea based inhibitors [3]. Inspection of the changes allows the following direct conclusions:

1. Although there is some overlap in resistance profiles, each inhibitor type selects for its own pattern of resistance, with the Roche compound having the most distinct set. This suggests (although there is already controversy) that the Roche compound can be paired with one of the others, either simultaneously or sequentially to suppress mutant viruses.

Table 1. Changes associated with resistance to HIV protease inhibitors

Structure	Fold-resistance (Virus)	Changes in protease
Saquinavir (Roche)	40	G48V, L90M
Ritonavir (Abbott)	10	M46I, L63P, A71V, V82F, I84V
Indinavir (Merck)	18	L10R, M46I, L63P, V82T, I84V
DMP 450 (ref. 3)	120	V82F, I84V

2. Many of the changes come from simple one nucleotide substitutions, lie at the substrate/inhibitor binding sites and are easily understood. However, there are others, eg leucine 63 or alanine 71, which are more difficult to interpret because of lack of known involvement in substrate binding. There are some data available to suggest that changes at positions distant from the substrate binding regions may operate to make the enzyme more efficient, although these claims are not altogether strongly supported.
3. Generally, the virus prefers to accommodate changes leading to insertion of smaller amino acids in the protease thus reducing the interactions with inhibitor by increasing contact distances, implying that the parental enzyme was optimized with respect to substrate binding distances. Smaller residues can be inserted with some loss of contacts, but larger ones are impractical, because of overly close packing when substrate is bound. 4. Single base changes lead to double amino acid alterations because the enzyme is a homo-dimer. This should reflect greater fold resistance to symmetrical inhibitors, and so in principle non-symmetric inhibitors may be preferred. However, non-symmetry may lead to other problems (eg in difficulty of syntheses), which may cancel out any advantages regarding resistance.

Not shown in Table 1, but as a direct result of alterations in protease coding sequence and enzyme structure, is that often the mutant enzyme (and the virus which contains it) are less robust than the drug-sensitive parent in terms of Km/Kcat and replication in cultured lymphocytes. The question remains open at this point as to whether a less fit virus will cause disease with reduced severity. There is practically no animal model for HIV disease, and experimental infection of man with mutant HIV cannot be contemplated. Therefore, as the treatment of individuals with protease inhibitors continues for months and years, and resistant viruses are selected, comparisons will have to be drawn epidemiologically as to the virulence of the "new" viruses which have emerged with alterations in protease sequence.

4. A CELLULAR PROTEIN SUBSTRATE FOR HIV PROTEASE

The essential role of HIV pr in processing viral precursor proteins has been under intense study, but also reported are cleavages of various cellular polypeptides by the viral protease (reviewed in 4). Those proteolytic events are generally poorly understood with respect to their relevance in virus replication.

Interest in cellular proteins as HIV pr substrates grew from our efforts to construct mammalian cell-based assays dependent on the action of the viral protease. We reported previously the splitting of firefly luciferase by HIV pr (5). The luciferases are potentially ideal reporters for cell-based assays, because the light-emitting reactions they catalyze are performed extremely efficiently (quantom yield approaching 1.0 for the firefly enzyme), and because measurement is quantitative, rapid and specific, with negligible background activity in animal cells.

During construction of cell-based assays, we found that the cytocidal action of HIV pr (6) was based on induction of apoptosis or programmed death in protease expressing cell (7). We searched for potential substrates and found that the cytoprotective protein bcl-2 was selectively cleaved by HIV pr, and propose below a model for HIV replication based upon proteolysis-initiated suicide of HIV infected lymphocytes.

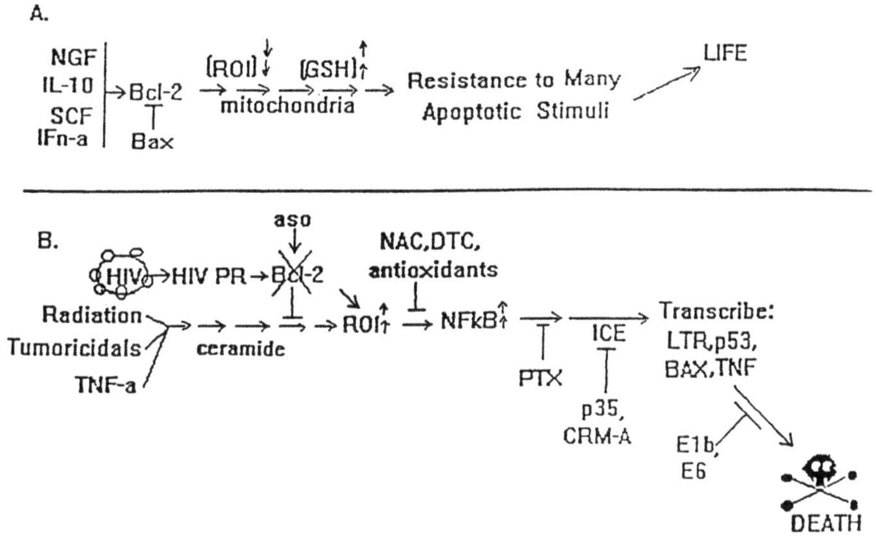

Figure 2. Model for regulation of HIV replication by bcl-2.

Cell death initiated by HIV closely parallels the course of disease in infected people. The mechanism is unclear, but likely involves a direct effect of one or more viral products on the cells rather than immuno-modulation or growth factor deprivation (see 7 for a recent review). We observed the direct splitting of the cytoprotective protein bcl-2 in cells or cell-free extracts exposed to small quantities of HIV-pr, primarily at a site between F112 and A113 located in a sequence in bcl-2 which is similar to an optimal HIV pr processing site. Although this could be incidental, it subsequently leads to a marked increase in oxidative stress in the cells and a lowering of intracellular pH; both conditions favoring HIV replication. Moreover, these conditions are also antecedents of death of cells by apoptosis or programmed suicide.

Why should a virus adopt such a lethal strategy? We propose that it accomplishes two goals of the virus. First, it activates transcription of viral RNA, and second it provides an additional way for the virus to ensure its spread in the human host, since apoptotic cell fragments, after being scavenged, lead to infection of macrophages, and transport to distant organs of the host, including possibly the nervous system (Figure 2).

5. CONCLUSIONS

The recent introduction of three new HIV pr inhibitors as antiviral treatments for AIDS is an important milestone in treating this catastrophic infectious disease. Because the viral protease may be the central actor in the virus strategy to kill host cells as it replicates, the new drugs are predicted to be highly effective at protecting the immune and nervous systems after HIV infection, as well as suppressing "viral load," ie, the quantity of infectious virus in circulation. A separate option we are pursuing is to modify the gene for bcl-2 so that the protein is less susceptible to cleavage by the HIV pr. This is not completely straightforward; altered bcl-2 must also retain its cytoprotective properties, and there are many amino acids in bcl-2 conserved throughout evolution from nematode to man, in particular in a portion of the molecule where the primary HIV cleavage site is located. However, if the proper mutein can be identified and its gene delivered in a suitable vector, it may make available an antiviral therapy to many HIV-positive people who cannot obtain the current anti-HIV drugs.

6. ACKNOWLEDGMENTS

Key contributions to this study were made by P. Strack, M. Frey, and Z. Lu using support provided by the DuPont Merck postdoctoral fellowship program. J. Corman carried out protein sequence analysis.

7. REFERENCES

1. A. Lalvani and J.S. Shastri. HIV Epidemic in India: Opportunity to learn from the past, Lancet 347 (1996), 1349–50.
2. M. Otto et al. In vitro isolation and identification of HIV variants with reduced sensitivity to a C-2 symmetrical inhibitor of HIV-1 protease, Proc. Natl. Acad. Sci. U.S.A. 90 (1993), 7543–7547.
3. C.N. Hodge et al. Improved cyclic urea inhibitors of the HIV-1 protease: synthesis, potency, resistance profile, human pharmacokinetics and x-ray crystal structure of DMP 450, Chemistry and Biology 3 (1996), 301–314.
4. A. Tomasselli and R. Heinrikson. Specificity of retroviral protease: An analysis of viral and non-viral protein substrates, Meth. Enzymol. **241**, (1994), 279–301.
5. B. Korant, Z. Lu, P. Strack and C. Rizzo. HIV protease mutations leading to reduced inhibitor sensitivity, Adv. Exp. Med. Biol., **389** (1996), 241–246.
6. H-G. Kraüsslich. Human immunodeficiency virus proteinase dimer as a component of the viral polyprotein prevents particle assembly and viral infectivity, Proc. Natl. Acad. Sci. USA **88**, (1991), 3213–3217
7. P. Strack et al. Apoptosis mediated by the HIV protease is preceded by cleavage of bcl-2, Proc. Natl. Acad. Sci. USA **93**, (1996), 9571–9576.

38

LEUKODIAPEDESIS, FUNCTION, AND PHYSIOLOGICAL ROLE OF LEUCOCYTE MATRIX METALLOPROTEINASES

Harald Tschesche

University Bielefeld
Faculty of Chemistry and Biochemistry
Universitätsstr. 25, D-33615 Bielefeld, Germany

1. INTRODUCTION

Any inflammatory process is accompanied by the invasion of polymorphonuclear neutrophils (PMN) which emigrate from the vascular bed through the blood capillary walls into the surrounding tissue towards the site of inflammation (Fig. 1).

This process designated as leukodiapedesis involves a cascade of events that start with adhesion of PMN to vascular endothelium initiated by an inflammatory stimulus [1]. The initial adhesion, which is transient and reversible induces a phenomenon called rolling of the cells and involves L-selectin constitutively functional on the PMN surface [2]. After rolling along the vessel wall PMN firmly adhere to the vascular endothelium. Although the precise mechanism responsible for such PMN activation has not yet been identified, it clearly involves the family of β_2 or leucocyte integrins for increased adhesiveness[3, 4]. After firm adhesion of PMN to the vascular endothelium transendothelial migration takes place and is followed by subendothelial migration through the basement membrane barrier of the vessel wall constituted from type IV collagen [5].

2. LEUKODIAPEDESIS

The morphological details of the process of leukodiapedesis, especially the penetration of PMN through basement membrane, have been studied by a scanning electron microscopic investigation in a model based on a modified Boyden chamber assembly (Fig. 2 and 3).

The two chamber compartments were separated by a micropore filter with an overlaying human amnion membrane serving as a basement membrane with epithelial cell

Cellular Peptidases in Immune Functions and Diseases, edited by Ansorge and Langner
Plenum Press, New York, 1997

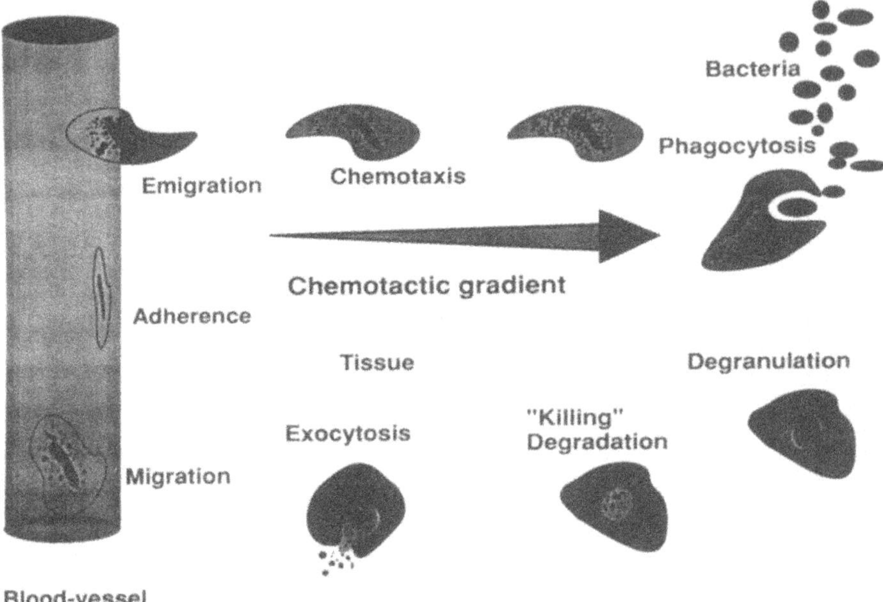

Figure 1. Scheme of leucocyte migration (leukodiapedesis) from the capillary blood vessels towards the site of inflammation.

Figure 2. Scanning electron micrograph of the three-layered structure of human amnion membrane: Epithelial cell layer; basement membrane and stroma tissue [6].

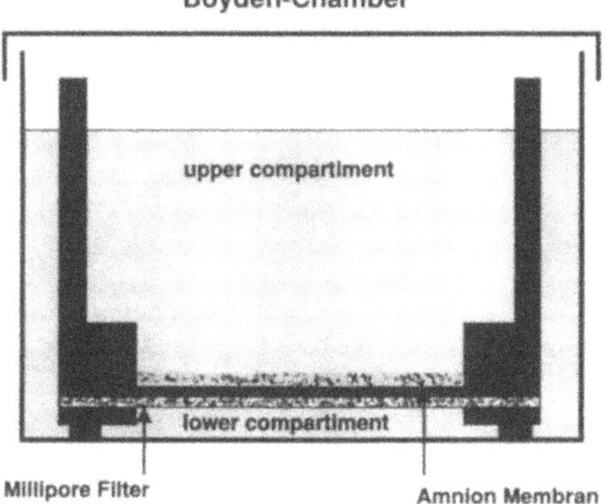

Figure 3. Schematic representation of a Boyden chamber.

layer and underlying stroma tissue [6]. When the lower compartment was filled with a solution of formylpeptides, e.g. 10^{-7} M FMLP, and the upper part with a steril suspension of freshly collected PMN migration of PMN through the membrane was induced. Thus, the amnion membrane was exposed to the cells migrating towards the developing FMLP gradient. This in vitro system allowed to study the individual steps of extravasation and provided a three-dimensional impression of PMN during penetration of basement membrane. Neutrophil activation via FMLP receptors and the processes involved are well characterized [7, 10]. After transendothelial migration, cells adhere to the basement membrane by four types of extracellular matrix receptors for laminin, C 3 bi / fibrinogen, fibronectin and vitronectin (Fig. 4). These receptors are located on the inner surface of special granules ("adhesomes") that fuse with the cell membrane and expose these receptors on the surface of PMN and allow attachment to the extracellular matrix. The second step of penetration involves partial degradation of the dense scaffold of the type IV collagen network of the membrane. A polarized cell which seems to begin lysing the matrix fibres by limited proteolysis in its immediate environment is seen in Fig. 5.

A partial degradation of fibres closely surrounding the cell can be observed which leads to local loss in fibre density and enables the cells to slip into the tissue. Local partial lysis of the basement membrane barrier obviously allows facilitated penetration [6].

Wright and Gallin [11] demonstrated that, during migration, PMN release the content of specific granules by exocytosis in response to chemotactic stimuli. These findings were confirmed by our investigations of the FMLP-stimulated dose-dependent release of collagenase and gelatinase B [9], whereby gelatinase B is the first enzyme to be secreted and detected in the extracellular environment of PMN. The in vitro investigations of Vissers [12] and Uitto [13] et al. demonstrated the gelatinolytical degradation of basement membranes. Involvement of metalloproteinases, expecially gelatinase B, in leukodiapedesis was further supported by inhibition experiments. The amnion membranes were preincubated for 15 min with a five-fold excess of TIMP-1 with regard to the total content of gelatinase. No local lysis of tissue fibres could be observed. The cell seemed to be unable to

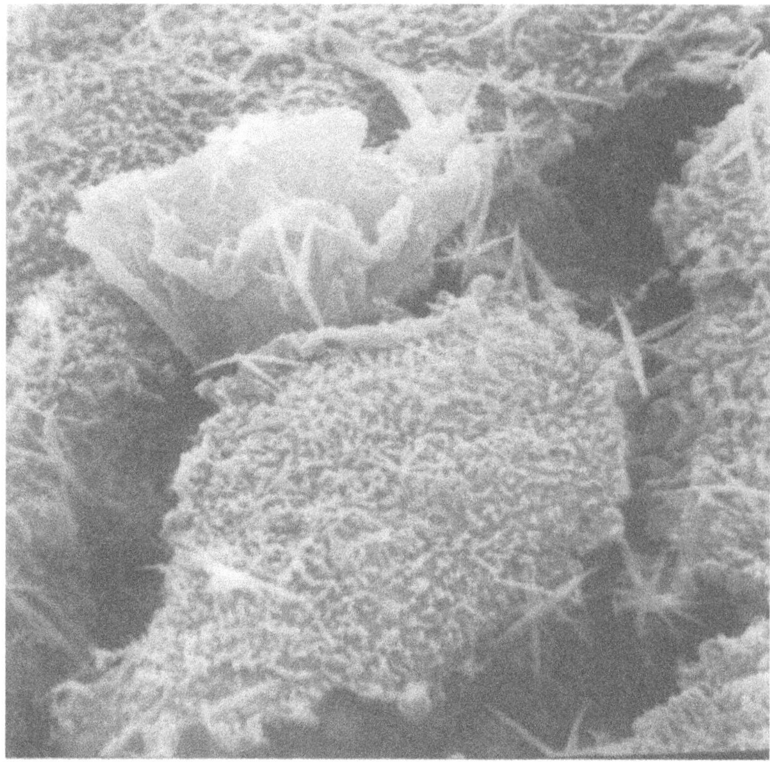

Figure 4. Scanning electron micrograph of a PMN in transendothelial migration [6].

enter into the membrane. Investigations of the reverse sides of the amnion membrane showed that only a few cells crossed the membrane (Fig.6).

In general, proteolytic degradation is controlled by proteinase inhibitors. The most important inhibitors in this case are the tissue inhibitors of metalloproteinases TIMP-1[14,15] and/or TIMP-2.[16, 17] Thus, tissue destruction could be limited to a region close to the cell environment, as can be seen in Fig. 5 by scanning electron micrography. An excess of TIMP lead to an inhibition of cell migration. This provided strong evidence for the participation of metalloproteinases in the process of leukodiapedesis.

In vivo, penetration of basement membranes must proceed in the presence of proteinase inhibitors, which are produced by a variety of cells. [14, 15] Campbell and Campbell[18] proposed a mechanism by which lysis of matrix is limited to the pericellular space between the inflammatory cell and matrix. Thus, cells are able to degrade matrix components in the presence of proteinase inhibitors by locally high concentrations of proteinases in their micro-environment.

Though an enzymatic proteolysis of basement membrane components seemed to be a prerequisite for migration, it could not be excluded that PMNL use mechanical forces for penetration.

After crossing the basement membrane barrier PMNL appeared in the loose meshwork of the stroma tissue, Fig 6, which did not seem to be an obstacle in the further process of emigration of PMNL through the loose stroma tissue to the site of infection. Our

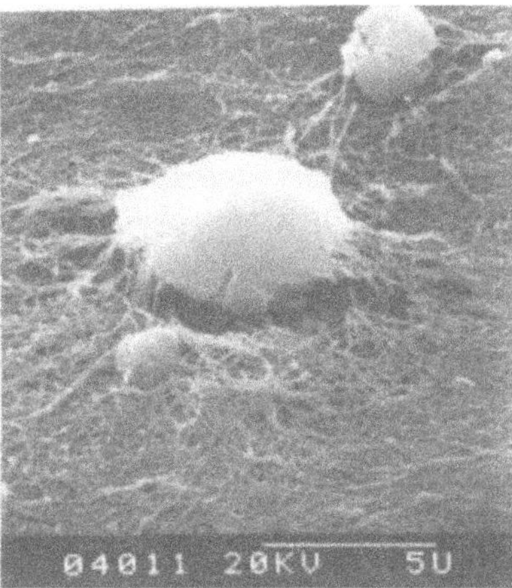

Figure 5. Scanning electron micrograph of a polarized PMN on basement membrane with tissue fibres damaged in the immediate environment of the cell [6].

observations indicated that little or no destruction of type I collagen fibres of stroma tissue took place.

These findings are in agreement with the results of Brown [19], who proposed two mechanisms by which the locomotion of PMNL can be explained. The two-dimensional locomotion is adherence-dependent. This mechanism can be suitable for the process of emigration from blood vessels and penetration of basement membranes. Motility in a three-dimensional collagen gel is adherence-independent, which is adequate for locomotion in the extracellular stroma tissue of type I collagen fibres. The invasion of tissue takes place without proteolytic degradation of collagen fibres [19], which is in accordance with our observations.

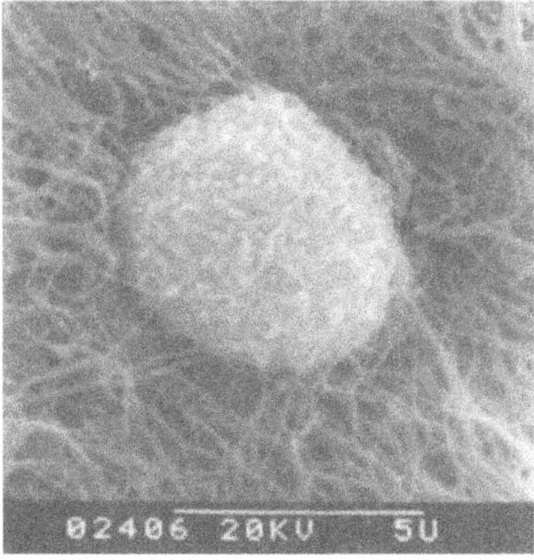

Figure 6. Scanning electron micrograph of a PMN after crossing the basement membrane and invading the stroma tissue [6].

3. MATRIX METALLOPROTEINASES

As documented, the PMN is well equipped with enzymes that are capable of degrading extracellular matrix components, such as types IV and V collagen and interstitial types I, II and II collagens. These enzymes belong to the family of matrix metalloproteinases (MMPs), that are composed as multi-domain proteins (Fig. 7).

The PMN contain at least two of these proteinases, the leucocyte interstitial collagenase (MMP-8) [20] stored as proenzyme in the specific (secondary) granules and a type IV collagenase, i.e. gelatinase B (MMP-9) [21] stored in the specific (tertiary) granules. Secretion of the proenzymes by degranulation of PMN is induced by a great number of agents, including formylpeptides, anaphylatoxins, leukotrien B_4, prostaglandin F_2, interleukin 8 a.o. [20]. This is a special feature of PMN since all other cells producing MMPs like fibroblasts, chondrocytes, synovial cells a.o. are unable to store their enzymes in granules, but rather export their proenzymes after biosynthesis. The induction of secretion from PMN by the various agents is an immediate and dose-dependent phenomenon [9, 10], see Fig. 8.

Degranulation is obviously a regulated process and prevented in the vascular bed by various inhibitory active proteins, such as angiogenin [22] and/or complement factor D[23] (Fig. 9).

The leucocyte MMPs are released from their granules as proenzymes and require extracellular activation. Activation in vitro can be achieved by various proteinases, e.g. trypsin, chymotrypsin, tissue kallikrein, cathepsin G but not by leucocyte elastase and not by

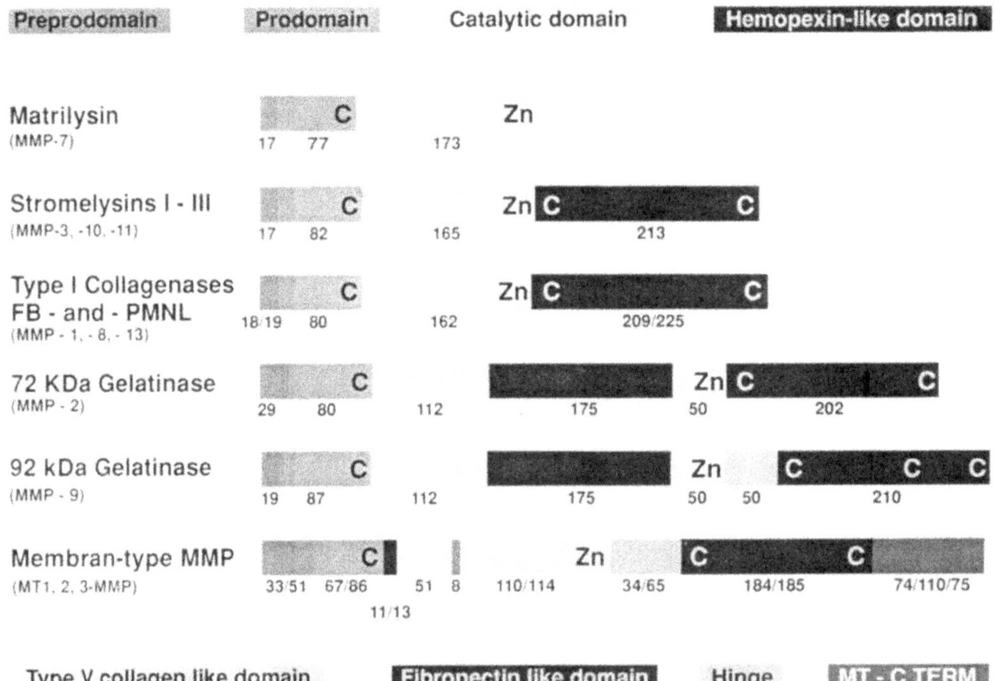

Figure 7. Schematic representation of the domain structure of MMPs.

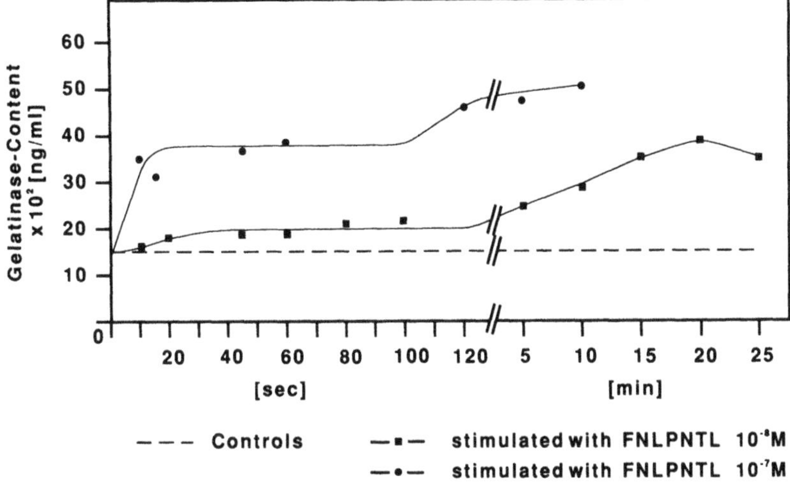

Figure 8. Dose-dependent secretion of gelatinase B from PMN stimulated with the peptide FNLPNTL [9].

plasmin [20, 21, 24–27]. Plasmin derived from the cascade plasminogen activator plasminogen has been made responsible for pro MMP-activation in general, but is only capable to activate the fibroblast interstitial collagenase (MMP-1) and prostromelysin (MMP-3) [26, 27]. Activation requires proteolytic removal of the propeptide domain, that is coordinated by its single cysteine residue within the propeptide consensus sequence PRCGVPD to the catalytic site zinc ion and masks the enzyme's reactive site.

As a consequence of the cysteine-zinc coordination activation of the proMMPs can also be achieved by mercury compounds, other sulfhydryl reagents [27] and by oxidative agents [27, 28], which obviously lead to disengagement of the propeptide domain from the

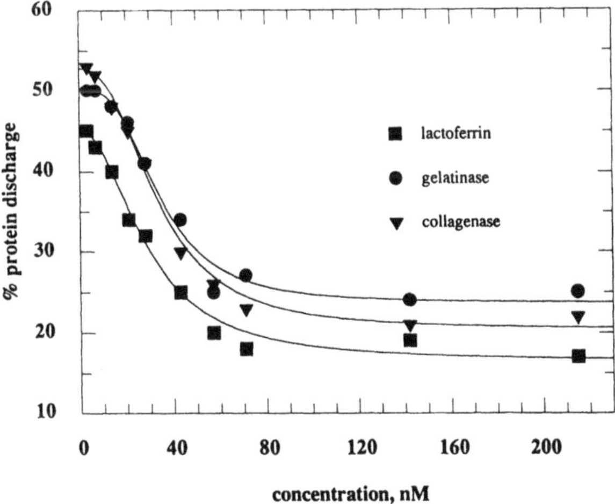

Figure 9. Inhibition of secretion of lactoferrin, gelatinase B, and collagenase by nM-concentrations of angiogenin[22].

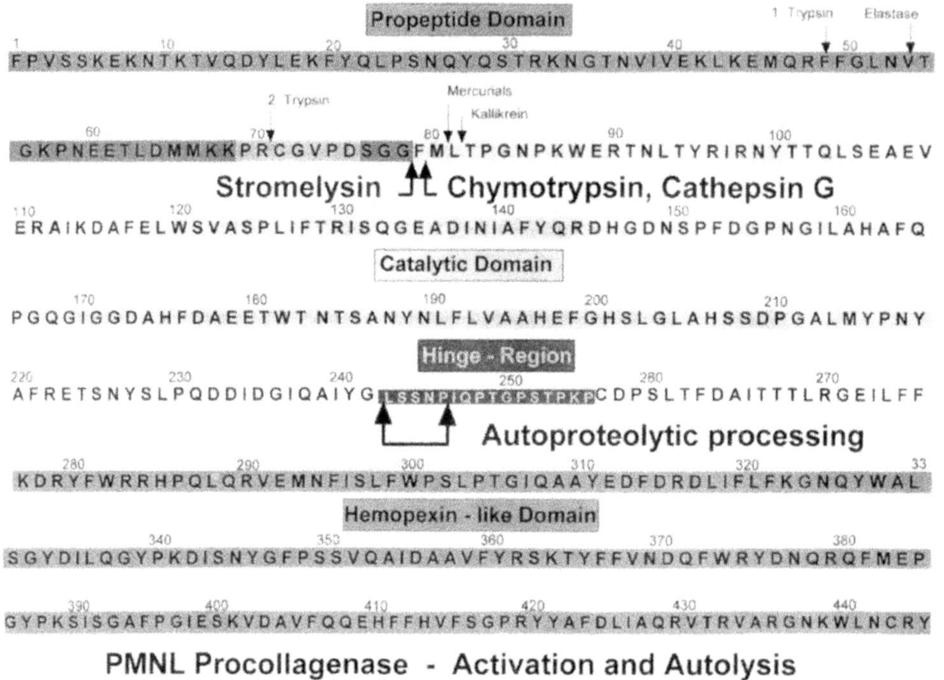

Figure 10. Amino acid sequence of leucocyte procollagenase MMP-8 with proteolytic activation and autoproteolytic processing sites [24].

zinc ion followed by autocatalytic cleavage. This activation mechanism was therefore designated as "cysteine switch mechanism" [29] (Fig. 10 and 11).

The three-dimensional structure of the catalytic domain of human leucocyte interstitial collagenase was solved at 2.0Å after crystallisation of the recombinant protein expressed in *Escherichia coli* [30]. It contains all the essential features of the catalytic sites of the members of the MMP-family. The spherical molecule contains a flat active-site cleft separating the smaller C-terminal part from the larger N-terminal part, which is built of a central, highly twisted five-stranded β-sheet, flanked by an S-shaped double loop and two additional bridging loops on its convex side and two long α-helices on its concave side. The catalytic zinc ion is located at the bottom of the active site cleft and is coordinated by the N atoms of the three His within the His^{197}-Glu^{198}-X-X-His^{201}-Y-X-Gly^{204}-X-X-His^{207} zinc binding consensus sequence. The active site helix contains His^{197}, Glu^{198}, and His^{201} and extends to Gly^{204}, where the polypeptide chain turns away from the helix axis toward the third zinc ligand, His^{207} (Fig. 12).

Besides the "catalytic" zinc ion, a second "structural" zinc ion is sandwiched between the surface S-shaped double loop Arg^{145}-Leu^{160} and the surface of the β-sheet. It is tetrahedrally coordinated by His^{147}, Asp^{149}, His^{175}, while a structural calcium ion is octahedrally coordinated by Asp^{154}, Gly^{155}, Asn ^{157}, Ile^{159}, Asp^{177}, and Glu^{180}. A second structural calcium ion is located on the convex side of the β-sheet. It is also octahedrally coordinated by Asp^{137}, Gly^{169}, Gly^{171}, Asp^{172}, and two water molecules [30].

The full length MMP-8 proenzyme as well as the catalytic domains (Met 80- Gly 242 and Phe 79- Gly 242) have been cloned and expressed in E.coli and the X-ray crystal

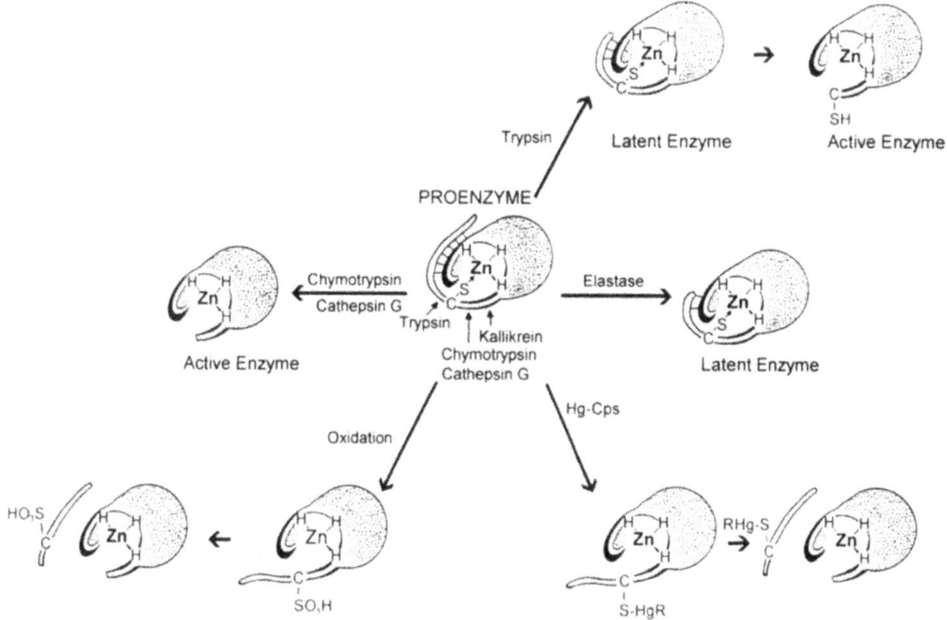

Figure 11. Schematic representation of the various modes of MMP activation.

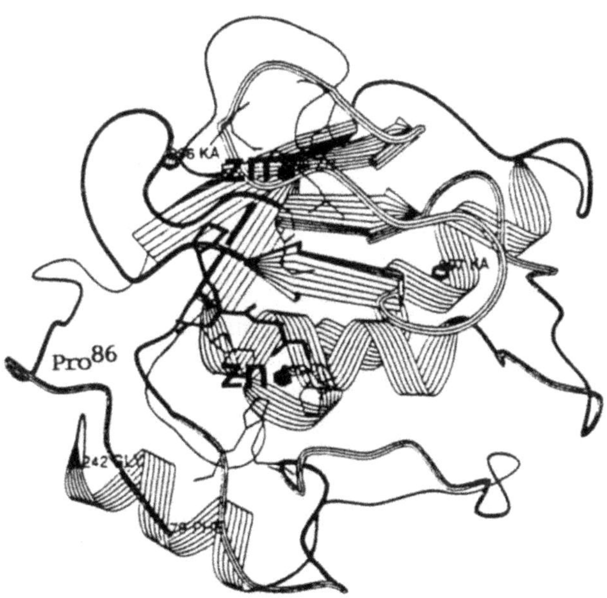

Figure 12. Ribbon plot representation of the catalytic domain of human leucocyte collagenase (MMP-8).[20,30]

structures of the catalytic domain with Met 80 and Phe 79 as N-terminus have been determined at 2.0 Å and 2.5 Å resolution, respectively [31].

The activated forms of the enzymes may vary in their specific activities depending on the respective N-terminus formed upon cleavage of the propeptide domain, as has been demonstrated for the interstitial collagenases (MMP-1 and MMP-8) [32 - 35]. Activation, e.g. by trypsin, chymotrypsin, or cathepsin G lead to the Met 80-form [26, 34] (counting from the N-terminus of the MMP-8 propeptide sequence), while activation by stromelysin resulted in the Phe 79-form, [34], which was a "superactivated" form having activity about four to five times increased, compared to the Met 80-form. X-ray crystallographic structure determinations of both forms revealed a significant difference in the arrangement of the first six N-terminal residues [31]. These are firmly associated with the globular structure of the enzyme in the Phe 79-form. In this form the N-terminal positively charged α-amino group forms an ion pair with the negatively charged side chain of the (in all MMPs) strongly conserved aspartate 232. Thus, the catalytic site or the transition state of substrate catalysis seems to be more favourable. This is also indicated by the inhibition constants of tissue inhibitor of metalloproteinases 2 (TIMP-2) which forms a more stable complex with the Phe 79-enzyme than with the Met 79-form [unpublished results].

The significance of the hemopexin-like domains in MMPs has so far clearly been demonstrated only for the interstitial collagenases (MMP-1 and MMP-8). In these two enzymes the hemopexin-like domains have been shown to be responsible for the helicase activity, i.e. substrate cleavage of intact triple-helical type I collagen into one quarter and three quarter fragments [35,36]. The catalytic domain alone cleaves a great variety of other protein and peptide substrates, such as aggrecan [37], serpins [38, 39], substance P and angiotensin I [40], but not triple helical types I, II and III-collagen. The molecular basis for this phenomenon has not yet been clarified. It is, however, anticipated that the hemopexin-like domain, perhaps together with the proline-rich hinge region plays a critical role in substrate binding and orientation and a clamping model has been hypothesized [20].

4. PHYSIOLOGICAL AND PATHOPHYSIOLOGICAL SIGNIFICANCE

The various matrix metalloproteinases are able to cleave most, if not all extracellular matrix proteins (Table 1).

Consequently, they have a fundamental role in many physiological events, such as leukodiapedesis, wound healing, growth and differentiation, embryo implantation, cervix ripening, uterus involution a.o. and also in pathophysiological situations such as inflammation, rheumatoid arthritis, paradontosis, Sjörgren's syndrome, emphysema, cholesteatomas, tumor invasion and metastasis a.o. Some recent examples are:

In cervix ripening e.g. it could be shown that a great number of PMN accumulate in the extracellular matrix of the cervix tissue prior to birth [41, 42]. With the onset of labour the cells degranulate and release their collagenases into the surrounding tissue where collagen degradation takes place and finally allows dilatation to open up the birth chanel. Interleukin 8 has been shown to increase dramatically during this process and seems to be responsible for triggering this event [43]. The enzyme release could be followed and detected in the circulation where a significant increase in leucocyte col-

Table 1. Substrate specificities of MMPs

Enzyme Name	M (kDa)	Matrix Substrates
Collagenase		
Interstitial Collagenase (MMP-1)	52	Collagen I, II, III, VII, X
Neutrophil Collagenase (MMP-8)	85	Collagen I, II, III
Collagenase 3 (MMP-13)	54	Collagen I, II, III, Gelatin
Gelatinases		
Gelatinase A (MMP-2)	72	Collagen IV, V, Gelatin, Fibronectin, Elastin
Gelatinase B (MMP-9)	92	Collagen IV, V, Gelatin, Elastin
Stromelysines		
Stromelysine-1 (MMP-3)	57	Collagen III, IV, V, IX, Gelatin, Proteoglycan, Laminin, Fibronectin
Stromelysine-2 (MMP-10)	53	analogous MMP-3
Stromelysine-3 (MMP-11)	28	Collagen IV, Gelatin, Proteoglycan, Fibronectin, Laminin
Membrane-type MMPs		
MT1-MMP (MMP-14)	66	Gelatin, activates MMP-2
MT2-MMP (MMP-15)	?	unknown
MT3-MMP (MMP-16)	?	unknown
MT4-MMP (MMP-17)	?	unknown
Other		
Matrilysin (MMP-7)	55	Collagen IV, Gelatin, Proteoglycan, Fibronectin, Elastin
Metalloelastase (MMP-12)	54	Fibronectin, Elastin

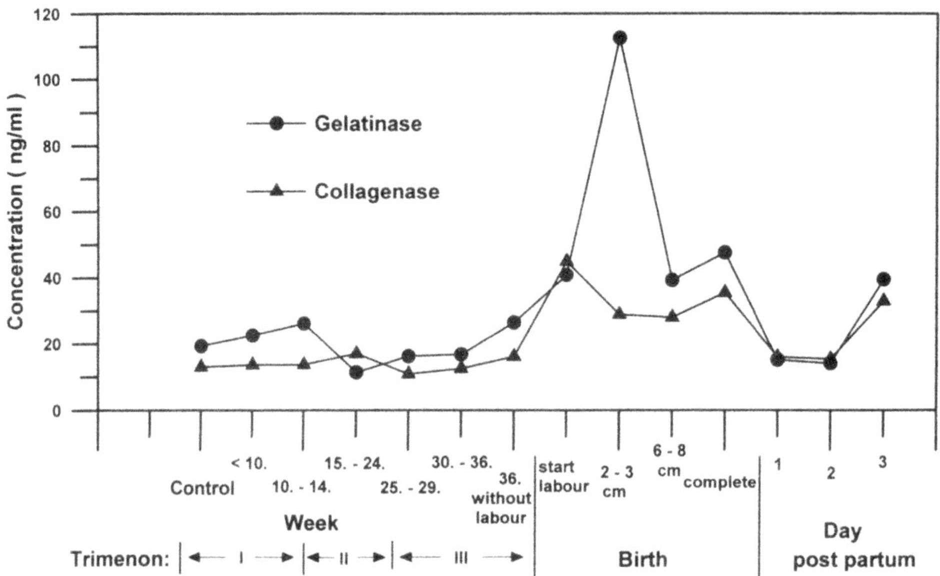

Figure 13. Concentrations of leucocyte collagenase (MMP-8) and gelatinase (MMP-9) in plasma during pregnancy, birth, and post partum [41, 42].

lagenases (MMP-8 and MMP-9) could be detected by specific ELISA-measurements, see Fig. 13.

Interestingly, the leucocytes are the only cells that store these enzymes in enzyme packages, their granules, and are equipped for the process of immediate triggered release.

The easy release of the leucocyte proteinases (MMP-8 and MMP-9) from PMN by non-biocompatible materials obviously also accounts for problems encountered with loosening of endoprotheses. Low TIMP-1 levels combined with high collagenolytic and gelatinolytic activity in the interface tissue has been made responsible for the weakening of periprosthetic connective tissue. Weakening combined with cyclic mechanical loading may lead to loosening of total hip arthroplasty endoprotheses [44].

Also in periodontal diseases both leucocyte collagenases are found in dental plaque rather than bacterial-derived collagenases [45]. But, in periodontally healthy individuals the enzymes exist in latent forms (proforms) and were found activated only in the dental plaque of periodontitis patients, probably reflecting proteolytic activation by potent perodontopathogenic bacterial proteases such as T. denticola or others [45]. The migration of cells through basement membranes as in leukodiapedesis [46] or of tumour cells in intravasation and extravasation obviously requires a type IV collagen degrading proteinase[47]. PMN contain a type IV-collagenase, i.e. gelatinase B, MMP-9, see Fig. 1, which is also produced from many transformed malignant cells. This enzyme cleaves type IV collagen also into one quarter and three quarter fragments at a single locus in the molecule (Fig. 14).

Besides gelatinas B (MMP-9) another type IV-collagenase, gelatinase A (MMP-2) is produced from fibroblast-lineage cells, either constitutively from benign cells and/or from

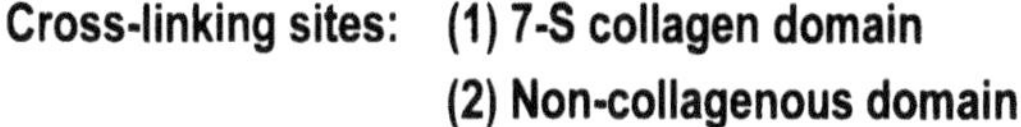

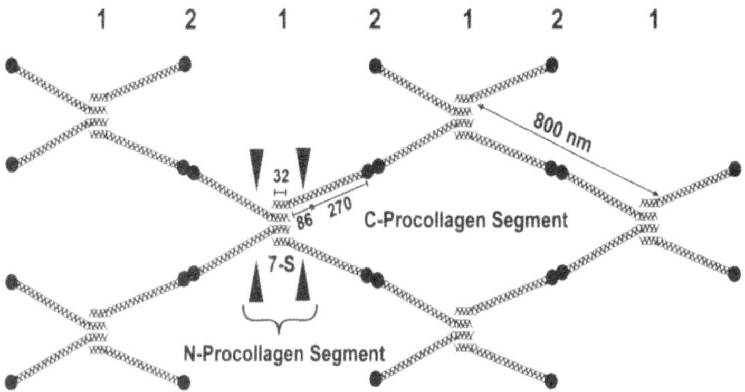

Figure 14. Schematic representation of the single cleavage sites in basement membrane type IV collagen by leucocyte gelatinase B (type IV-collagenase) [24].

tumour cells. The amounts of type IV-collagenase found in malignant cells has been correlated with their metastatic potential [47, 48]. Furthermore, the amounts found in tumour tissues have been used for prognostic relevance of patients outcome. Sophisticated statistical analyses, such as Kaplan-Meier investigations, have been performed for follow ups. In some cancers a high statistical relevance for prognosis of outcome was found, e.g. in colon cancer and the adeno-carcinoma of the esophagus (Fig. 15).

The recurrence of death from a number (n=54) of cancer patients was followed over a period of sixty months. Those patients with values of MMP-9 above 1280 ng/mg protein and MMP-8 above 214 ng/mg protein in their adenocarcinomas had a bad statistical prognosis with outcome of death after about 40 months, while about fifty percent of the patients with lower values survived.

5. IN VIVO ACTIVATION PROCESSES

All soluble MMPs so far known are secreted as proenzymes as are gelatinases A and B. Several proteinases have shown to be activated in vitro [20, 21, 24, 29, 34, 46], but the mechanisms of in vivo activations of most, if not all MMPs, are still debated or unclear. For in vivo activation of gelatinase A the recently discovered membrane-type MMP-14 and/or MMP-15, also designated as MT1-MMP and MT2-MMP respectively, have been made responsible [49–52]. However, the catalytic activation by these MT-MMPs, studied with recombinant enzymes in vitro is a rather slow process [53, 54]. The in vivo activation of gelatinase B ist still more abscure since a corresponding membrane-type MMP has not yet been described. Newly discovered cDNAs for membrane-type proteinases, MT3-MMP [51] and MT4-MMP [54], have not yet been expressed and characterized as enzymes, but are possible candidates for surface associated proenzyme activation.

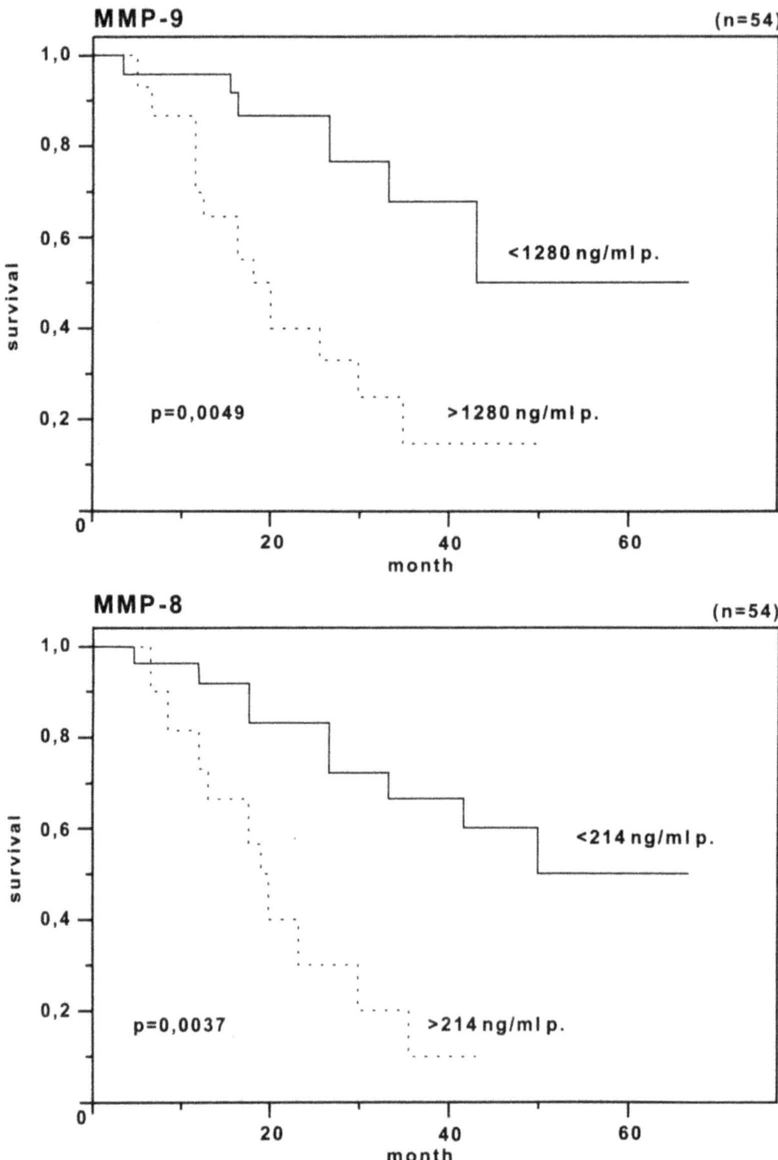

Figure 15. Kaplan-Meier statistics of follow ups of leucocyte collagenase (MMP-8) and gelatinase B (MMP-9) over sixty months of patients (N-54) suffering from adenocarcinoma of the oesophagus (unpublished results).

REFERENCES

1. T. J. Williams and P. G. Hellewell, Endothelial cell biology. Adhesion molecules involved in the microvascular inflammatory response, *Am. Rev. Respir. Dis.* **146,** 45–50 (1992)
2. M. P. Bevilacqua and R. M. Nelson, Selectins, *J. Clin. Invest.* **91**, 379–387 (1993)
3. E. C. Butcher, Leukocyte-endothelial cell recognition: Three (or more) steps to specificity and diversity, *Cell* **67**, 1033–1036 (1991)
4. R. Pardi, L. Inverardi, and J. R. Bender, Regulatory mechanisms in leukocyte adhesion: flexible receptors for sophisticated travelers, *Immunol. Today* **13**, 224 (1993)

5. T. M. Carlos and J. M. Harlan, Leukocyte-endothelial adhesion molecules, *Blood* **84**, 2068–2101 (1994)
6. B. Bakowski and H. Tschesche, Migration of polymorphonuclear leukocytes through human amnion membrane - A scanning electron microscopic study *Biol. Chem. Hoppe-Seyler* **373**, 529–546 (1992)
7. I. I. Singer, S. Scott, D. W. Kawda, D. M. Kazazis, Adhesomes: Specific granules containing receptors for laminin, C3bi/fibrinogen, fibrinectin, and vitronectin in human polymorphonuclear leucocytes and monocytes, *J. Cell Biol.* **109**, 3169–3182 (1989)
8. R. Snyderman and R. J. Uhing, Phagocytic cells: stimulus-response coupling mechanisms, in: *Inflammation: Basic Principles and Clinical Correlates*, 309–332 (J. I. Gallin, I. M. Goldstein, and R. Snyderman, eds.), Raven Press, New York (1988)
9. H. Tschesche, A. Schettler, H. Thorn, B. Bakowski, V. Knäuper, H. Reinke, and B. M. Jockusch, Chemotaxis, proteinase secretion and activation of collagenase of PMN leucocytes in: *Proteinases and Their Inhibitors - Recent Developments*, Proceedings of the 8th Winter School, (E. Auerswald, H. Fritz, V. Turk, eds.) 31–36, KFA Jülich GmbH, (1989)
10. A. Schettler, H. Thorn, B. M. Jockusch, and H. Tschesche, Release of proteinases from stimulated polymorphonuclear leukocytes: Evidence for subclasses of the main granule types and their association with cytoskeletal components, *Eur. J. Biochem.* **197**, 197–202 (1991)
11. D. G. Wright and J. I. Gallin, Secretory responses of human neutrophils: Exocytosis of specific (secondary) granules of human neutrophils during adherence in vitro and during exudation in vivo, *J. Immunol.* **123**, 285–294 (1979)
12. M. C. M. Vissers, C. C. Winterbourn, and J. S. Hunt, Degradation of glomerular basement membrane by human neutrophils in vitro, *Biochim. Biophys. Acta*, **804**, 154–160 (1984)
13. V.-J. Uitto, D. Schwartz, and A Veis, Degradation of basement-membrane collagen by neutral proteases from human leukocytes, *Eur. J. Biochem.* **105**, 409–417 (1980)
14. R. A. D. Bunning, G. Murphy, S. Kumar, P. Phillips, and J. J. Reynolds, Metalloproteinase inhibitors from bovine cartilage and body fluids, *Eur. J. Biochem.* **139**, 75–80 (1984)
15. Y. A. De Clerck, T.-D. Yean, B. J. Ratzkin, H. S. Lu, and K. E. Langley, Purification and characterization of two related but distinct metallo-proteinase inhibitors secreted by bovine aortic endothelial cells, *J. Biol Chem.* **264**, 17445–17453, (1989)
16. G. I. Goldberg, B. L. Marmer, G. A. Grant, A. Z. Eisen, and S. Wilhelm, Human 72-kilodalton type IV collagenase forms a complex with a tissue inhibitor of metalloproteases designated TIMP-2, *Proc. Natl. Acad. Sci. USA* **86**, 8207–8211 (1989)
17. W. G. Stetler-Stevenson, H. C. Krutzsch, and L. A. Liotta, Tissue Inhibitor of Metalloproteinase (TIMP-2), *J. Biol. Chem.* **264**, 17374–17378 (1989)
18. E. J. Campbell and M. A. Campbell, Pericellular proteolysis by neutrophils in the presence of proteinase inhibitors: Effects of substrate opsonization, *J. Cell Biol.* **106**, 667–676 (1988)
19. A. F. Brown, Neutrophil granulocytes: Adhesion and locomotion on collagen substrata and in collagen matrices, *J. Cell Sci.* **58**, 455–467 (1982)
20. H. Tschesche, Human neutrophil collagenase in: *Methods in Enzymology*, **248**, (A.J.Barrett, ed.), 431–449, Academic Press, San Diego (1995)
21. G. Murphy and T. Crabbe, Gelatinases A and B, *Methods in Enzymol.* **248**, 470–484 (1995)
22. H. Tschesche, C. Kopp, W. H. Hörl, and U. Hempelmann, Inhibition of degranulation of polymorphonuclear leukocytes by angiogenin and its tryptic fragment, *J. Biol.Chem.* **269**, 30274–30280 (1994)
23. N. Balke, U. Holtkamp, W.H. Hörl, and H. Tschesche, Inhibition of degranulation of human polymorphonuclear leukocytes by complement factor D, *FEBS Lett.* **371**, 300–302 (1995)
24. H. Tschesche, V. Knäuper, S. Krämer, J. Michaelis, R. Oberhoff, and H. Reinke, Latent collagenase and gelatinase from human neutrophils and their activation, in: *MATRIX Supplement* **1**, 245–255 (H. Birkedal-Hansen, Z. Werb, H. Welgus and H. Van Wart eds.), Gustav-Fischer Verlag, Stuttgart, New York, (1992)
25. G. Murphy, R. Ward, J. Gavrilovic, and S. Atkinson, Physiological mechanisms for metalloproteinase activation, in: *MATRIX Supplement* **1**, 224–230 (H. Birkedal-Hansen, Z. Werb, H. Welgus and H. Van Wart eds.), Gustav Fischer Verlag, Stuttgart, New York (1992)
26. H. Nagase, K. Suzuki, T. Morodomi, J. J. Enghild, and G. Salvesen, Activation Mechanisms of the precursors of matrix metalloproteinases 1, 2 and 3, in: *MATRIX Supplement* **1**, 237–244 (H. Birkedal-Hansen, Z. Werb, H. Welgus and H. Van Wart eds.), Gustav Fischer Verlag, Stuttgart, New York (1992)
27. E. B. Springman, E. L. Angleton, H. Birkedal-Hansen, and H. E. van Wart, Biochemical basis for multiple modes of activation of human fibroblast collagenase, in: *MATRIX Supplement* **1**, 76–77(H. Birkedal-Hansen, Z. Werb, H. Welgus and H. Van Wart eds.), Gustav Fischer Verlag, Stuttgart, New York (1992)
28. S. J. Weiss, G. Peppin, X. Ortiz, C. Ragsdale, and S. T. Test, Oxidative activation of latent collagenase by human neutrophils, *Science* **227**, 747- 749 (1985)

29. E. B. Springman, E. L. Angleton, H. Birkendal-Hansen, and H. E. Van Wart, Multiple modes of activation of latent human fibroblast collagenase: Evidence for a role of a Cys[73] active-site zinc complex in latency and a "cysteine switch" mechanism for activation, *Proc. Natl. Acad. Sci. USA* **87,** 364–368 (1990)
30. W. Bode, P. Reinemer, R. Huber, T. Kleine, S. Schnierer, and H. Tschesche, The X-ray crystal structure of the catalytic domain of human neutrophil collagenase inhibited by a substrate analogue reveals the essentials for catalysis and specificity, *EMBO J.* **6,** 1263–1269 (1994)
31. P. Reinemer, F. Grams, R. Huber, T. Kleine, S. Schnierer, M. Pieper, H. Tschesche, and W. Bode, Structural implications for the role of the N-terminus in the "superactivation" of collagenases. A crystallographic study, *FEBS Lett.* **338**, 227–233 (1994)
32. V. Knäuper, S. Krämer, H. Reinke, and H. Tschesche, Characterization and activation of procollagenase from human polymorphonuclear leucocytes - N-terminal sequence determination of the proenzyme and various proteolytically activated forms, *Eur. J. Biochem.* **189**, 295–300 (1990)
33. J. M. Clark and T. E. Cawston,Fragments of human fibroblast collagenase purification and characterization, *Biochem. J.* **263**, 201–206 (1989)
34. V. Knäuper, S. M. Wilhelm, P. K. Seperack, Y. A. DeClerck, K. E. Langley,A. Osthues, and H. Tschesche. Direct activation of human neutrophil procollagenase by recombinant stromelysin, *Biochem. J.* **295**, 581–586(1993)
35. S. Schnierer, T. Kleine, T. Gote, A. Hillemann, V. Knäuper, and H. Tschesche, The recombinant catalytic domain of human neutrophil collagenase lacks type I collagen substrate specificity, *Biochem. Biophys. Res. Comm.* **191**, 319–326 (1993)
36. E. J. Miller, E. D. Harris, Jr., E. Chung, D. E. Finch, Jr., P. A. McCroskery, and W. T. Butler, Cleavage of type II and III colagens with mammalian collagenase: Site of cleavage and primary structure at the NH_2-terminal portion of the smaller fragment released from both collagens, *Biochemistry* **15**, 787 (1976)
37. A. J. Fosang, K. Last, P. J. Neame, G. Murphy, V. Knäuper, H. Tschessche, C. E. Hughes, B. Caterson, and T. E. Hardingam. Neutrophil collagenase (MMP-8) cleaves at the aggrecanase site E^{373}-A^{374} in the interglobular domain of cartilage aggrecan, *Biochem. J.* **304**, 347–351 (1994)
38. V. Knäuper, H. Reinke and H. Tschesche,Inactivation of Human Plasma 1-Proteinase Inhibitor by Human PMN Leucocyte Collagenase, *FEBS Lett.* **263**, 355–357 (1990)
39. V. Knäuper, S. Triebel, H. Reinke and H. Tschesche,Inactivation of human plasma C1-inhibitor by human PMN leucocyte matrix metalloproteinases, *FEBS Lett.* **290**, 99–102 (1991)
40. O. Diekmann and H. Tschesche, Degradation of kinins, angiotensins and substance P by polymorphonuclear matrix metalloproteinases MMP 8 and MMP 9, *Braz. J. Med. Biol. Res.* **27**, 1877–1883 (1994)
41. I. Walter, I. Wölker and W. Kuhn, Serum collagenase levels during pregnancy and parturition. R. Osmers, M. A. Pflanz, W. Rath, M. Szeverényi, V. Süwer, H. Tschesche, *Eur.J. Obstet. Gynecol. Reprod. Biol.* **53**, 55–57 (1994)
42. R. G. W. Osmers, B. C. Adelmann-Grill, W. Rath, H. W. Stuhlsatz, H. Tschesche, and W. Kuhn, Biochemical events in cervical ripening dilatation during pregnancy and parturition, J. Obstet. Gynaecol. **21**, 185–194 (1996)
43. R. Osmers, H. Tschesche, J. Bläser, Th. Cunze, B. Lefhalm, and W. Kuhn, Bedeutung von Il-1 und Il-8 während der Geburt, *110. Congress Norddeutsche Gesellschaft für Gynäkologie und Geburtshilfe*, **Abstract No. 89**, 142–143 (1995)
44. M. Takagi, Y. Konttinen, P. Kemppinen, T. Sorsa, H. Tschesche, J. Bläser, A. Suda, and S. Santavirta, Tissue inhibitor of metalloproteinase (TIMP)-1 and collagenolytic and gelatinolytic potential in loose THR endoprostheses, *J. Rheumatol.* **22,** 2285–2290, (1995)
45. T. Sorsa, Y.-L. Ding, T. Ingman, T. Salo, U. Westerlund, M. Haapasalo, H. Tschesche, and Y.T. Konttinen, Cellular source, activation and inhibition of dental plaque collagenase, *J. Clin. Periodontol.* **22**, 709–717 (1995)
46. H. Tschesche, B. Bakowski, A. Schettler, V. Knäuper, and H. Reinke, Leukodiapedesis, release of PMN leucocyte proteinases and activation of PMNL procollagenase, *Biomed. Biochim. Acta* **50**, 755–761 (1991)
47. L. A. Liotta, U. P. Thorgeirsson, and S. Garbisa, Role of collagenases in tumor cell invasion. *Cancer Metastasis Rev.* **1,** 277–288 (1982)
48. W. G. Stetler-Stevenson, Type IV collagenases in tumour invasion and metastasis, *Rev.* **9**, 289–303 (1990)
49. H. Sato, T. Takino, Y. Okada, J. Cao, A. Shinagawa, E. Yamamoto, and M. Seiki, A matrix metalloproteinase expressed on the surface of invasive tumour cells, *Nature* **370**, 61–65 (1994)
50. A. Y. Strongin, I. Collier, G. Bannikov, B. L. Marmer, G. A. Grant, and G I. Goldberg, Mechanism of cell surface activation of 72-kDa type IV collagenase, *J. Biol. Chem.* **270**, 5331–5338 (1995)
51. H. Sato and M. Seiki, Membrane-type matrix metalloproteinases (MT-MMPs) in tumor metastasis, *J. Biochem.* **119**, 209–215 (1996)

52. R. V. Ward, S. J. Atkinson, J. J. Reynolds, and G. Murphy, Cell surface- mediated activation of progelatinase A: Demonstration of the involvement of the C-terminal domain of progelatinase A in cell surface binding and activation of progelatinase A by primary fibroblasts, *Biochem J.* **304**, 263- 269 (1994)
53. H. Will, S. J. Atkinson, G. S. Butler, B. Smith, and G. Murphy, The soluble catalytic domain of membrane type I matrix metalloproteinase cleaves the propeptide of progelatinase A and initiates autoproteolytic activation, *J. Biol. Chem.* **271**, 17119–17123 (1996)
54. A. Lichte, H. Kolkenbrock, and H. Tschesche, The recombinant catalytic domain of membrane-type matrix metalloproteinase-1 (MT-MMP) induces activation of progelatinase A and progelatinase A complexed with TIMP-2, *FEBS Lett.* **397**, 277–282 (1996)
55. X. S. Puente, A. M. Pendás, E. Llano, G. Velasco, and C. López-Otin, Molecular cloning of a novel membrane-type matrix metalloproteinase from a human breast carcinoma, *Cancer Research* **56**, 944–949 (1996)

MATRIX METALLOPROTEINASES IN EXPERIMENTAL AUTOIMMUNE ENCEPHALOMYELITIS

Bernd C. Kieseier and Hans-Peter Hartung

Department of Neurology
Neuroimmunology Branch and Clinical Research Group for Multiple Sclerosis
Julius-Maximilians-Universität
Würzburg, Germany

1. MATRIX METALLOPROTEINASES - STRUCTURE AND REGULATION

The matrix metalloproteinases (MMPs) belong to a large subgroup of Zn^{2+}-dependent neutral endoproteinases, which includes the collagenases, gelatinases, and the stromelysins. The number of MMPs known is growing rapidly and they all share at least four common features [1]: (a) they all display proteolytic activity, (b) they are functionally active in the extracellular space, and on the cDNA level (c) protein sequences for the cysteine switch mechanism (PRCGxPD), which is important for their activation, and (d) protein sequences for the binding of catalytic zinc (HExGHxxGxxHS/T) can be found.

MMPs are secreted into the extracellular space by a wide range of cell types as latent pro-enzymes that undergo proteolytic cleavage of an amino-terminal domain during activation. The recently discovered membrane type MMPs (MT-MMP-1, -2, -3, and -4) are bound to the cellular surface [2].

The regulation of MMP activity is strictly controlled at different levels [3–5]: At the transcriptional level, different cytokines, such as tumor necrosis factor (TNF)-α, Interleukin-1, transforming growth factor-β, can directly induce or suppress MMP expression. After secretion the activation of the latent proenzymes is modulated by other proteinases, such as plasmin, a serin proteinase. Furthermore, certain MMPs are capable to activate others: MT-MMP-1, for example, can activate the 72 kDa gelatinase (MMP-2) [6], whereas stromelysin-1 (MMP-3) activates collagenases [7]. Another regulatory mechanism is the interaction with specific tissue inhibitors of metalloproteases (TIMP-1, -2, and -3), which are expressed ubiquitous in the extracellular milieu and form a complex of 1:1 stoichiometry with the endoproteinases.

Cellular Peptidases in Immune Functions and Diseases, edited by Ansorge and Langner
Plenum Press, New York, 1997

2. MMPs – POTENTIAL INVOLVEMENT

MMPs can degrade all protein components of the extracellular matrix (ECM). Furthermore, they have been shown to be capable of processing TNF-α precursor to its mature form [8, 9]. ECM degradation is an important step in many physiological and pathological processes, in which MMPs are believed to be of critical importance [1]. Normal processes include endometrial cycling, pregnancy and parturition, wound healing, and bone remodeling. The probably best known pathologic condition MMPs are believed to be involved relates to tumor biology: Tumor invasion and metastasis seem to be highly dependent on their proteolytic activity. Clinical trials using synthetic MMP-inhibitors as a potential treatment for cancer are currently underway. In rheumatoid arthritis, gelatinases seem to be crucial factors in the degradation of collagen type IV, a structural element of cartilage. And finally, there is an emerging body of evidence that MMPs might be implicated in the pathogenesis of inflammatory demyelinating disorders of the central and peripheral nervous system, such as multiple sclerosis (MS) and the Guillain-Barré syndrome.

3. EAE AS A MODEL FOR HUMAN DEMYELINATING DISEASES

Multiple sclerosis is a common neurologic disease of unknown etiology. Present consensus holds that damage to the nervous system results from aberrant immune response to myelin and possibly non-myelin self-antigens [10]. Autoreactive T-lymphocytes, specific for different antigens within the central nervous system (CNS) including myelin basic protein (MBP), are thought to play an important role in the pathogenesis of this disorder. To reach the target structure within the CNS, circulating T-cells have to cross the blood-brain barrier (BBB).

Experimental autoimmune encephalomyelitis (EAE) is an inflammatory disease of the CNS and is commonly used as an animal model for MS [11]. In Lewis rats it can be actively induced by immunization with CNS myelin proteins or can be adoptively transferred (AT-EAE) by injection of activated encephalitogenic T-cells specific for these antigens [12]. EAE is an acute paralytic disease, which begins 3 to 4 days in AT-EAE and 10 to 12 days in active EAE after immunization, respectively. Clinically animals start to loose weight and to develop loss in tail tonicity, and progress further to paraplegia. Most rats recover completely by 20 days after immunization.

Perivascular inflammatory infiltrates are the histomorphologic hallmark in EAE [13]. The most important effector cells in EAE and MS are T-cells and macrophages, which mainly form the inflammatory infiltrates within the CNS. The mechanism of homing and transmigration of circulating autoreactive T lymphocytes through the blood vessels includes the interaction of a variety of adhesion molecules, such as ICAM-1, VCAM-1, VLA-4, and different cytokines [10], However, the understanding of the exact pathomechanism is only incomplete. Some of these adhesion molecules and cytokines are also known to regulate MMP expression themselves.

4. MMPs IN NEUROINFLAMMATION

Evidence for a potential involvement of proteases in inflammatory diseases of the nervous system was strengthened when Cuzner and colleagues [14] demonstrated increased neutral proteolytic activity in the cellular fraction of the cerebrospinal fluid (CSF) in acute MS. Gijbels et al. [15] detected increased gelatinase activity in the CSF of patients with MS and other neuroinflammatory diseases and proposed a potential involvement of gelatinases in blood-brain barrier (BBB) breakdown. Particularly the 92 kDa gelatinase (MMP-9) and the 72 kDa gelatinase might be involved in BBB breakdown: Increased CSF levels of MMP-9 in MS patients were recently shown to be associated with a leaky BBB on magnetic resonance imaging [16]. Histomorphologic studies of MS lesions revealed astrocytes and microglia to be immunoreactive for the 72 kDa and 92 kDa gelatinase [17]. Also macrophages, neutrophils [18], monocytes [19], and T-cells [20] are known to secrete MMP-9. Moreover, in vitro studies did show that T-cell migration is mediated by gelatinases, and in vivo injection of gelatinases into the brain resulted in disruption of the BBB [21].

4.1. MMPs in EAE

The hypothesis of MMPs being potentially involved in the pathogenesis of EAE received support when it was shown that the application of an unspecific hydroxamate MMP inhibitor (GM 6001) could suppress and even reverse ongoing disease [22]. In another study, MMP-9 was detected in the CSF of animals with EAE [23]. In vitro studies revealed that 72 kDa and 92 kDa gelatinases, but also matrilysin, stromelysin-1, and interstitial collagenase are capable to degrade MBP [24]. All these data indicate a potential involvement of MMPs in EAE, however there are no data about the temporospatial regulation of MMP-2, MMP-9 or other MMPs during the time course of neuroinflammatory diseases available.

Therefore we investigated the expression pattern of 72 kDa and 92 kDa gelatinases during the clinical course of AT-EAE. Animals were immunized with encephalitogenic T-cells, and groups of animals were sacrified at different time points during the course of the ongoing disease. RNA was obtained out of the spinal cord and MMP expression was measured using an RT-PCR assay as described elsewhere [25]. To correlate mRNA expression with proteolytic activity, gelatin zymography was performed. We could not detect any regulation of MMP-2 during the course of AT-EAE, neither on the mRNA level nor on the proteolytic activity level. However, MMP-9 was upregulated during the early phase of the disease. Increased mRNA expression and proteolytic activity were found. Our data coroborate the hypothesis that MMPs and especially MMP-9 might have a role in the pathogenesis of neuroinflammatory diseases. How far these data can be transferred from the animal model to human diseases is difficult to answer. The DNAs that encode MMPs, like collagenases, are highly preserved with 97% identity between rat and mouse, however, there are significant differences between other species - 55% identity to humans, rabbits, pigs, and cows [26]. One might expect that these differences on the DNA level are also reflected on a functional level. Nevertheless, increased knowledge about the expression patterns of different MMPs during the course of neuroinflammatory disease might contribute to a better understanding of the basic pathomechanisms causing these disorders.

REFERENCES

1. Woessner Jr JF. The family of matrix metalloproteinases. Ann NY Acad Sci 1994;732:11–21
2. Takino T, Sato H, Yamamoto E, Seiki M. Cloning of a human gene potentially encoding a novel matrix metalloproteinase having a C-terminal transmembrane domain. Gene 1995;155:293–298
3. Ries C, Petrides PE. Cytokine regulation of matrix metalloproteinase activity and its regulatory dysfunction in disease. Biol Chem Hoppe-Seyler 1995;376:345–355

4. Kleiner Jr DE, Stetler-Stevenson WG. Structural biochemistry and activation of matrix metalloproteases. Curr Opin Cell Biol 1993;5:891–897
5. Murphy G, Willenbrock F, Crabbe T, O'Shea M, Ward R, Atkinson S, O'Connell J, Docherty A. Regulation of matrix metalloproteinase activity. Ann NY Acad Sci 1994;732:31–41
6. Vassalli J-D, Pepper MS. Membrane proteases in focus. Nature 1994;370: 14–15
7. Krane SM. Clinical importance of matrix metalloproteinases and their inhibitors. Ann NY Acad Sci 1994;732:1–10
8. Gearing AJH, Beckett P, Christodoulou M, Churchill M, Clements J, Davidson AH, Drummond AH, Galloway WA, Gilbert R, Gordon JL, Leber TM, Mangan M, Miller K, Nayee P, Owen K, Patel S, Thomas W, Wells G, Wood LM, Wooley K. Processing of tumor necrosis factor-α precursor by metalloproteinases. Nature 1994;370:555–557
9. Mohler KM, Sleath PR, Fitzner JN, Cerretti DP, Alderson M, Kerwar SS, Torrance DS, Otten-Evans C, Greenstreet T, Weerawarna K, Kronheim SR, Petersen M, Gerhart M, Kozlosky CJ, March CJ, Black RA. Protection against lethal dose of endotoxin by an inhibitor of tumor necrosis factor processing. Nature 1994;370:218–220
10. Hartung HP. Pathogenesis of inflammatory demyelination: implications for therapy. Curr Opin Neurol 1995;8:191–199
11. Schwanborg RH. Esperimental autoimmune encephalomyelitis in rodents as a model for human demyelinating disease. Clin Immunol Immunopathol 1995;77:4–13
12. Jung S, Toyka K, Hartung H-P. Suppression of experimental autoimmune encephalomyelitis in Lewis rats by antibodies against CD2. Eur J Immunol 1995;25:1391–1398
13. Lassmann H, Zimprich F, Rössler K, Vass K. Inflammation in the nervous system. Basic Mechanisms and immunological concepts. Rev Neurol (Paris) 1991;147:763–781
14. Cuzner ML, Davison AN, Rudge P. Proteolytic enzyme activity of blood leukocytes and cerebrospinal fluid in multiple sclerosis. Ann Neurol 1978;4:337–344
15. Gijbels K, Masure S, Carton H, Opdenakker G. Gelatinase in the cerebrospinal fluid of patients with multiple sclerosis and other inflammatory neurological disorders. J Neuroimmunol 1992,41:29–34
16. Rosenberg GA, Dencoff JE, Correa N, Reiners M, Ford CC. Effect of steroids on CSF matrix metalloproteinases in multiple sclerosis: relation to blood-brain barrier injury. Neurology 1996;46:1626–1632
17. Maeda A, Sobel RA. Matrix metalloproteinases in the normal human central nervous system, microglial nodules, and multiple sclerosis lesions. J Neuropathol Exp Neurol 1996;55:300–309
18. Nielsen BS, Timshel S, Kjeldsen L, Sehested M, Pyke C, Borregaard N, Dano K. 92kDa type IV collagenase (MMP-9) is expressed in neutrophils and macrophages but not in malignant epithelial cells in human colon cancer. Int J Cancer 1996;65:57–62
19. Welgus HG, Campbell EJ, Cury JD, Eisen AZ, Senior RM, Wilhelm SM, Goldberg GI. Neutral metalloproteinases produced by human mononuclear phagocytes. J Clin Invest 1990;86:1496–1502
20. Leppert D, Waubant E, Galardy R, Bunnett NW, Hauser SL. T cell gelatinases mediate basement membrane transmigration in vitro. J Immunol 1995;154:4379–4389
21. Rosenberg GA, Dencoff JE, McGuire PG, Liotta LA, Stetler-Stevenson WA. Injury-induced 92-kilodalton gelatinase and urokinase expression in rat brain. Lab Invest 1994;71:417–422
22. Gijbels K, Galardy RE, Steinman L. Reversal of experimental autoimmune encephalomyelitis with a hydroxamate inhibitor of matrix metalloproteinases. J Clin Invest 1994;94:2177–2182
23. Gijbels K, Proost P, Carton H, Billiau A, Opdenakker G. Gelatinase B is present in the cerebrospinal fluid during experimental autoimmune encephalomyelitis and cleaves myelin basic protein. J Neurosci Res 1993;36:432–440
24. Chandler S, Coates R, Gearing A, Lury J, Wells G, Bone E. Matrix metalloproteinases degrade myelin basic protein. Neurosci Lett 1995;201: 223–226
25. Wells GMA, Catlin G, Cossins JA, Mangan M, Ward GA, Miller KM, Clements JM. Quantitation of matrix metalloproteinases in cultured rat astrocytes using the polymerase chain reaction with a multi-competitor cDNA standard. Glia 1996;in press
26. Henriet P, Rousseau GG, Eeckhout Y. Cloning and sequencing of mouse collagenase cDNA. FEBS Lett 1992;310:175–178

40

INTERACTION OF TRANSFORMING GROWTH FACTOR ß (TGFß) WITH PROTEINASE 3

J. Kekow,[1] E. Csernok,[2] C. Szymkowiak,[2] and W. L. Gross[2]

[1]Clinic of Rheumatology
Otto von Guericke University of Magdeburg
D 39245 Vogelsang, Germany
[2]Department of Clinical Rheumatology and
Rheumaklinik Bad Bramstedt GmbH
Medical University of Lübeck
Ratzeburger Allee 160, D 23538 Lübeck, Germany

SUMMARY

TGFß is a multifunctional cytokine modulating onset and course of autoimmune diseases as shown in experimental models. Aim of this study was to investigate possible interactions of TGFß with lysosomal enzymes identified as ANCA autoantigens (e.g. proteinase 3, PR3). This included TGFß effects on the translocation the lysosomal enzymes to the cell surface of polymorphonuclear cells (PMN), and the presumabe activation of non bioactive, latent TGFß by these enzymes. Flow cytometry analysis showed TGFß1 to be a potent translocation factor for PR3 comparable with other neutrophil activating factors such as interleukin 8 (IL8). The PR3 membrane expression on primed PMN increased by up to 51% after incubation with TGFß1. PR3 itself was revealed as a potent activator of latent TGFß, thus mediating bioeffects of this cytokine. Patients with various types of systemic vasculitis (SV) showed marked TGFß overexpression correlating with disease. Mean TGFß1 plasma levels in the ANCA associated vasculitis (AAV) patients ranged from 8.9 (Wegeners granulomatosis, WG) to 13.3 ng/ml (Churg-Strauss syndrome, CSS) (control: 4.2 ng/ml, $p<0.01$) while TGFß2 levels were not elevated. Our findings, together with other features of TGFß's such as induction of angiogenesis and its strong chemotactic capacity, indicate that TGFß might serve as a proinflammatory factor in SV, especially in AAV.

1. INTRODUCTION

TGFß shows with multiple actions on various cells and tissues (1,2). In animal models of human autoimmune diseases, local or systemic administration of TGFß modulates

Cellular Peptidases in Immune Functions and Diseases, edited by Ansorge and Langner
Plenum Press, New York, 1997

onset and course of inflammatory processes (3,4). Since no data concerning the expression of TGFß were available in AAV such as WG, and autoantigens in AAV display proteolytic activities and were expected to be able to activate latent TGFß, we conducted a study on TGFß expression in different types of SV, and on the interaction of TGFß with lysosomal enzymes. For comparison with another inflammatory disease, TGFß expression was studied in rheumatoid arthritis (RA). TGFß1 and TGFß2 isoform determinations in plasma specimens were done using a newly established ELISA (5). Since other cytokines are capable of translocating antigens such as PR3 to the cell membrane (6,7), we investigated the effect of TGFß on membrane expression of various lysosomal enzymes present in PMN.

Collectively, our data revealed marked elevations of TGFß1 in AAV patients, but also in RA. In the given setting, TGFß was found to be a potent translocation factor for PR3 on PMN, comparable with other PMN activating factors (e.g. PMA, IL8). PR3 also activated latent TGFß which was comparable with previous reports on plasmin (8).

2. MATERIAL AND METHODS

2.1. Patients

59 vasculitis patients and 37 healthy donors were included in this study. The diagnosis of the patients selected included different types of systemic vasculitides (CSS, Churg-Strauß syndrome; MPA, microscopic polyangiitis; PAN, polyarteritis nodosa; WG, Wegeners granulomatosis) and another inflammatory rheumatic disease (rheumatoid arthrtis: RA, n=15) (9). All patients were included with active, generalized disease and irrespective of current treatment.

2.2. Isolation of Cells for Flow Cytometry Analysis

Human neutrophils were prepared from EDTA-anticoangulated blood from healthy donors using Ficoll-Paque centrifugation (400g, 35 min, 20°C). PMN-containing erythrocyte pellets were subsequently mixed with two volumes of polyvinylalcohol. After sedimentation for 20 min, the PMN-containing supernatant was collected. Contaminating erythrocytes were removed by hypotonic lysis. PMN purity was more than 98% as determined by Giemsa/Wright staining.

2.3. Cytokine Assays

TGFß determination was performed by a newly established ELISA specific for TGFß1 and TGFß2, respectively. The samlpes were tested with and without transient acidification to activate latent TGFß (10,11). For details refer to Szymkowiak et al. (5).

2.4. Flow Cytometric Analysis of PMN Surface Expression of Lysosomal Proteins

Flow cytometric analysis were used to investigate the effect of rTGFß1 on membrane expression of lysosomal proteins (PR3, human leukocyte elastase (HLE) and myeloperoxidase (MPO) on PMN. Monoclonal antibodies against PR3 (designated WGM2, IgG1) (12), HLE and MPO (both IgG1 isotype) (Dakopatts, Hamburg, Germany) were used to identify cell surface expression. As negative control an murine monoclonal IgG1 "irrelevant" has been used (Dianova, Hamburg, Germany). PMN (2 x 10^6/ml) from healthy donors were incubated for 30 min either with TGFß1 (1 ng/ml), PMA (1μg/ml), or with buffer alone at 37°C. Incubations were stopped by ice-cold buffer and cell washing. All experiments were performed in duplicate. Prior to exposure to TGFß1, PMA or buffer alone, PMN were incubated with tumor necrosis factor α (TNFα) (150 pg/ml, 10 min), thus priming them to better respond to TGFß1 and other cytokines, as established elsewhere (13). Flow cytometry was performed as described elsewhere (13).

2.5. Effect of PR3 on the Activation of Latent TGFß

Enzymatically active PR3 was isolated from azurophilic granules of PMN as described by Leid et al. (14). The PBMC culture supernatants (pH 7.4) containing latent TGFß1 were incubated with PR3 (1 μg/ml), Plasmin (55 U/ml, Sigma), HLE (1 μg/ml, Calbiochem, Bad Soden, Germany) and cathepsin G (1 μg/ml, Calbiochem) at 37°C for 2 h (8). The protease activity was then inhibited by the addition of PMSF and aprotinin to final concentration of 2mM and 0.55 TIU/ml. Complete activation of latent TGFß was achieved by transient acidification.

2.6. Statistics

Significance was determined applying the Wilcoxon rank sum test.

3. RESULTS

3.1. TGFß1 Expression in Plasma

The vasculitis patients showed elevated TGFß1 plasma levels of between 8.9 and 13.3 ng/ml, all in the latent form (Figure 1). The values measured in patients with active RA (mean value: 13.8 ng/ml) indicate that high TGFß1 levels are associated with inflammation in general and not with vasculitis per se. TGFß2 levels in plasma ranged about 3.6 ng/ml in controls and were not found to be elevated in the vasculitis patients nor in RA respectively.

3.2. Translocation of Lysosomal Enzymes by TGFß

In contrast to the well-known anti-inflammatory TGFß profile, incubation of PMN with rTGFß1 revealed an increase (up to 51%) in PR3 membrane expression (Figure 2). This experimental approach included a TNFα priming step which mimicks in vivo conditions since high TNFα levels are present in active SV (15,16). The ommission of PMN priming by TNFα resulted in a 60% reduction in PR3 translocation. Antibodies against other proteases present in PMN (e.g. HLE, MPO) did not show significant changes in their membrane expression.

3.3. Effect of PR3 on the Activation of Latent TGFß

Figure 3 shows typical examples of the activation experiments of latent TGFß from pooled supernatants (SN) from monunuclear cells of 4 WG patient. PR3 showed the high-

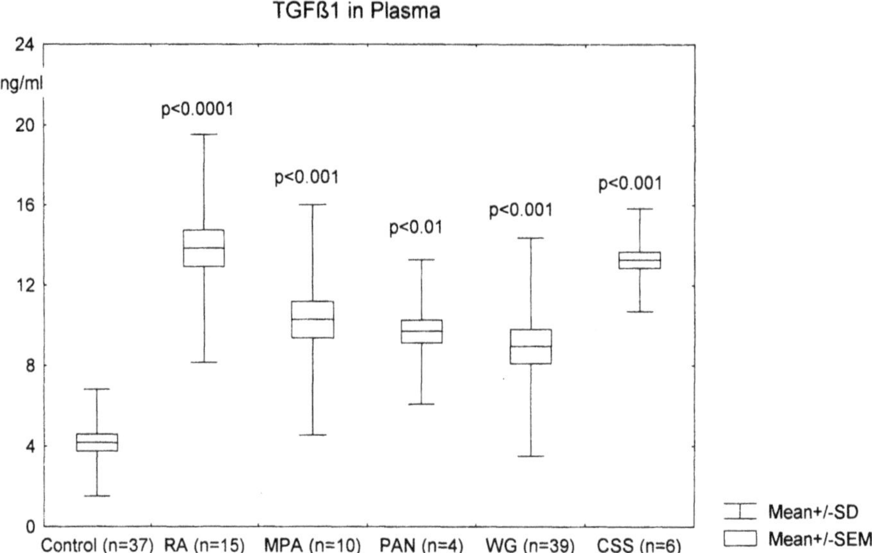

Figure 1. Plasma levels of TGFß1 in patients with RA, MPA, PAN, WG, and CSS as measured by ELISA.

est level of activation reaching up to 51% of total activation achievable by transient acidification of the samples. PR3 was even more effective than plasmin (32%) in TGFß activation. Two other lysosomal proteases, HLE and cathepsin G, showed a weak activation of latent TGFß (8% and 23% respectively).

4. DISCUSSION

The present study of cytokine expression in different types of vasculitic disorders revealed signs of moderate to strong overexpression of TGFß1 in plasma. All TGFß1 detected in plasma of our patients was biologically inactive, which may raise doubts as to the physiological meaning of these findings. In vitro and in vivo studies, however, have clearly indicated constituitive TGFß activation (2). So far TGFß overexpression has only been described in autoimmune diseases, such as diabetes, in malignancies, and in infectious diseases (10,11). Some pathogenetic aspects described in these groups may, however, play a role in SV as well.

The studies on TGFß's effects on the translocation of lysosomal enzymes revealed a new proinflammatory feature of TGFß by making the autoantigen PR3 accessible to the corresponding autoantibody (cANCA). The observed TGFß overexpression may be included into the established hypothetical model for the development of AAV (17). In this model, the interaction between autoantigen (e.g. PR3) and the corresponding antibody (e.g. cANCA) results in PMN degranulation, release of lysosomal proteases, generation of oxygen radicals, and subsequent endothelial cell injury.

Factors activating latent TGFß have been described and it was expected that a wide range of granular proteases, cleave the 'latency associated peptide' (2,8). This is especially likely at inflammatory sites. We found that PR3 is a potent activator of latent TGFß. Further experiments have to clarify whether cell membrane associated PR3 also activates TGFß as shown for free PR3. The fact that free PR3 in circulation is physiologically in-

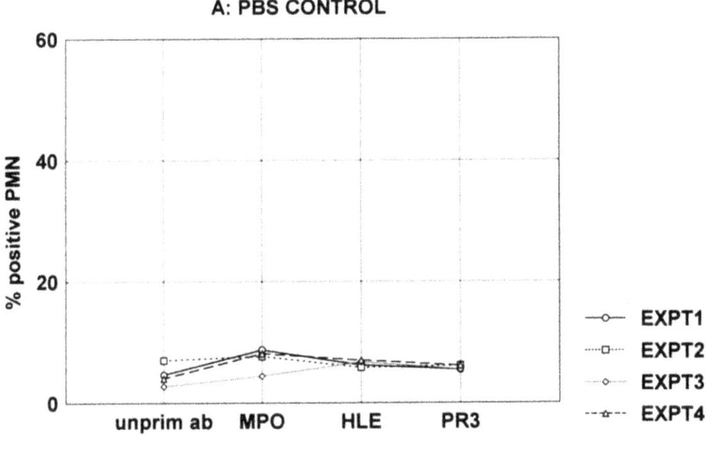

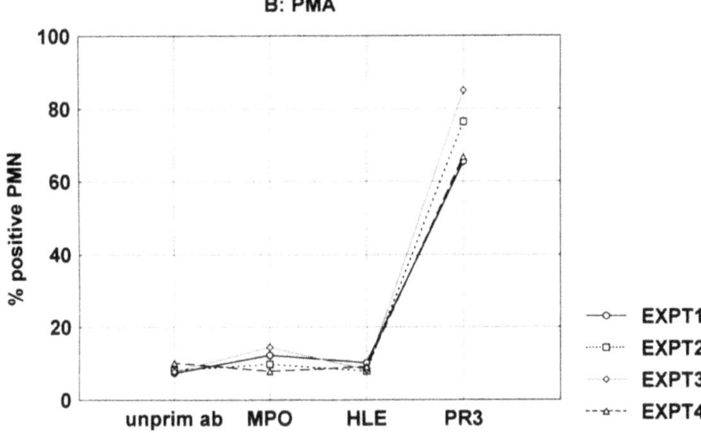

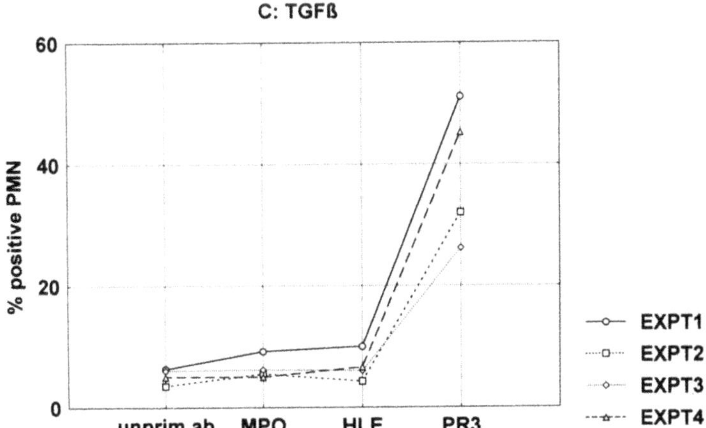

Figure 2. Effect of TGFß and PMA on membrane expression of PR3, HLE, and MPO. Results are shown as percentage of PMN positively stained for the respective enzyme. Panel A: control experiment using PBS, panel B: PMN stimulated with PMA (1μg/ml), panel C: PMN stimulated with TGFß1 (1 ng/ml).

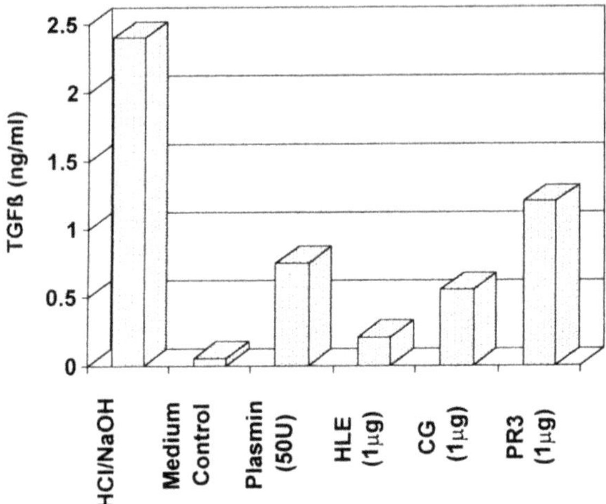

Figure 3. TGFß activation by lysosomal enzymes. The experiment was conducted as described in Materials and Methods. The plot shows the results from 4 individual WG patients. HCl/NaOH shows the complete activation of TGFß1 by acidification. The control consists of untreated SN (medium control) with no TGFß1 activity in the ELISA.

hibited by α_1 antitrypsin explains the observation that no bioactive TGFß was found in our plasma samples. Our results confirmed the proinflammatory type effects of TGFß which have been described for this cytokine before (3).

Regarding the effect of TGFß on different cell types, no other report has implicated TGFß in PMN activation. Since enzymes such as PR3 released by PMN can activate latent TGFß, they may also contribute to elevated TGFß levels because TGFß augments its own expression (2). Further studies on TGFß expression in autoimmune diseases, including AAV, are necessary for understanding their pathogenesis and to enable new therapeutic interventions, as already anticipated in experimental studies using TGFß antibodies or TGFß-binding proteins such as decorin (1,18,19). This will also include studies to elucidate the role of other proteases in cytokine activation.

ACKNOWLEDGMENTS

Work was supported by different grants (Bundesministerium für Forschung und Technologie, autoimmunity grant 01 KD 89030, Bonn; Deutsche Forschungsgemeinschaft grant GR 609/7-1, Bonn; Verein zur Förderung der Erforschung und Bekämpfung rheumatischer Erkrankungen Bad Bramstedt e.V., Bad Bramstedt, Germany).

PR3 was as kind gift from Dr. M.R. Daha, Department of Nephrology, University Hospital Leiden, The Netherlands.

REFERENCES

1. Wahl, S.M. 1994. Transforming growth factor beta: the good, the bad, and the ugly. *J. Exp. Med.* 180:1587–1590.
2. Kekow, J. and G.J. Wiedemann. 1995. Transforming growth factor β: A cytokine with multiple actions in oncology and potential clinical applications (Review). *Int. J. Oncol.* 7:177–182.

3. Allen, J.B., C.L. Manthey, A.R. Hand, K. Ohura, L. Ellingsworth, and S.M. Wahl. 1990. Rapid onset synovial inflammation and hyperplasia induced by transforming growth factor β. *J. Exp. Med.* 171:231–247.
4. Brandes, M.E., J.B. Allen, Y. Ogawa, and S.M. Wahl. 1991. TGF-β1 suppresses leukocyte recruitment and synovial inflammation in experimental arthritis. *J. Clin. Invest.* 87:1108–1113.
5. Szymkowiak, C., I. Mons, W.L. Gross, and J. Kekow. 1995. Determination of transforming growth factor β 2 determination in human blood samples by ELISA. *J. Immunol. Methods* 184:263–271.
6. Gross, W.L., E. Csernok, and B.K. Flesch. 1993. 'Classic' anti-neurophil cytoplasmic autoantibodies (cANCA), 'Wegener's autoantigen' and their immunopathogenic role in Wegener's granulomatosis. *Autoimmunity* 6:171–184.
7. Gross, W.L., W.H. Schmitt, and E. Csernok. 1993. ANCA and associated diseases: Immunodiagnostic and pathogenic aspects. *Clin. Exp. Immunol.* 91:1–12.
8. Lyons, R.M., L.E. Gentry, A.F. Purchio, and H.L. Moses. 1990. Mechanism of activation of latent recombinant transforming growth factor beta 1 by plasmin. *J. Cell Biol.* 110:1361–1367.
9. Kekow, J. and W.L. Gross. 1997. Immunologic assessments. In: Diagnostics of vascular diseases. Priciples and technology. P. Lanzer and M. Lipton, editors. Berlin, Springer. 226–236.
10. Kekow, J., W. Wachsman, J.A. Mc Cutchan, W.L. Gross, M. Zachariah, D.A. Carson, and M. Lotz. 1991. TGF-β and suppression of humoral immune responses in HIV infection. *J. Clin. Invest.* 87:1010–1016.
11. Kekow, J., W. Wachsman, J.A. MacCutchan, M. Cronin, D.A. Carson, and M. Lotz. 1990. Transforming growth factor β and noncytopathic mechanisms of immunodeficiency in human immunodeficiency virus infection. *Proc. Natl. Acad. Sci. U. S. A.* 87:8321–8325.
12. Csernok, E., J. Lüdemann, W.L. Gross, and D.F. Bainton. 1990. Ultrastructural localization of proteinase 3, the target antigen of anti-cytoplasmic antibodies circulating in Wegener's granulomatosis. *Am. J. Pathol.* 137:1113–1120.
13. Csernok, E., M. Ernst, W. Schmitt, D.F. Bainton, and W.L. Gross. 1994. Activated neutrophils express proteinase 3 on their plasma membrane in vitro and in vivo. *Clin. Exp. Immunol.* 95:244–250.
14. Leid, R.W., B.E. Ballieux, I. van der Heijden, C. Kleyburg van der Keur, E.C. Hagen, L.A. van Es, F.J. Van der Woude, and M.R. Daha. 1993. Cleavage and inactivation of human C1 inhibitor by the human leukocyte proteinase, proteinase 3. *Eur. J. Immunol.* 23:2939–2944.
15. Kekow, J., C. Szymkowiak, and W.L. Gross. 1992. Involvement of cytokines in granuloma formation within primary systemic vasculitis. In: New advances on cytokines. S. Romagnani, T.R. Mosmann, and A.K. Abbas, editors. New York, Raven Press. 341–348.
16. Deguchi, Y., N. Shibata, and S. Kishimoto. 1990. Enhanced expression of the tumour necrosis factor/cachectin gene in peripheral blood mononuclear cells from patients with systemic vasculitis. *Clin. Exp. Immunol.* 81:311–314.
17. Gross, W.L., S. Hauschild, and N. Mistry. 1994. The clinical relevance of ANCA in vasculitis. *Clin. Exp. Immunol.* 93:7–11.
18. McCartney Francis, N.L. and S.M. Wahl. 1994. Transforming growth factor beta: a matter of life and death. *J. Leukoc. Biol.* 55:401–409.
19. Border, W.A., N.A. Noble, T. Yamamoto, J.R. Harper, Y. Yamaguchi, M.D. Pierschbacher, and E. Ruoslahti. 1992. Natural inhibitor of transforming growth factor-β protects against scarring in experimental kidney disease. *Nature* 360:361–364.

41

LIVER CYSTEINE PROTEINASES IN MACROPHAGE DEPRESSION INDUCED BY GADOLINIUM CHLORIDE

T. Korolenko,[1] I. Svechnikova,[1] K. Urazgaliyev,[1] G. Vakulin,[2] and S. Djanaeva[1]

[1]Laboratory of Cellular Biochemistry, Institute of Physiology RAMS
630117, Novosibirsk, Timakov St. 4, Russia
[2]Novosibirsk Medical Institute, Krasny Prospekt 52
630091 Novosibirsk, Russia

ABSTRACT

Liver lysosomal enzymes during macrophage depression (gadolinium chloride, 7 mg/kg, intravenously) and macrophage stimulation (zymosan, 100 mg/kg, intravenously) have been studied. It was shown that gadolinium chloride treatment of rats reduced the rate of carbon particles phagocytosis at 24 and 48 h after the single administration. Decreased endocytic capacity of Kupffer cells was confirmed also by electron microscopy. Gadolinium chloride induced labilization of liver lysosomes (increased free activity of cathepsins B and L); there was no changes of specific activity of liver cysteine proteinases (24 h). Gadolinium chloride prevented death of rats after administration of non-sonicated particular zymosan particles, resulting to 70% survival, compare with the 17% survival in group with zymosan alone. We can summarize that macrophages depression by gadolinium chloride abolish symptoms of inflammation in zymosan-model, influencing on cysteine proteinases of Kupffer cells.

1. INTRODUCTION

Recently the new model of selective depression of liver macrophage function in vivo was introduced with help of injection of gadolinium chloride. This compound is commonly used for the deletion of Kupffer cells to study the contribution of these cells in liver physiology and pathology. Intravenous injection of gadolinium chloride inhibits receptor-mediated endocytosis and phagocytosis of Kupffer cells and selectively eliminates large Kupffer cells. Accumulation of this lysosomotropic compound inside of lysosomes can induce significant changes of lysosomes leading to deletion of Kupffer cells.

Cellular Peptidases in Immune Functions and Diseases, edited by Ansorge and Langner
Plenum Press, New York, 1997

The main source of liver cysteine proteinases are macrophages (Kupffer cells) endowed by cathepsin B and aspartic proteinase cathepsin D. In this work we studied the liver lysosomal changes in model of macrophage depression induced by gadolinium chloride with special attention to cysteine proteinases and lysosome stability.

2. METHODS

Male Wistar rats weighing 180–200 g and CBA/C57Bl mice were used in this experiment. Gadolinium chloride (kindly provided by Dr. Bouma, The Netherlands) was administered to rats intravenously into tail vein in a dose of 7 mg/kg b. w. (diluted in 0.15 M NaCl at pH 3.5). Animals were killed at 4, 6, 8, 12, 24, 48, and 72 hours after for evaluation of carbon particle phagocytosis. As a positive control macrophage stimulator zymosan A (Sigma, USA, kind gift of Dr. Janssen) was administered in a dose of 100 mg/kg b.w. in rats intravenously. In other group of rats gadolinium chloride was administered in the same dose as mentioned above 24 h before zymosan. Animals were killed 24 h and 5 days after the last administration of agents.

Phagocytosis was evaluated with carbon clearance test (Gunther-Wagner). For biochemical analysis liver was homogenized in 0.25 M sucrose (1:9, w/v), pH 7.4, containing 1 mM EDTA, as was described earlier. The specific activity was measured with of 0.1% Triton X-100 (final concentration), free activity -in presense of 0.25 M sucrose solution (incubation time 10 min). The activity of cathepsin B was measured against fluorogenic substrate Z-Arg-Arg-NMec, and cathepsin L - against Z-Phe-Arg-NMec with inhibitors. Determination of other lysosomal enzymes was described. The statistical significance was calculated using the Student's t-test.

For electron microscopy liver specimens were fixed in glutaraldehyde solution in 0.1 M phosphate buffer, pH 7.4. Tissue dehydrated in graded alcohols were embedded in Epon-812. Ultrathin sections were obtained on ultratome LKB-8800, contrasted by saturated water solution of uranyl acetate and lead citrate and were studied in electron microscopes JEM-7A and JEM-100S.

3. RESULTS

3.1. Survival Animals after Treatment

Intravenous administration of non-sonicated particular zymosan in a dose of 100 mg/kg to rats resulted to death at 12–24 h (survival 17 %). Gadolinium chloride alone didn't cause mortality of rats. Pretreatment by gadolinium chloride prevented significantly (survival 70 %) the death of animals caused by zymosan.

3.2. Electron Microscopy

3.2.1. Controls. A normal ultrastucture of liver cells was observed in the control animals which were given saline.

3.2.2. Gadolinium Chloride. 24 h after gadolinium chloride treatment ultrastructure of Kupffer cells showed signs of significant decrease of phagocytic function: almost total deletion of worm-like processes in Kupffer cells, rare filopodia and invaginates formed by plasma membrane; limited number of heterophagosomes were noted.

In hepatocytes lipid infiltration and appearance of large lipid-containing lipolysosomes and secretory vacuoles with VLDL particles were noted. Mitochondrial matrix was more electron-dense with dilatation of cristae.

Taken together the results of electron microscopic investigation suggest that gadolinium chloride induces deletion of endocytic capacity of liver macrophages and exerts some injurious effect on parenchymal cells.

3.2.3. Zymosan. On the contrast to treatment by gadolinium chloride macrophage stimulation by zymosan (5th day) induced formation of numerous worm-like processes, invaginates, filopodia and heterophagolysosomes in Kupffer cells (signs of increased phagocytic activity). Increased uptake of erythrocytes at different stages of digestion inside of lysosomes and appearance of large ring-like lysosomes were noted. Increased proliferative activity was shown according to appearance of centrioles and nuclear changes typical for mitosis. In hepatocytes numerous small primary lysosomes with homogenic osmiophylic matrix and large autophagosomes were noted.

3.2.4. Gadolinium Chloride and Zymosan. Comparatively to gadolinium chloride pretreatment by this agent of zymosan-stimulated rats (5th day) was followed by some activation of phagocytosis in Kupffer cells and discharge of hepatocytes from lipids, probably, thanks to the increased exocytosis of such material by Kupffer cells. Typical for zymosan ring-like lysosomes were shown. In hepatocytes the significant decrease of lipid droplets and glycogen occured.

3.3. Carbon Particle Phagocytosis

Comparatively to the control animals received only saline gadolinium chloride administration induced decrease of index of phagocytosis (24 h - 0.027 + 0.005, 48 h - 0.025 + 0.002 versus control 0.065 + 0.008, $p < 0.05$). Zymosan treatment increased twice the index of phagocytosis and pretreatment by gadolinium chloride resulted to its normalization.

3.4. Enzyme Assays in Liver

Gadolinium chloride administration 24 h after induced increase of free activity of cathepsin B (Fig.1) and cathepsin L (Fig.2) calculated as enzyme activity or as a percentage from the total activity. In macrophage stimulation by zymosan increased both free (Fig.1) and specific (Fig.3) activities of cathepsin B were noted 5 days after. So, increased free activity of cathepsin B as evidence of lysosomal membrane labilization of macrophages was observed both in liver macrophage stimulation and depression.

Preliminary administration of gadolinium chloride (24 h before) to rats with zymosan injection did not cause any changes of free and total activities of cathepsin B (Fig. 1,3), but free activity of cathepsin L remained elevated (Fig.2). Specific activity of cathepsin L was not changed (Fig.4). There was no significant changes in free and specific activity of acid phosphatase in all of groups of animals studied (data are not shown).

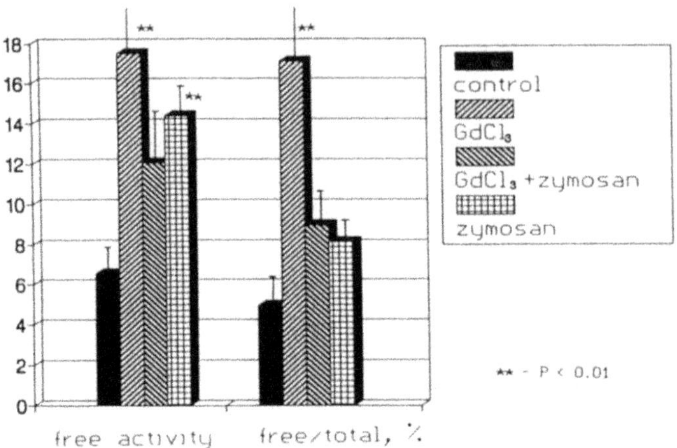

Figure 1. Free activity of cathepsin B in liver of rats received saline (control), gadolinium chloride ($GdCl_3$), gadolinium chloride and zymosan ($GdCl_3$ + zymosan) or only zymosan. Results are expressed as nmol MCA/g of protein per min (free activity) or as a percentage from the total activity of enzyme. Notes: ** $p < 0.01$ vs control. Data shown represent the mean values + SEM for 9–12 animals.

3.5. Spleen Weight and Protein Content

In group of animals received gadolinium chloride plus zymosan a significant increase of relative spleen weight was noted (0.0078 + 0.0012 versus 0.0027 + 0.00011 in control, $p < 0.001$), as well as protein content (157 + 18 versus control 112 + 10 mg/g $p < 0.05$). Zymosan injection was followed also by increase of spleen weight (0.0044 + 0.00016, $p < 0.001$ versus control) and protein content (160 + 5, $p < 0.01$ versus control).

3.6. Enzyme Assays in Spleen, Lung and Kidney

Spleen specific cathepsin B activity increased during administration of gadolinium chloride plus zymosan (Fig.3). Zymosan increased cathepsin B activity in lung (Fig.5), similar tendency was observed in spleen (Fig.3). Pretreatment by gadolinium chloride did not prevent this increase in zymosan-group in lung (Fig. 5). There was no changes in specific activity of cathepsin B in kidney, tissue not so enriched by macrophages.

4. DISCUSSION

System of mononuclear phagocytes includes three main pools of macrophages from liver (Kupffer cells), lung and spleen, having common precursor and related unctionally. The liver is largest reticuloendothelial organ, however liver macrophages are in close relationship with pools from other sources of macrophages. Gadolinium chloride in vivo was used in many investigations as a model of selective deletion of Kupffer cell function. It is not clear yet the exact fate and proteolytic function of Kupffer cells after such intervention. The question arises about the toxicity of this compound for liver macrophages and possibly for hepatocytes. One can suggest that gadolinium chloride can be selectively accumulated inside of lysosomes of macrophages and, possibly, of hepatocytes (lysosomotropic action), like some heavy metals and lantanoids. Decreased phagocytic activity of

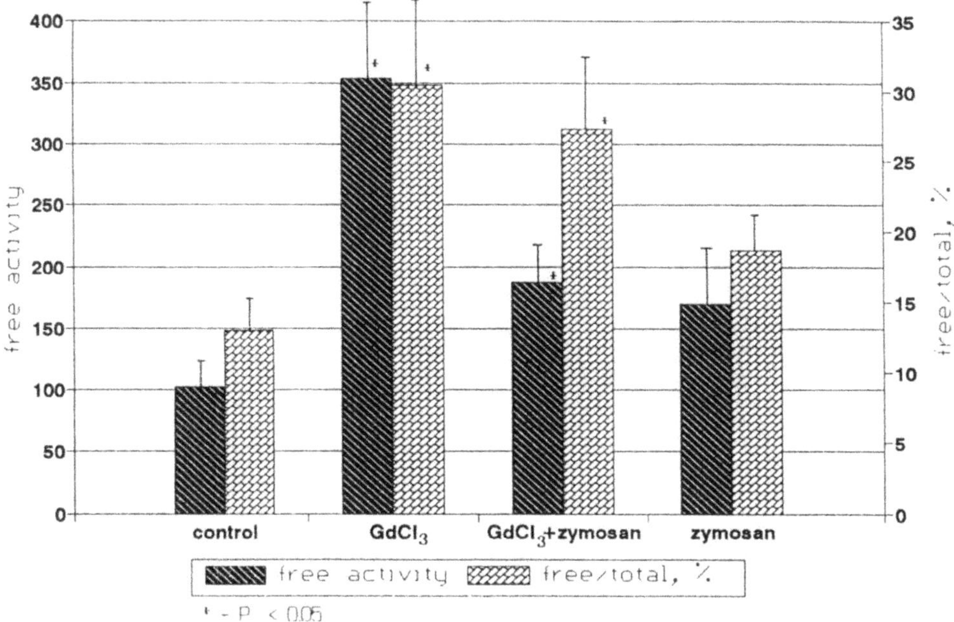

Figure 2. Free activity of cathepsin L in rat liver (here and below the subscriptions as was mentioned for Fig.1). Notes: * $p < 0.05$

liver macrophages and disappearance of staining for Kupffer cell-specific antigen were noted after the treatment by gadolinium chloride in mice and rats. In pathology gadolinium chloride was shown to prevent the release of inflammatory cytokines produced by Kupffer cells and limit the production of toxic oxygen radicals.

Earlier in our study the increased serum secretion of lysosomal enzymes N-acetyl-β-glucosaminidase and N-acetyl-β-galactosaminidase, marker enzymes for liver macrophages, in mice 24 h after gadolinium chloride administration in the same dose was shown. Similar data were obtained during macrophage activation by zymosan. However, it is still unclear wether this elevation of serum level of lysosomal enzymes was a result of secretion from Kupffer cells or other macrophages.

According to our electron microscopic data gadolinium chloride exerted some deleterious effect on liver parenchymal cells. The most significant were lipid infiltration, accumulation of VLDL-containing vesicles and increased auto- and heterophagy in hepatocytes.

The injurious effect of gadolinium chloride was confirmed in biochemical studies by lysosomal labilization test. Free activity of cathepsin B (main cysteine proteinase of macrophages) and cathepsin L (more typical for hepatocytes) was elevated in experiment with gadolinium chloride. So, there was labilization of lysosomal membranes of Kupffer cells and hepatocytes. However, the mechanism of labilization in macrophage depression and stimulation was different, as was suggested. In macrophage stimulation lysosomal labilization was accompanied by increase of the specific activity cathepsin B. It reflected the input of new monocytes/macrophages or proliferation of resident ones. In macrophage depression loading of lysosomes by gadolinium chloride and formation of secondary lysosomes were responsible for labilization of lysosomal membranes. Increasing of free activity of cathepsin L in gadolinium chloride model was related to disturbances of

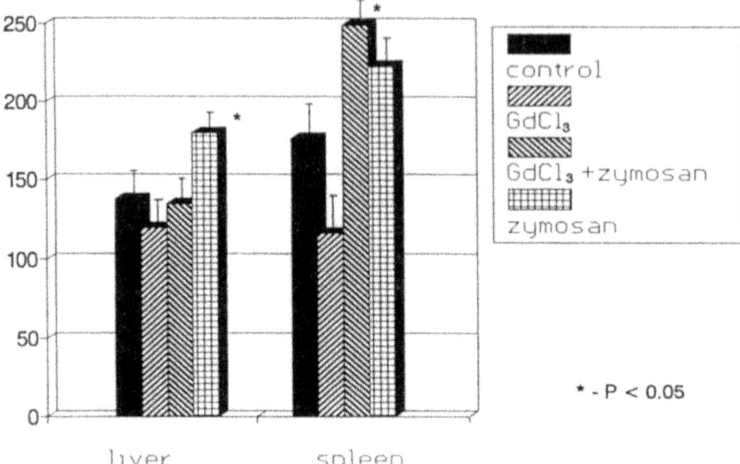

Figure 3. The specific activity of cathepsin B in liver and spleen of rats (nmol MCA/g of protein per min). Groups of animals and notes are as in Fig. 1.

lysosomes of hepatocytes. So, according to these data one can conclude that gadolinium chloride exert its harmful effect not only on liver macrophages, but also on hepatocytes.

Surprisingly, the finding of drastic increase of spleen weight and protein content in group with gadolinium chloride plus zymosan was noted, followed by increased specific cathepsin B activity. Possibly, it reflects significant input of monocyte/macrophage into spleen during stimulation of different pools of macrophages by zymosan which was not

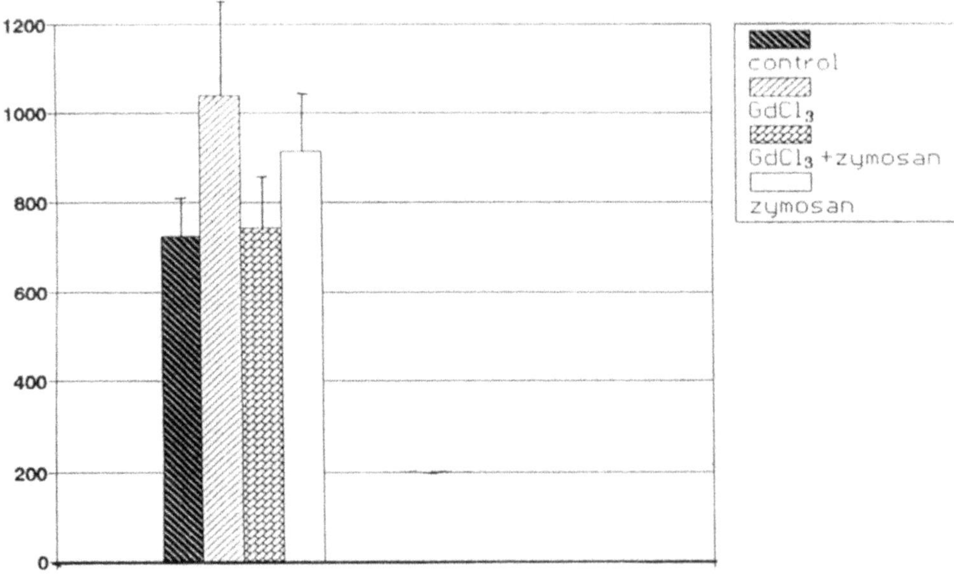

Figure 4. The specific activity of cathepsin L in rat liver (nmol MCA/g of protein per min).

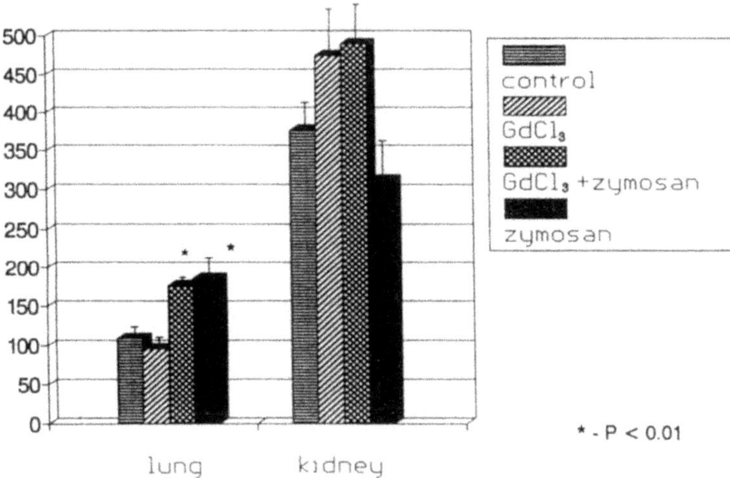

Figure 5. The specific activity of cathepsin B in lung and kidney (nmol MCA/g of protein per min). Notes: * $p < 0.01$

abolished by gadolinium chloride pretreatment. Similar results were obtained for lung cathepsin B specific activity, but not for kidney. We can summarize that gadolinium chloride effect was targeted predominantly on liver macrophages, and preliminary administration of gadolinium chloride attenuated harmful effect of zymosan, influencing on liver cysteine proteinases.

REFERENCES

1. Bouma, J.M.W., and M.J.Smit (1989) Gadolinium chloride selectively blocks endocytosis by Kupffer cells. In : Cells of the Hepatic Sinusoid. Vol. 2. E.Wisse, D.L.Knook, K.Decker, eds. The Kupffer Cell Foundation, Rijswijk, 132–133
2. Lazar, G., G.Lazar, Jr., J.Kazaki, J.Olah, and E.Husztik (1995) Protection against anaphylactic shock and the release of anaphylactic mediators by gadolinium chloride-induced Kupffer cell blockade. In: Cells of the Hepatic Sinusoid. Vol. 5. E.Wisse, D.L.Knook and K.Wake, eds. The Kupffer Cell Foundation, Leiden, 58–59
3. Hardonk, M.J., F.W.J.Dijkhuis, and A.M.Jonker (1995) Selective depletion of Kupffer cells by gadolinium chloride attenuates both acute galactosamine-induced hepatitis and carbon tetrachloride toxicity in rats. In: Cells of the Hepatic Sinusoid. Vol.5. E.Wisse, D.L.Knook and K.Wake eds. The Kupffer Cell Foundation, Leiden, 29–32
4. Kamei,T., M.P.Callery, and M. W. Flye (1990) Kupffer cell blockade prevents induction of portal venous tolerance in rat cardiac allograft transplantation. Journal of Surgical Research, 48: 393–396
5. Korolenko, T.A., E.V.Rukavishnikova, A.F.Safina, M.I.Dushkin, and G.I.Mynkina (1992) Endocytosis by liver cells during suppression of intralysosomal proteolysis. Biol. Chem. Hoppe-Seyler, 373: 573–580
6. Barrett, A.J. and H.Kirschke (1981) Cathepsin B, cathepsin H and cathepsin L. In: Methods in Enzymology. 80: 535–561
7. Safina, A.,T.Korolenko, G.Mynkina, M.Dushkin, and G.Krasnoselskaya (1992) Liver and serum lysosomal enzymes activity during zymosan-induced inflammationin mice. Agents and Actions Supplements, 38/III: 370–375
8. Ii, K., Hizawa, K., Kominami, E., Bando, Y., Katunuma, N. (1985) Different immunolocalization of cathepsins B, H, and L in the liver. J. Histochem. Cytochem, 33: 1173–1175

CONTRIBUTORS

C. A. Abbott
Centenary Institute for Cancer Medicine and Cell Biology
Royal Prince Alfred Hospital
Sydney, Australia

Magnus Abrahamson
Department of Clinical Chemistry
University of Lund
Lund, Sweden

Karin Agha-Amiri
Institute of Medical Neurobiology
Otto von Guericke University
Magdeburg, Germany

Siegfried Ansorge
Institute of Experimental Internal Medicine
Otto von Guericke University
Magdeburg, Germany

M. Arndt
Institute of Experimental Internal Medicine
Otto von Guericke University
Magdeburg, Germany

Patrick Auberger
Faculté de Médecine
CJF INSERM 9605
Nice, France

U. Bank
Institute of Experimental Internal Medicine
Otto von Guericke University
Magdeburg, Germany

Brigitte Bauvois
Institut Curie
Paris, France

Julià Blanco
Institut Pasteur
Paris, France

Lothar Bock
Department of Immunology and Cell Biology
Research Centre Borstel
Borstel, Germany

Judith S. Bond
Department of Biochemistry and Molecular Biology
The Pennsylvania State University College of Medicine
Hershey, Pennsylvania

Ernst Brandt
Department of Immunology and Cell Biology
Research Centre Borstel
Borstel, Germany

Wolfgang Brandt
Fachbereich Biochemie/Biotechnologie
Martin Luther University
Halle, Germany

Frank Bühling
Institute of Experimental Internal Medicine
Otto von Guericke University
Magdeburg, Germany

Christian Callebaut
Institut Pasteur
Paris, France

Frédéric Checler
Institut de Pharmacologie Moléculaire et Cellulare du DNRS
Valbonne, France

W. Conca
Institute of Rheumatology and Clinical Immunology
University of Freiburg
Freiburg, Germany

E. Csernok
Department of Clinical Rheumatology and Rheumaklinik Bad Bramstedt
Medical University of Lübeck
Lübeck, Germany

I. De Meester
Department of Clinical Biochemistry
University of Antwerp
Wilrijk, Belgium

S. Djanaeva
Institute of Physiology RAMS
Novosibirsk, Russia

Rui-Ping Dong
Dana-Farber Cancer Institute and Department of Medicine
Harvard Medical School
Boston, Massachusetts

Eileen Ebert
Thoraxklinik
Heidelberg, Germany

Werner Ebert
Thoraxklinik
Heidelberg, Germany

Jan-Erik Ehlert
Department of Immunology and Cell Biology
Research Centre Borstel
Borstel, Germany

Jürgen Faust
Department of Biochemistry and Biotechnology
Martin Luther University
Halle, Germany

Hans-Dieter Flad
Department of Immunology and Cell Biology
Forschungszentrum
Borstel, Germany

Bernhard Fleischer
Bernhard Nocht Institute for Tropical Medicine
Hamburg, Germany

Rafael Franco
Departamento de Bioquimica i Biologica Molecular
Universidad de Barcelona
Barcelona, Spain

K. Frank
Institute of Experimental Internal Medicine
Otto von Guericke University
Magdeburg, Germany

H. G. Gassen
Institute for Biochemistry
Technical University of Darmstadt
Darmstadt, Germany

Johannes Gerdes
Department of Immunology and Cell Biology
Forschungszentrum
Borstel, Germany

F. Goossens
Department of Clinical Biochemistry
University of Antwerp
Wilrijk, Belgium

M. D. Gorrell
Centenary Institute for Cancer Medicine and Cell Biology
Royal Prince Alfred Hospital
Sydney, Australia

W. L. Gross
Department of Clinical Rheumatology and Rheumaklinik Bad Bramstedt
Medical University of Lübeck
Lübeck, Germany

Sandrine Guérin
Faculté de Médecine
CJF INSERM 9605
Nice, France

Margrit Hahn
Department of Immunology and Cell Biology
Forschungszentrum
Borstel, Germany

Walter Halangk
Department of Experimental Surgery
University Magdeburg
Magdeburg, Germany

Luc Härter
Department of Immunology and Cell Biology
Research Centre Borstel
Borstel, Germany

Hans-Peter Hartung
Department of Neurology
Julius-Maximilians-University
Würzburg, Germany

Martin Hegen
Dana-Farber Cancer Institute and Department of Medicine
Harvard Medical School
Boston, Massachusetts

Gesine Heinemann
Department of Immunology and Cell Biology
Forschungszentrum
Borstel, Germany

D. Hendriks
Department of Clinical Biochemistry
University of Antwerp
Wilrijk, Belgium

Uta Hermanns
Department of Biochemistry and Biotechnology
Martin Luther University
Halle, Germany

Nigel M. Hooper
Department of Biochemistry and Molecular Biology
University of Leeds
Leeds, United Kingdom

Ara G. Hovanessian
Institut Pasteur
Paris, France

Jochen Hühn
Bernhard Nocht Institute for Tropical Medicine
Hamburg, Germany

Ralph J. Hyde
Department of Biochemistry and Molecular Biology
University of Leeds
Leeds, United Kingdom

A. Ittenson
Institute of Experimental Internal Medicine
Otto von Guericke University
Magdeburg, Germany

Etienne Jacotot
Institut Pasteur
Paris, France

Weiping Jiang
Department of Biochemistry and Molecular Biology
The Pennsylvania State University College of Medicine
Hershey, Pennsylvania

L. Juliano
Biophysics Department
Escola Paulista de Medicina
Sao Paulo, Brasil

M. A. Juliano
Biophysics Department
Escola Paulista de Medicina
Sao Paulo, Brasil

Britta Jülke
Thoraxklinik
Heidelberg, Germany

Thilo Kähne
Institute of Experimental Internal Medicine
Otto von Guericke University
Magdeburg, Germany

Junichi Kameoka
Dana-Farber Cancer Institute and Department of Medicine
Harvard Medical School
Boston, Massachusetts

Astrid Kehlen
Institute of Medical Immunology
Martin Luther University
Halle, Germany

J. Kekow
Clinic of Rheumatology
Otto von Guericke University
Vogelsang, Germany

Bernd C. Kieseier
Department of Neurology
Julius-Maximilians-University
Würzburg, Germany

Heidrun Kirschke
Institute of Physiological Chemistry
Martin Luther University
Halle, Germany

Regine Koelsch
Institute of Physiological Chemistry
Martin Luther University
Halle, Germany

Klaus Kokholm
Department of Medical Biochemistry and Genetics
University of Copenhagen
Copenhagen, Denmark

Natasa Kopitar-Jerala
Department of Biochemistry and Molecular Biology
Jozef Stefan Institute
Ljubljana, Slovenia

Bruce D. Korant
DuPont Merck Pharmaceutical Company
Wilmington, Delaware

T. Korolenko
Institute of Physiology RAMS
Novosibirsk, Russia

Janko Kos
Department of Biochemistry and Molecular Biology
Jozef Stefan Institute
Ljubljana, Slovenia

Bernard Krust
Institut Pasteur
Paris, France

Dagmar Kunz
Institute of Clinical Chemistry
University Magdeburg
Magdeburg, Germany

Tamara Lah
National Institute of Biology
Ljubljana, Slovenia

Jürgen Langner
Institute of Medical Immunology
Martin Luther University
Halle, Germany

Jürgen Lasch
Institute of Physiological Chemistry
Martin Luther University
Halle, Germany

Uwe Lendeckel
Institute of Experimental Internal Medicine
Otto von Guericke University
Magdeburg, Germany

M. T. Levy
Centenary Institute for Cancer Medicine and Cell Biology
Royal Prince Alfred Hospital
Sydney, Australia

Jaeseung Lim
Department of Biochemistry and Molecular Biology
University of Leeds
Leeds, United Kingdom

Hans Lippert
Department of Experimental Surgery
University Magdeburg
Magdeburg, Germany

Carmen Lluis
Departamento de Bioquímica i Biologia Molecular
Universidad de Barcelona
Barcelona, Spain

Matthias Löhn
Institute of Medical Immunology
Martin Luther University
Halle, Germany

Susan Lorey
Department of Biochemistry and Biotechnology
Martin Luther University
Halle, Germany

Andreas Ludwig
Department of Immunology and Cell Biology
Research Centre Borstel
Borstel, Germany

U. Machein
Institute of Rheumatology and Clinical Immunology
University of Freiburg
Freiburg, Germany

Philippe Marambaud
Institut de Pharmacologie Moléculaire et Cellulaire du CNRS
Valbonne, France

Jean Martinez
URA CNRS 1845
Montpellier, France

Taila Mattern
Department of Immunology and Cell Biology
Forschungszentrum
Borstel, Germany

Rainer Matthias
Department of Experimental Surgery
University Magdeburg
Magdeburg, Germany

Oleg Mayboroda
Institute of Medical Neurobiology
Otto von Guericke University
Magdeburg, Germany

G. W. McCaughan
Centenary Institute for Cancer Medicine and Cell Biology
Royal Prince Alfred Hospital
Sydney, Australia

Anne Micouin
Institut Curie
Paris, France

Chikao Morimoto
Dana-Farber Cancer Institute and Department of Medicine
Harvard Medical School
Boston, Massachusetts

Sylke Moschner
Institute of Physiological Chemistry
Martin Luther University
Halle, Germany

Christoph Mueller
Institute of Medical Immunology
Martin Luther University
Halle, Germany

Klaus Neubert
Department of Biochemistry and Biotechnology
Martin Luther University
Halle, Germany

Ove Norén
Department of Medical Biochemistry and Genetics
University of Copenhagen
Copenhagen, Denmark

Jørgen Olsen
Department of Medical Biochemistry and Genetics
University of Copenhagen
Copenhagen, Denmark

Frank Petersen
Department of Immunology and Cell Biology
Research Centre Borstel
Borstel, Germany

Dirk Reinhold
Institute of Experimental Internal Medicine
Otto von Guericke University
Magdeburg, Germany

Christoph Rethfeldt
Institute of Medical Neurobiology
Otto von Guericke University
Magdeburg, Germany

Dagmar Riemann
Institute of Medical Immunology
Martin Luther University
Halle, Germany

François Rieunier
URA CNRS 1845
Montpellier, France

Y. Rivière
Institut Pasteur
Paris, France

Christopher J. Rizzo
DuPont Merck Pharmaceutical Company
Wilmington, Delaware

S. Scharpé
Department of Clinical Biochemistry
University of Antwerp
Wilrijk, Belgium

K. Schatteman
Department of Clinical Biochemistry
University of Antwerp
Wilrijk, Belgium

Dagmar Scheel-Toellner
Department of Rheumatology
University of Birmingham
Birmingham, United Kingdom

Stuart F. Schlossman
Dana-Farber Cancer Institute and Department of Medicine
Harvard Medical School
Boston, Massachusetts

Abidat Schneider
Institute of Clinical Chemistry
University Magdeburg
Magdeburg, Germany

Walter Schubert
Institute of Medical Neurobiology
Otto von Guericke University
Magdeburg, Germany

Hans-Ulrich Schulz
Department of Experimental Surgery
University Magdeburg
Magdeburg, Germany

Ulrike Seitzer
Department of Immunology and Cell Biology
Forschungszentrum
Borstel, Germany

Hans Sjöström
Department of Medical Biochemistry and Genetics
University of Copenhagen
Copenhagen, Denmark

Eberhard Spiess
Deutsches Krebsforschungszentrum
Heidelberg, Germany

Christiane Steeg
Bernhard Nocht Institute for Tropical Medicine
Hamburg, Germany

Beate Stiebitz
Department of Biochemistry and Biotechnology
Martin Luther University
Halle, Germany

Angela Stöckel
Department of Biochemistry and Biotechnology
Martin Luther University
Halle, Germany

Jörg Stürzebecher
Centre of Vascular Biology and Medicine
University Jena
Erfurt, Germany

I. Svechnikova
Institute of Physiology RAMS
Novosibirsk, Russia

C. Szymkowiak
Department of Clinical Rheumatology and Rheumaklinik Bad Bramstedt
Medical University of Lübeck
Lübeck, Germany

Michael Täger
Institute of Experimental Internal Medicine
Otto von Guericke University
Magdeburg, Germany

Katja Thiele
Institute of Medical Immunology
Martin Luther University
Halle, Germany

Harald Tschesche
Faculty of Chemistry and Biochemistry
University of Bielefeld
Bielefeld, Germany

Anthony J. Turner
Department of Biochemistry and Molecular Biology
University of Leeds
Leeds, United Kingdom

K. Urazgaliyev
Institute of Physiology RAMS
Novosibirsk, Russia

G. Vakulin
Novosibirsk Medical Institute
Novosibirsk, Russia

Agustín Valenzuela
Departamento de Bioquímica i Biologia Molecular
Universidad de Barcelona
Barcelona, Spain

Gret Vanhoof
Department of Clinical Biochemistry
University of Antwerp
Wilrijk, Belgium

Lotte K. Vogel
Department of Biochemistry
Panum Institute
Copenhagen, Denmark

Arne von Bonin
Bernhard Nocht Institute for Tropical Medicine
Hamburg, Germany

Bernd Werle
Thoraxklinik
Heidelberg, Germany

Thomas Wex
Institute of Experimental Internal Medicine
Otto von Guericke University
Magdeburg, Germany

Sherwin Wilk
Department of Pharmacology
Mount Sinai School of Medicine
New York

Sabine Wolf
Institute for Biochemistry
Technical University of Darmstadt
Darmstadt, Germany

Sabine Wrenger
Institute of Experimental Internal Medicine
Otto von Guericke University
Magdeburg, Germany

INDEX

www.ingramcontent.com/pod-product-compliance
Ingram Content Group UK Ltd.
Pitfield, Milton Keynes, MK11 3LW, UK
UKHW051130260726
13967UKWH00010B/2970